The Economic Aspect of Geology

C. K. Leith

Writat

This edition published in 2022

ISBN : 9789356393165

Contents

CHAPTER I

INTRODUCTION

SURVEY OF FIELD

In adapting ourselves to physical environment it has been necessary to learn something about the earth. Mainly within the last century has this knowledge been organized into the science of geology, and only within the last few decades have the complex and increasing demands of modern civilization required the applications of geology to practical uses, resulting in the development of the science generally known as *economic geology*. This science is not sharply marked off from the science of geology proper; almost any phase of geology may at some time or some place take on its economic aspect.

The usefulness of economic geology was first recognized in relation to mineral resources,—and particularly in relation to metallic resources, their discovery and development,—but the science has been found to have much wider practical application. The practice of the economic geologist in recent years has taken on many new phases.

The geologist is called upon to study the geologic features of mineral deposits, their occurrence, structure, and origin. The basic information thus acquired is useful in estimating reserves and life of mineral deposits. This leads naturally to considerations of valuation. Because valuation plays such a large part in any tax program, the geologist is being used by tax boards of the federal and state governments.

Both in the formulation of laws relating to mineral resources, and in the litigation growing out of the infraction of these laws, the economic geologist plays a part.

One cannot go very far with the study of mineral resources without consideration of the question of conservation. Geologists are called on not only for broad surveys of the mineral reserves, but for the formulation of general principles of conservation and their application to specific mines and minerals.

The geologist's familiarity with the distribution and nature of mineral resources has given him a part in coping with broad questions of international use of natural resources. War conditions made it necessary to use new sources of supply, new channels of distribution, and new methods of utilization. The economic geologist came into touch with questions of international trade, tariffs, and shipping.

But economic geology is not solely confined to mineral resources. In relation to engineering enterprises of the greatest variety—canals, aqueducts, tunnels, dams, building excavations, foundations, etc.—geology now figures largely, both in war and in peace.

The nature, amount, and distribution of underground water supplies are so involved with geologic considerations that a considerable number of geologists give up their time wholly to this phase of the subject.

It might seem from this list of activities that geology is spreading too far into the fields of engineering and commerce, but there are equally rapid extensions of other fields of knowledge toward geology. The organization of these intermediate fields is required both in the interest of science and in the interest of better adaptation of the race to its environment. The geologist is required to do his part in these new fields, but not to abandon his traditional field.

It is proposed in this volume to discuss the economic aspects of geology without exhaustive discussion of the principles of geology which are involved. Practically the whole range of geologic science has some sort of economic application, and it would be futile to attempt in one volume even a survey of the science of geology as a whole. Our purpose is rather to indicate and illustrate, in some perspective, the general nature of the application of geology to practical affairs.

In professional preparation for the practice of economic geology there is no easy short-cut. Students sometimes think that a smattering of geological principles, combined with a little business and economic information, may be sufficient. Analysis of professional successes should make it clear that economic geologists are most effective and in most demand, not primarily because of business aptitude, though this helps, but because of their proficiency in the science of geology itself. In short, to enter successfully the field of economic geology one should first become a scientist, if only in a limited field.

The traditional conception of the geologist as a musty and stooped individual, with a bag, hammer, and magnifying glass, collecting specimens to deposit in a dusty museum, will doubtless survive as a caricature, but will hardly serve to identify the economic geologist in his present-day work. In writing this book, it is hoped in some measure to convey an impression of the breadth and variety in this field. Few other sciences offer so wide a range of opportunity, from the purely scientific to the practical and commercial, coupled with travel, exploration, and even adventure.

ECONOMIC APPLICATIONS OF THE SEVERAL BRANCHES OF GEOLOGY AND OF OTHER SCIENCES

There is no phase of geology which at some time or place does not have its economic application. Many references to these applications are made in other chapters. It is proposed here to indicate briefly some of the phases of geologic science which are most necessary to the practice of economic geology. The student in his preparation cannot afford to eliminate any of them on the ground that they are merely "scientific" or "academic" or "theoretical."

MINERALOGY AND PETROLOGY

Mineralogy, the study of minerals, and petrology, the study of rocks (aggregations of minerals), are of course elementary requisites in preparation. There must be familiarity with the principal minerals and rocks, and especially with the methods and processes of their identification, with their nature, and with their origin. This involves a study of their crystallography, chemical composition, physical qualities, and optical properties as studied with the microscope. In recent years the microscopical study of polished and etched surfaces of ores has proved a valuable tool.

Stratigraphy and Paleontology

Stratigraphy and paleontology are concerned with the sedimentary and life history of the earth. The determination of the ages of the earth's strata and of the conditions of their deposition is required in the practice of economic geology. For example, a detailed knowledge of the succession of rocks and their ages, as determined by fossils and other stratigraphic evidence, is vital to the interpretation of conditions in an oil or coal field, and to the successful exploration and development of its deposits. The success of certain paleontologists and stratigraphic specialists in oil exploration is an evidence of this situation. Certain iron ores, phosphates, salts, potash, and other minerals, as well as many of the common rocks used for economic purposes, are found in sedimentary deposits, and require for their successful exploration and development the application of stratigraphic and paleontologic knowledge.

Closely related to stratigraphy (as well as to physiography, see pp. 6-10) is the study of sedimentation,—*i. e.*, the study of the physical, chemical, climatic, and topographic conditions of the deposition of sediments. This is coming to play an increasingly large part in geologic work, and is essential to the interpretation of many mineral deposits, particularly those in which stratigraphic and physiographic questions are involved.

Still another aspect of the problem of stratigraphy and sedimentation is covered by the study of ***paleogeography***, or the areal distribution of the faunas and sediments of geologic periods caused by the alternating submergence and emergence of land areas. In the search for the treasures of sedimentary deposits, a knowledge of ancient geographies and of ancient faunas makes it possible to eliminate certain regions from consideration. From a study of the faunas of eastern Kansas and Missouri, and of those along the eastern part of the Rocky Mountains, it has been inferred that a ridge must have extended across eastern Kansas during early Pennsylvanian time,—a conclusion which is of considerable economic importance in relation to oil exploration.

Structural Geology

Structural geology is the study of the physical forms and relations of rocks which result mainly from deformation by earth forces. If rocks remained in their original forms the structural problem would be a comparatively easy one, but usually they do not. Often they are faulted and folded and mashed to such an extent that it is difficult to go behind the superposed structural features to the original conditions in order to work out the geologic history. Not only is structural study necessary for the interpretation of geologic history, but it is often more directly applicable to economic problems,—as when, for instance, ore deposits have been formed in the cracks and joints of rocks, and the ore deposits themselves have been faulted and folded. Water resources are often located in the cracks and other openings of rocks, and are limited in their distribution and flow because of the complex attitude of deformed rocks. Oil and gas deposits often bear a well-defined relation to structural features, the working out of which is almost essential to their discovery.

It is not desirable to stop with the merely descriptive aspects of structural geology, as is so often done; for much light can be thrown on the economic applications of this subject by consideration of the underlying principles of mechanics,—involving the relations of earth stresses to rock structures. The mere field mapping and description of faults and joints is useful, but in some cases it is necessary to go a step further and to ascertain the mechanical conditions of their origin in order to interpret them clearly. If, for illustration, there are successive groups of mineralized veins in a mining camp, the later ones cutting the earlier ones, these might be treated as separate structural units. But if it can be shown that the several sets of veins have formed from a single movement, that there is no sharp genetic separation between the different sets and that they are a part of a single system, this interpretation throws new light on exploration and development, and even on questions of ownership and extralateral rights (Chapter XVI).

PHYSIOGRAPHY

Physiography is a phase of geology which investigates the surface features of the earth. It has to do not only with the description and classification of surface forms, present and past (physical geography or geomorphology), but with the processes and history of their development. The subject is closely related to geography, climatology, sedimentation, and hydrology. As one of the latest phases of geology to be organized and taught, its economic applications have been comparatively recent and are not yet widely recognized. Because of this fact its economic applications may be summarized at somewhat greater length than those of the other branches of geology above mentioned, which are to be more or less taken for granted.

The central feature of physiography is the so-called erosion cycle or topographic cycle. Erosion, acting through the agencies of wind, water, and ice, is constantly at work on the earth's surface; the eroded materials are in large part carried off by streams, ultimately to be deposited in the ocean near the continental margins. The final result is the reduction of the land surface to an approximate plain, called a *peneplain*, somewhere near sea level. Geological history shows that such peneplains are often elevated again with reference to sea level, by earth forces or by subsidence of the sea, when erosion again begins its work,—first cutting narrow, steep gulches and valleys, and leaving broad intervening uplands, in which condition the erosion surface is described as that of *topographic youth*; then forming wider and more extensive valleys, leaving only points and ridges of the original peneplains, in which stage the surface is said to represent *topographic maturity*; then rounding off and reducing the elevations, leaving few or none of the original points on the peneplain, widening the valleys still further and tending to reduce the whole country to a nearly flat surface, resulting in the condition of *topographic old age*. The final stage is again the peneplain. This cycle of events is called the *erosion cycle* or *topographic cycle*. Uplift may begin again before the surface is reduced to base level; in fact, there is a constant oscillation and contest between erosion and relative uplift of the land surface.

The action of the erosion cycle on rocks of differing resistance to erosion and of diverse structure gives rise to the great variety of surface forms. The physiographer sees these forms, not as heterogeneous units, but as parts of a definite system and as stages in an orderly series of events. He is able to see into the topographic conditions beyond the range of immediate and direct observation. He is able to determine what these forms were in the past and to predict their condition in the future. He is able to read from the topography the underground structure which has determined that topography. A given structure may in different stages of topographic development give quite diverse topographic forms. In such a case it is important to realize that the diversity is only superficial. On the other hand, a slight local divergence from the usual topographic forms in a given region may reflect a similar local divergence in the underground structure. Thus it is that an appreciation of the physiographic details may suggest important variations in the underground structure which would otherwise pass undiscovered.

Many mineral deposits owe their origin or enrichment to weathering and other related processes which are preliminary to erosion. These processes vary in intensity, distribution, and depth, with the stage of erosion, or in relation to the phase of the erosion cycle. They vary with the climatic conditions which obtain on the erosion surface. Mineral deposits are therefore often closely related to the topographic features, present and past, in kind, shape, and distribution. A few illustrative cases follow.

Many of the great copper deposits of the western United States owe their values to a secondary enrichment through the agency of waters working down from the surface. When this fact of secondary enrichment was discovered, it was naturally assumed that the process was related to the present erosion surface and to present climatic and hydrologic conditions. Certain inferences were drawn, therefore, as to depth and distribution of the enriched ores. This conception, however, proved to be too narrow; for evidences were found in many cases that the copper deposits had been concentrated in previous erosion cycles, and therefore in relation to erosion surfaces, now partly buried, different from the present surface. The importance of this knowledge from an exploring and development standpoint is clear. It has made it possible to find and follow rich ores, far from the present erosion surface, which would otherwise have been disclosed solely by chance. Studies of this kind in the copper camps are yet so recent that much remains to be learned. The economic geologist advising exploration and development in copper ores who does not in the future take physiographic factors into account is likely to go wrong in essential ways, as he has done in some cases in the past.

Not only is it necessary to relate the secondary enrichment of copper deposits to the erosion surface, present or past, but by a study of the conditions it must be ascertained how closely erosion has followed after the processes of enrichment. In some cases erosion has followed so slowly as to leave large zones of secondary enrichment. In other cases erosion has followed up so closely after the processes of secondary enrichment as to remove from the surface important parts of the secondarily enriched deposits.

The iron ores of the Lake Superior region are the result of the action of waters from the surface on so-called iron formations or jaspers. Here again it was at first supposed that the enrichment was related to the present erosion surface; but upon further studies the fact was disclosed that the concentration of the ores took place in the period between the deposition of Keweenawan and Cambrian rocks, and thus a new light was thrown on the possibilities as to depth and distribution of the ores. The old pre-Cambrian surface, with reference to which the concentration took place, can be followed with some precision beneath the present surface. This makes it possible to forecast a quite different depth and distribution of the ores from that which might be inferred from present surface conditions. Present surface conditions, of low relief, considerable humidity, and with the water table usually not more than 100 feet from the surface, do not promise ore deposits at great depth. The erosion which formed the old pre-Cambrian surface, however, started on a country of great relief and semi-arid climate, conditions which favored deep penetration of the surface waters which concentrated the ores.

The iron ores of eastern Cuba are formed by the weathering of a serpentine rock on an elevated plateau of low relief, where the sluggish streams are unable rapidly to carry off the products of weathering. Where streams have cut into this plateau and where the plateau breaks down with sharp slopes to the ocean, erosion has removed the products of weathering, and therefore the iron ore. An important element, then, in iron ore exploration in this country is the location of regions of slight erosion in the serpentine area. One of the largest discoveries was made purely on a topographic basis. It was inferred merely from a study of topography that a certain large unexplored area ought to carry iron ore. Subsequent work in the thick and almost impenetrable jungle disclosed it.

Bauxite deposits in several parts of the world require somewhat similar conditions of concentration, and a study of the physiographic features is an important factor in their location and interpretation.

A physiographic problem of another sort is the determination of the conditions surrounding the origin of sedimentary ores. Certain mineral deposits, like the "Clinton" iron ores, the copper ores in the "Red Beds" of southwestern United States and in the Mansfield slates of Germany, many salt deposits, and almost the entire group of placer deposits of gold, tin, and other metals, are the result of sedimentation, from waters which derived their materials from the erosion of the land surface. It is sometimes possible from the study of these deposits to discover the position and configuration of the shore line, the depth of water, and the probable continuity and extent of the deposits. Similar questions are met in the study of coal and oil.

This general problem is one of the phases of geology which is now receiving a large amount of attention, not only from the standpoint of ore deposition, but from a broader geologic standpoint. In spite of the fact that sedimentary processes of great variety can be observed in operation today, it is yet extremely difficult to infer from a given sedimentary deposit the precise conditions which determined its deposition and limited its distribution. For instance, sedimentary iron formations furnish a large part of the world's iron ore. The surface distribution, the structure, the features of secondary enrichment, are all pretty well understood; likewise the general conditions of sedimentation are reasonably clear,—but the close interpretation of these conditions, to enable us to predict the extent of one of these deposits, or to explain its presence in one place and absence in another, is in an early and sketchy stage.

An understanding of the principles and methods of physiography is also vital to an intelligent application of geology to water resources, to soils, to dam and reservoir construction, and to a great variety of engineering undertakings, but as these subjects involve the application of many other phases of geology, they are considered in separate chapters. (Chapters V, VI, and XX.)

Rock Alterations or Metamorphism

This is one of the newer special phases of geology which for a long time was regarded as the plaything of the petrographer or student of rocks. With the systematic development of the subject, however, it was found that the extremely numerous and complex alterations of rocks and minerals may be definitely grouped, and that they are controlled by broad principles. It became apparent also that these principles apply both to the economic and non-economic minerals and rocks,—in other words, that the segregation of economic minerals is a mere incident in pervasive cycles of the alterations which affect all rocks. Metamorphic geology, therefore, for some geologists becomes a convenient approach to the subject of economic geology. It has the great advantage that it tends to keep all minerals and all processes of ore deposition in proper perspective with relation to rocks and rock processes in general. It is not argued that this is the only approach or that it is the best for all purposes. A brief account of this phase of geology is given in Chapter II.

Application of Other Sciences

Geology is sometimes defined as the application of other sciences to the earth. Considered broadly, there is no phase of science which is not involved in economic geology. In other chapters in this book many references are made to applications of engineering, mathematics, physics, chemistry, metallurgy, biology, and economics.

At different times and places the requirements for earth materials are quite different. In the Stone Age there was little use for metals; in later ages the use of metals broadened. The multiplicity of demands of modern civilization, the increasing knowledge of processes of metallurgy, chemistry and physics, better transportation, better organization of commercial life, and many other factors, tend to bring new earth materials into use,—and, therefore, into the field of economic geology. A comparatively few years ago alumina, one of the most common and abundant substances of the earth's crust, was in no general demand except for very limited use as an ornament. Little attention was paid to it by economic geologists as a commercial product; now, however, aluminum is in great demand, and the raw materials which produce it have become the subjects of intensive study by economic geologists.

In short, economic geology includes the consideration of man in reaction to his physical environment. There are some earth materials and some conditions of the earth environment which do not yet come within the field of economic geology. But so large a proportion of them do, that the "complete economic geologist" should indeed be almost omniscient. When one considers what an insignificantly small portion of this field can be covered by any individual, it is apparent that the title of economic geologist implies no mastery of the entire field. There is yet no crowding.

TREATMENT OF THE SUBJECT IN THIS VOLUME

In scope and manner of treatment this volume follows somewhat the writer's presentation of the subject in university teaching. The purpose is to explain the nature of the economic demands for the science of geology, and to discuss something of the philosophy of the finding and use of raw materials.

Somewhat generalized statistics are used as a means of gaining perspective. No effort has been made for detailed accuracy or for completeness. So far as possible the quantitative features are expressed in general proportions, and where specific figures are given they are meant to indicate only such general proportions. The thought has been not to be so specific that the figures would soon be out of date. All standard statistical sources have been drawn on, but the principal sources have been the results of the various special investigations called out by the war, in which the writer had a part.

On the geologic side many sources have been drawn on outside of the writer's own experience. For the most part, no specific references or acknowledgments are made, on the ground that the book aims to present the general features which are now the more or less common knowledge of economic geologists. To make the references really adequate for exhaustive study would not only burden the text, but would require a specificity of treatment which it has been hoped to avoid.

The illustrative cases chosen for discussion are often taken from the writer's field of experience. This field has been principally the Lake Superior region, but has included also the principal mineral deposits of North America, Cuba, and limited areas in South America and Europe. Thus the Lake Superior iron and copper region might seem to be brought forward more than is warranted by its scientific or economic importance. For this, the writer offers no apology. An author's perspective is largely determined by his background of training and experience, and a frank recognition of this fact may aid in determining the weight to be given to his conclusions. It might even add to scientific efficiency if each writer were to confine his discussion almost solely to matters within his own range of observation and study.

The writer's indebtedness for information derived from the printed page and for personal discussion and advice is of wide range. He would express his warm appreciation of the friendly spirit of coöperation and advice with which this effort has been aided—a spirit which he likes to think is particularly characteristic of the profession of economic geology. In particular he would acknowledge the efficient aid of Mr. Julian D. Conover in preparation and revision of the manuscript.

CHAPTER II

THE COMMON ELEMENTS, MINERALS, AND ROCKS OF THE EARTH AND THEIR ORIGINS

A list of the solid substances of the earth making up the so-called lithosphere (or rock sphere) in order of their abundance, does not at all correspond to a list made in order of commercial importance. Some of the most valuable substances constitute such a small proportion of the total mass of the lithosphere that they hardly figure at all in a table of the common substances.

RELATIVE ABUNDANCE OF THE PRINCIPAL ELEMENTS OF THE LITHOSPHERE

When reduced to the simplest terms of elements the outer ten miles of the lithosphere consists of:[1]

PERCENTAGE OF PRINCIPAL ELEMENTS IN THE LITHOSPHERE

Oxygen	47.33
Silicon	27.74
Aluminum	7.85
Iron	4.50
Calcium	3.47
Magnesium	2.24
Sodium	2.46
Potassium	2.46
	98.05

The remainder of the elements exist in quantities of less than 1 per cent. None of these principal elements occur separately in nature and none of them are mined as elements for economic purposes.

RELATIVE ABUNDANCE OF THE PRINCIPAL MINERALS OF THE LITHOSPHERE

Minerals exceptionally consist of single elements, but ordinarily are combinations of two or more elements; for instance, quartz consists of a chemical combination of silicon and oxygen. The proportions of the common minerals in the outer ten miles of the lithosphere are in round numbers as follows:

PERCENTAGE OF COMMON MINERALS IN LITHOSPHERE

Feldspar	49
Quartz	21
Augite, hornblende, and olivine	15
Mica	8
Magnetite	3
Titanite and ilmenite	1
Kaolin, limonite, hematite, dolomite, calcite, chlorite, etc.	3
	100

In making up this table it is assumed that the rocks to a depth of ten miles are about 95 per cent of igneous type, that is, crystallized from molten magma, and about 5 per cent of sedimentary type, that is, formed from the weathering and erosion of igneous rocks or preëxisting sediments, and deposited in beds or layers, either by water or by air (see pp. 16-17).

More reliable figures for the relative abundance of the minerals are available for each of the two classes of rocks, igneous and sedimentary. The igneous rocks contain minerals in about the following proportions:

PERCENTAGE OF COMMON MINERALS IN IGNEOUS ROCKS

Feldspar	50
Quartz	21
Augite, hornblende, olivine, etc.	17
Mica	8
Magnetite	3
Titanite and ilmenite	1
	100

The sedimentary rocks contain minerals in about the following proportions:

PERCENTAGE OF COMMON MINERALS IN SEDIMENTARY ROCKS

Quartz	35
Feldspar	16
White mica	15
Kaolin (clay)	9
Dolomite	9
Chlorite	5
Calcite	4
Limonite	4
Gypsum, carbon, rutile, apatite, magnetite, etc.	3
	100

The sedimentary rocks comprise three main divisions: (1) The muds and clays, with their altered equivalents, shale, slate, etc.;

(2) the sands, with their altered equivalents, sandstone, quartzite, quartz-schist, etc.; (3) the marls, limestones, and dolomites, with their altered equivalents, marble, talc-schist, etc. For brevity these groups are referred to respectively as shale, sandstone, and limestone. The proportions of minerals in each of these groups of rocks are as follows:

PERCENTAGE OF COMMON MINERALS IN SHALE, SANDSTONE, AND LIMESTONE

	Average shale	*Average sandstone*	*Average limestone*
Quartz	31.91	69.76	3.71
Kaolin	10.00	7.98	1.03
White mica	18.40		
Chlorite	6.40	1.15	
Limonite	4.75	.80	
Dolomite	7.90	3.44	36.25[1]
Calcite		7.21	56.56
Gypsum	1.17	.12	.10
Feldspar	17.60	8.41	2.20
Magnetite		.58	
Rutile	.66	.12	.06
Ilmenite		.25	
Apatite	.40	.18	.09
Carbon	.81		
Total	100.00	100.00	100.00

[1]Includes small amount of $FeCO_3$.

In comparing the mineral composition of igneous and sedimentary rocks, it will be noted that the most abundant single mineral of the igneous rocks, and the most abundant mineral of the lithosphere as a whole, is *feldspar*; that next in order is *quartz*; and that third comes a group of dark green minerals typified by augite and hornblende, commonly called *ferro-magnesian silicates* because they consist of iron and magnesia, with other bases, in combination with silica. The sedimentary rocks, which are ultimately derived from the destruction of the igneous rocks, contrast with the igneous rocks mainly in their smaller proportions of feldspars and ferro-magnesian minerals, their higher proportions of quartz and white mica (sericite or muscovite), and their content of kaolin, dolomite, calcite, chlorite, limonite, etc., which are nearly absent from the unaltered igneous rocks. Evidently the development of sediments from igneous rocks has involved the destruction of much of the feldspars and ferro-magnesian silicates, and the building from the elements of these destroyed minerals of more quartz, white mica, clay, dolomite, calcite, chlorite and limonite. The composition of the minerals of the sedimentary rocks is such as to indicate that the constituents of the air and water have been added in important amounts to accomplish this change of mineral character. For instance, carbon dioxide of the atmosphere has been added to lime and magnesia of the igneous rocks to make calcite and dolomite, water has been added to some of the alumina and silica of the igneous rocks to make kaolin or clay, and both oxygen and water have been added to the iron of the igneous rocks to make limonite.

RELATIVE ABUNDANCE OF THE PRINCIPAL ROCKS OF THE LITHOSPHERE

Just as elements combine chemically to form minerals, so do minerals combine mechanically, either loosely or compactly, to form rocks. For instance, quartz is a mineral. An aggregation of quartz particles forms sand or sandstone or quartzite. Most rocks contain more than one kind of mineral.

Sedimentary rocks occupy considerable areas of the earth's surface, but they are relatively superficial. It has been estimated that if spread evenly and continuously over the earth, which they are not, they would constitute a shell scarcely a half mile thick.[2] Igneous rocks are relatively more abundant deep below the surface. If the sediments be assumed to be limited to a volume equivalent to a half-mile shell, and the remainder of the rocks be assumed to be igneous, it is evident that to a depth of ten miles 95 per cent of the rocks are igneous. Our actual observation is confined to a shallow superficial zone in which sediments make up at least half of all the rocks.

Igneous rocks can be divided for convenience into two main types: (1) granite and allied rocks, containing a good deal of silica and therefore *acid* in a chemical sense, and (2) basalt and allied types, containing less silica and more lime, magnesia, iron, soda and potassa, and therefore *basic* in a chemical sense. The former are light-colored gray and pink rocks while the latter are dark-colored green and gray rocks. Granite and basalt as technically defined are very common igneous rocks,—so common that the names are sometimes used to classify igneous rocks in general into two great groups, the granitic and the basaltic. It has been estimated that about 65 per cent of the igneous rocks are of the granitic group and 35 per cent of the basaltic group.

Sedimentary rocks, as already indicated, consist principally of three groups, which for convenience are named shale, sandstone, and limestone. If we approximate the average composition of each group and the average composition of the igneous rocks from which they are ultimately derived, it can be calculated that sedimentary rocks must form in the proportions of 82 per cent shale, 12 per cent sandstone, and 6 per cent limestone. Only this combination of the three sediments will yield an average composition comparable with that of the parent igneous rocks. As actually observed in the field the sandstones and limestones are in relatively higher percentage than is here indicated, suggesting that part of the shales may have been deposited in deep seas where they cannot be observed, and that part may have been so changed or metamorphosed that they are no longer recognized as shales.

SOILS AND CLAYS

Weathered and disintegrated rocks at the surface form soils and clays. No estimate is made of abundance, but obviously the total volume of these products is small as compared with the major classes of earth materials above noted, and in large part they may be included with these major classes.

Water (Hydrosphere)

It has been estimated that all the water of the earth, including the ocean, surface waters, and underground waters, constitutes about 7 per cent of the volume of the earth to a depth of 10 miles.[3]

COMPARISON OF LISTS OF MOST ABUNDANT ROCKS AND MINERALS WITH COMMERCIAL ROCKS AND MINERALS

Of the common rocks and minerals figuring as the more abundant materials of the earth's crust, only a few are prominently represented in the tables of mineral resources. Of these water and soils stand first. Others are the common igneous and sedimentary rocks used for building and road materials. Missing from the lists of the most abundant minerals and rocks, are the greater part of the commercially important mineral resources—including such as coal, oil, gas, iron ore, copper, gold, and silver,—implying that these mineral products, notwithstanding their great absolute bulk and commercial importance, occur in relatively insignificant amounts as compared with the common rock minerals of the earth.

THE ORIGIN OF COMMON ROCKS AND MINERALS

The common rocks and minerals develop in a general sequence, starting with igneous processes, and passing through stages of weathering, erosion, sedimentary processes, and alterations beneath the surface. The commercial minerals are incidental developments under the same processes.

Igneous Processes

The earliest known rocks are largely igneous. Sedimentary rocks are formed from the breaking down of igneous rocks, and the origin of rocks therefore starts with the formation of igneous rocks. Igneous rocks are formed by the cooling of molten rock material. The ultimate source of this molten material does not here concern us. It may come from deep within the earth or from comparatively few miles down. It may include preëxisting rock of any kind which has been locally fused within the earth. Wherever and however formed, its tendency is to travel upward toward the surface. It may stop far below the surface and cool slowly, forming coarsely crystallized rocks of the granite and gabbro types. Igneous rocks so formed are called *plutonic* intrusive rocks. Or the molten mass may come well toward the surface and crystallize more rapidly into rocks of less coarse, and often porphyritic, textures. Such intrusive rocks are porphyries, diabases, etc. Or the molten mass may actually overflow at the surface or be thrown out from volcanoes with explosive force. It then cools quickly and forms finely crystalline rocks of the rhyolite and basalt types. These are called effusives or extrusives, or lavas or volcanics, to distinguish them from intrusives formed below the surface. The intrusive masses may take various forms, called stocks, batholiths, laccoliths, sills, sheets and dikes, definitions and illustrations of which are given in any geological textbook. The effusives or volcanics at the surface take the form of sheets, flows, tuffs, agglomerates, etc.

Some of the igneous rocks are themselves "mineral" products, as for instance building stones and road materials. Certain basic intrusive igneous rocks contain titaniferous magnetites or iron ores as original constituents. Others carry diamonds as original constituents. Certain special varieties of igneous rocks, known as pegmatites, carry coarsely crystallized mica and feldspar of commercial value, as well as a considerable variety of precious gems and other commercial minerals. Pegmatites are closely related to igneous after-effects, discussed under the next heading. As a whole, the mineral products formed directly in igneous rocks constitute a much less important class than mineral products formed in other ways, as described below.

Igneous after-effects. The later stages in the formation of igneous rocks are frequently accompanied by the expulsion of hot waters and gases which carry with them mineral substances. These become deposited in openings in adjacent rocks, or replace them, or are deposited in previously hardened portions of the parent igneous mass itself. They form "contact-metamorphic" and certain vein deposits. Pegmatites, referred to above, are in a broad sense in this class of "igneous after-effects," in that they are late developments in igneous intrusions and often grade into veins clearly formed by aqueous or gaseous solutions. Among the valuable minerals of the igneous after-effect class are ores of gold, silver, copper, iron, antimony, mercury, zinc, lead, and others. While mineral products of much value have this origin, most of them have needed enrichment by weathering to give them the value they now have.

Weathering of Igneous Rocks and Veins

No sooner do igneous rocks appear at or near the earth's surface, either by extrusion or as a result of removal by erosion of the overlying cover, than they are attacked vigorously by the gases and waters of the atmosphere and hydrosphere as well as by various organisms,—with maximum effect at the surface, but with notable effects extending as far down as these agents penetrate. The effectiveness of these agents is also governed by the climatic and topographic conditions. Under conditions of extreme cold or extreme aridity, weathering takes the form mainly of mechanical disintegration, and chemical change is less conspicuous. Under ordinary conditions, however, processes of chemical decomposition are very apparent. The result is definitely known. The rocks become softened, loose, and incoherent. Voids and openings appear. The volume tends to increase, if all end products are taken into account. The original minerals, largely feldspar, ferro-magnesian minerals, and quartz, become changed to clay, mixed with quartz or sand, calcite or dolomite, and iron oxide, together with residual particles of the original feldspars and ferro-magnesian minerals which have only partly decomposed. In terms of elements or chemical composition, water, oxygen, and carbon dioxide, all common constituents of the atmosphere and hydrosphere, have been added; and certain substances such as soda, potassa, lime, magnesia, and silica have in part been carried away by circulating waters, to be redeposited elsewhere as sediments, vein fillings, and cements. Figure 1 illustrates the actual mineral and volume changes in the weathering of a granite—one of the most common rocks. The minerals anorthite, albite, and orthoclase named in this figure are all feldspars; sylvite and halite are chlorides of potash and soda. The weathering processes tend to destroy the original minerals, textures, and chemical composition. They are collectively known as *katamorphic* alterations, meaning destructive changes. The zone in which these changes are at a maximum is called the *zone of weathering*. This general zone is principally above the surface or level of the ground-waters, but for some rocks it extends well below this level. In some regions the ground-water level may be nearly at the surface, and in others, especially where arid, it may be two thousand or more feet down. Disintegrated weathered rocks form a blanket of variable thickness, which is sometimes spoken of as the residual mantle, or "mantle rock."

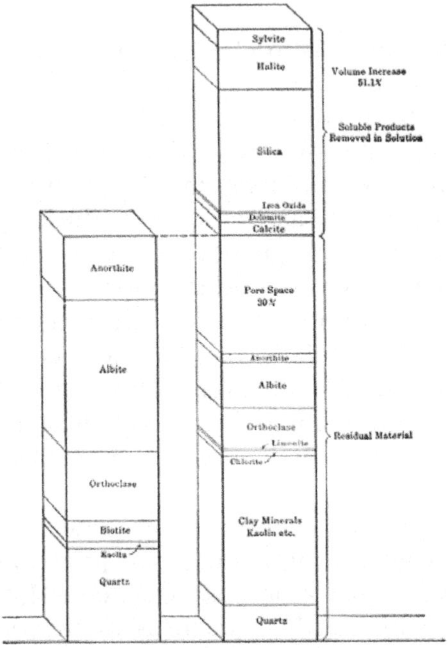

FIG. 1. Graphic representation of volume change in weathering of a Georgia granite.ToList

Mineral products formed by weathering from common igneous rocks include soils, clay, bauxite, and certain iron, chromite, and nickel ores. Again the commercial importance of this group is not large, as compared with products formed in other ways described below.

The same weathering processes described above for igneous rocks cause considerable changes of economic significance in deposits formed as igneous after-effects. In some cases they result in removing the less valuable minerals, thus concentrating the more valuable ones, as well as in softening the rock and making it easier to work; and in other cases they tend to remove the valuable constituents, which may then be redeposited directly below or may be carried completely out of the vicinity. The *oxide zones* of many ore bodies are formed by these processes.

Sedimentary Processes

Sedimentary rocks are formed by the removal and deposition of the weathered products of a land surface. Air, water, and ice, moving under the influence of gravity and other forces, all aid in this transfer. The broken or altered rock materials may be merely moved down slopes a little way and redeposited on the surface, forming one type of *terrestrial* or *subaërial deposits*, or they may be transferred and sorted by streams. When deposited in streams or near their mouths, they are known as *river*, *alluvial*, or *delta deposits*. When carried to lakes and deposited they form *lake deposits*. Ultimately the greater part of them are likely to be carried to the ocean and deposited as *marine sediments*.

Part of the weathered substances are carried mechanically as clay and sand, which go to make up the *shale* and *sandstone* sediments. Part are carried in solution, as for example lime carbonate and magnesium carbonate, which go to make up *limestone* and *dolomite*. Some of the dissolved substances are never redeposited, but remain in solution as salts in the sea, the most abundant of which is sodium chloride. Some of the dissolved substances of weathering, such as calcite, quartz, and iron oxide, are carried down and deposited in openings of the rocks, where they act as cements.

The sediments as a whole consist of three main types,—*shales* (kaolin, quartz, etc.), *sandstones* (quartz, feldspar, etc.), and *limestones* or *dolomites* (carbonates of lime and magnesia). Of these, the shale group is by far the most abundant. There are of course many sediments with composition intermediate between these types. There are also sediments made up of large undecomposed fragments of the original rocks, cemented to form *conglomerates*, or made up of small fragments of the original rocks cemented to form *arkoses* and *graywackes*. These, however, may be regarded as simply stages in the alteration, which in repeated cycles of weathering must ultimately result in producing the three main groups,—shales, sandstones, and limestones.

Mineral products formed by sedimentary processes include sandstones, limestones, and shales, used as building stone and road materials; certain sedimentary deposits of iron, like the Clinton ores of the southeastern United States and the Brazilian ores; important phosphate deposits; most deposits of salt, gypsum, potash, nitrates, etc.; comparatively few and unimportant copper deposits; and important placer deposits of gold, tin, and other metals, and precious stones. With the aid of organic agencies, sedimentary processes also account for the primary deposition of coal and oil.

Weathering of Sedimentary Rocks

After sedimentary rocks are formed, and in many cases covered by later sediments, they may be brought again by earth movements and erosion to the surface, where they in turn are weathered. The weathering of sedimentary rocks proceeds along lines already indicated for the igneous rocks. Residual mantles of impure clay and sand are commonly formed. The mineral composition of sedimentary rocks being different from that of igneous rocks to start with, the resulting products are in slightly different proportions; but the changes are the same in kind and tend merely to carry the general process of alteration farther in the same direction,—that is, toward the production of a few substances like clay, quartz, iron oxide, and calcite, which are transported and redeposited to form clay, sand, and limestone. Cycles of this kind may be repeated indefinitely.

By weathering of sedimentary rocks are produced some soils, certain commercial clays, iron ores, lead and zinc ores, and other valuable mineral products.

Consolidation, Cementation, and Other Subsurface Alterations of Rocks.

Cementation. No sooner are residual weathered mantles formed or sedimentary rocks deposited, whether under air or water, than processes of consolidation begin. Settling, infiltration of cementing materials, and new growths, or recrystallization, of the original minerals of the rock all play a part in the process. The mud or clay becomes a shale, the sand becomes sandstone or quartzite, the marl becomes limestone or marble. All the minute openings between the grains, as well as larger openings such as fissures and joints, may thus be filled. At the same time the cementing materials may replace some of the original minerals of the rock, the new minerals either preserving or destroying the original textures. This process is sometimes called *metasomatic replacement*. Igneous rocks as a rule are compact, and hence are not so much subject to the processes of cementation as sedimentary rocks; but certain of the more porous phases of the surface lavas, as well as any joints in igneous rocks, may become cemented. All of these changes may be grouped under the general term *cementation*.

A special phase of consolidation and cementation is produced near intrusive igneous rocks through the action of the heat and pressure and the expelled substances of the igneous rock. This is called *contact metamorphism* or *thermal metamorphism*. The processes are even more effective when acting in connection with the more intense metamorphism described under the next heading.

By cementation some of the common rocks, especially the sediments, become sufficiently compact and strong to be useful as commercial products, such as building stones and road materials.

More important as mineral products are the cementing materials themselves. These are commonly quartz, calcite, or iron oxide, of no especial value, but locally they include commercially valuable minerals containing gold, copper, silver, lead, zinc, and many other mineral products.

It is a matter of simple and direct observation, about which there is no controversy, that many minerals are deposited as cements in the openings in rocks or replacing rocks. As to the source of the solutions bringing in these minerals, on the other hand, there has been much disagreement. In general, the common cementing materials such as quartz and calcite, as well as some of the commercial minerals, are clearly formed as by-products of weathering, and are transported and redeposited by the waters penetrating downward from the surface. The so-called *secondary enrichment* of many valuable veins is merely one of the special phases of cementation from a superficial source. In other cases it is believed that deep circulation of ordinary ground-waters may pick up dispersed mineral substances through a considerable zone, and redeposit them in concentrated form in veins and other trunk channels. For still other cementing materials, it is suspected that the ultimate source is in igneous intrusions; in fact, deposits of this general character show all gradations from those clearly formed by surface waters, independently of igneous activity, to those of a contact-metamorphic nature and others belonging under the head of "igneous after-effects."

Hypothesis and inference play a considerable part in arriving at any conclusion as to the source of cementing materials,—with the result that there is often wide latitude for difference of opinion and of emphasis on the relative importance of the different sources of ore minerals.

Dynamic and contact metamorphism. Beneath the surface rocks are not only cemented, but may be deformed or mashed by dynamic movements caused by great earth stresses; the rocks may undergo rock flowage. The result is often a remarkable transformation of the character of the rocks, making it difficult to recognize their original nature. Also, igneous intrusions may crowd and mash the adjacent rocks, at the same time changing them by heat and contributions of new materials. This process may be called *contact metamorphism*, but in so far as it results in mashing of the rocks it is closely allied to *dynamic metamorphism*. The former term is also applied to less profound changes in connection with igneous intrusions, which result merely in cementation without mashing.

Dynamic and contact metamorphism may in some cases produce rocks identical in appearance with those produced by ordinary processes of cementation and recrystallization without movement. For instance, it is difficult to tell how much movement there has been in the production of a marble, because both kinds of processes seem to produce much the same result. Commonly, however, the effect of dynamic metamorphism is to produce a parallel arrangement of mineral particles and to segregate the mineral particles of like kind into bands, giving a *foliated* or *schistose* or *gneissic* structure, and the rocks then become known as slates, schists, or gneisses. Commonly they possess a capacity to part along parallel surfaces, called *cleavage*. The development of the schistose or gneissic structure is accompanied by the recrystallization of the rock materials, producing new minerals of a platy or columnar type adapted to this parallel arrangement. Even the composition of the rock may be substantially changed, though this is perhaps not the most common case. Whereas by weathering the rock is loosened up and disintegrated, substances like carbon dioxide, oxygen, and water are abundantly added, and light minerals of simple composition tend to develop,—by dynamic metamorphism on the other hand, carbon dioxide, oxygen, and water are usually expelled, the minerals are combined to make heavier and more complex minerals, pore space is eliminated, and altogether the rock becomes much more dense and crystalline. While segregation of materials is characteristic of the surficial products of weathering, the opposite tendency, of mixing and aggregation, is the rule under dynamic metamorphism, notwithstanding the minor segregation above noted.

Dynamic metamorphism is for the most part unfavorable to the development of mineral products. Ore bodies brought into a zone where these processes are active may be profoundly modified, but not ordinarily enriched. One of the exceptions to this general rule is the development of the cleavage of a slate, which enables it to be readily split and thereby gives it value. Contact metamorphism, on the other hand, may develop valuable mineral deposits (see pp. 20, 45-46).

THE METAMORPHIC CYCLE AS AN AID IN STUDYING MINERAL DEPOSITS

All of the chemical, mineralogical, and textural changes in rocks above described may be collectively referred to as *metamorphism*. The phase of metamorphism dealing with surficial weathering, similar changes below the surface, and the formation of sediments, is called *katamorphism* or destructive change. The phase of metamorphism dealing with the constructive changes in rocks, due to cementation, dynamic movements, and igneous influences, is called *anamorphism*. Some geologists confine the term metamorphism to the changes involved in contact and dynamic metamorphism, and call the resulting products *metamorphic rocks*.

The zone in which katamorphism is most active, usually near the surface, is called the *zone of katamorphism*. The deeper zone in which anamorphism is preponderant is called the *zone of anamorphism*. There are no definite limits of depth to these zones. A given rock may be undergoing katamorphism while rocks on either side at the same depth are suffering anamorphism.

By katamorphism rocks break down to produce the surficial rocks, and by anamorphism the surficial rocks are again consolidated and altered to produce highly crystalline rocks, which are not dissimilar in many of their characteristics to the igneous rocks from which all rocks trace their ultimate origin. In other words, anamorphism tends toward the reproduction of igneous rocks, though it seldom fully accomplishes this result. These two main groups of changes together constitute the *metamorphic cycle*. Some rocks go through all phases of the cycle, but others may pass directly from one phase to an advanced phase without going through the intermediate stages. For instance, an igneous rock may become a schist without going through the intermediate stage of sedimentation.

Rocks are not permanent in their condition, but at practically all times and places are undergoing some kind of metamorphism which tends to adapt them to their environment. The conception of rocks as representing phases or stages in a progressive series of changes called the metamorphic cycle aids greatly in correlating and holding in mind many details of rock nature and origin, and brings into some sort of perspective the conditions which have produced rocks. A schistose sediment comes to be regarded as an end product of a long series of alterations, beginning with igneous rocks and passing through the stages of weathering, sedimentation, cementation, etc., each of which stages has been responsible for certain mineralogical, chemical, and textural features now characterizing the rock. The alternation of constructive and destructive changes of the metamorphic cycle, and the repetitions of the cycle itself, periodically work over the earth materials into new forms. Usually the cycles are not complete, in the sense that they seldom bring the rock back to exactly the same condition from which it started. More sediments are formed than are changed to schists and gneisses, and more schists and gneisses are formed than are changed back to igneous rocks. Salts in the ocean continuously accumulate. The net result of the metamorphic cycle, is, therefore, the accumulation of materials of the same kinds. Incidental to these accumulations is the segregation of commercial mineral products.

The metamorphic cycle becomes a logical and convenient geologic basis for correlating, interpreting, and classifying mineral products. Because of the great variety of materials and conditions represented in mineral deposits, prodigious efforts are required to remember them as independent entities; but as incidents or stages in the well-known progress of the metamorphic cycle, their essential characteristics may be easily remembered and kept in some perspective.

Ores of certain metals, such as iron, occur in almost every phase of the metamorphic cycle,—as igneous after-effects, as weathered products, as sediments, and as schists. The ores of each of these several phases have group characteristics which serve to distinguish them in important particulars from ores belonging to other phases of the cycle. Having established the position of any particular ore in the metamorphic cycle, a number of safe inferences are possible as to mineralogical composition, shape, extent, and other conditions, knowledge of which is necessary for an estimate of commercial possibilities.

FOOTNOTES:

[1] Clarke, F. W., Data of geochemistry: *Bull. 695, U. S. Geol. Survey*, 1920, p. 35.

[2] Clarke, F. W., Data of geochemistry: *Bull. 695, U. S. Geol. Survey*, 1920, p. 33.

[3] Clarke, F. W., Data of geochemistry: *Bull. 695, U. S. Geol. Survey*, 1920, pp. 22-23.

CHAPTER III

SOME SALIENT FEATURES OF THE GEOLOGY AND CLASSIFICATION OF MINERAL DEPOSITS

VARIOUS METHODS OF CLASSIFICATION

Mineral products may be classified according to use, commercial importance, geographic distribution, form and structure, mineralogical and chemical composition, or origin. Each of these classifications is useful for some purposes. The geologist usually prefers a classification based on origin or genesis. In the following chapters on mineral resources, however, such a classification is not the primary one, because of the desire to emphasize economic features. The mineral commodities are treated as units and by group uses. Some mineral commodities have so many different kinds of origin in different regions that to distribute them among several genetic groups in description would make it impossible to preserve the unity necessary for consideration of the economic features.

While in the descriptive chapters many references are made to origin, it may be difficult for the reader to assemble them in perspective; for this reason we summarize at the outset some of the salient features of origin of mineral deposits and of their geologic classification.

To the layman the reason for emphasis on origin is often not clear. The "practical" man frequently regards this phase of the subject as merely incidental to the immediate economic questions—a playground for harmless theorists. The answer of the economic geologist is that in no other way than by a knowledge of origin is it possible to arrive at an understanding of conditions which so well enables one to answer many practical questions. In the exploration for mineral deposits, it is obvious that an understanding of the kinds of geologic conditions and processes under which a given type of deposit is known to develop results in the elimination of much unpromising territory, and the concentration of work on favorable localities. In forming any estimate of mineral deposits beyond the ground immediately opened up,—for instance, in estimating depth, form, change in values, mineralogical character, or interruptions due to faulting,—it is difficult to form any intelligent conception of the probabilities unless the history of the deposit is understood. If, for instance, the ore is known to be formed by hot waters, associated with the cooling of igneous rocks, different conditions are to be expected below the zone of observation than if the ore is formed by surface waters. If the ore body is formed as a single episode under simple geologic conditions, the interpretation of the possibilities in the situation may be quite different from the interpretation applied where the history has been more complex. If the surface conditions suggest possibilities of secondary enrichment of the ores, the interpretation of the conditions underground will be different from those applied where there is no evidence of such enrichment.

Where a mineral deposit is completely opened up in three dimensions, it is often possible to work out economic questions of tonnage, grade, shape, and values, without the aid of geology. Also, where conditions are comparatively simple and uniform throughout a district, the local knowledge of other mines may be a sufficient basis for answering these questions for any new property developed. Empirical methods may suffice. However, it is seldom that the conditions are so simple that some geological inference is not necessary. Even where problems are settled without calling in the geologist, geological inferences are required in the interpretation of, and projection from, the known facts. It is often the case that the practical man has in his mind a rather elaborate assortment of geologic hypotheses, based on his individual experience, which make the so-called theories of the geologist seem conservative in comparison. The geologist comes to the particular problem with a background of established geologic principles and observations, and his first thought is to ascertain all the local conditions which will aid in deciphering the complete history of the mineral deposit. There is no fact bearing on the history, however remote from practical questions, which may not be potentially valuable.

With this digression to explain the geologist's emphasis on origin of mineral products, we may return to a consideration of a few of the principles of rock and mineral genesis which have been found to be significant in the study of mineral products.

In the preceding chapters it has been indicated that mineral deposits are mere incidents in the mass of common rocks; that they are made by the same processes which make common rocks, that none of the processes affecting mineral deposits are unique for these minerals, and that most common rocks are on occasion themselves used as mineral resources. These facts are emphasized in order to make it clear that the study of mineral deposits cannot be dissociated from the study of rocks, and that the study of the latter is essential to bring mineral deposits into their proper perspective. Absorption in the details of a mineral deposit makes it easy for the investigator to forget or minimize these relations.

Nevertheless, in the study of mineral deposits, and especially deposits of the metallic minerals, certain geologic features stand out conspicuously against the common background indicated above. Our discussion of these features will follow the order of rock genesis indicated in the description of the metamorphic cycle.

NAMES

Any classification of mineral deposits on the basis of origin is more or less arbitrary. The sharp lines implied by the use of class names do not exist in nature. Mineral deposits are so complex and so interrelated in origin, that a classification according to genesis indicates only the essential and central class features; it does not sharply define the limits of the classes.

It is practically impossible for any geologist to present a classification which will be accepted without qualification by other geologists, although there may be agreement on essential features. Difficulties in reaching agreement are increased by the inheritance from the past of names, definitions, and classifications which do not exactly fit present conceptions based on fuller information,—but which, nevertheless, have become so firmly established in the literature that it is difficult to avoid their use. In the progress of investigation many new names are coined to fit more precisely the particular situation in hand, but only in fortunate cases do these new names stand up against the traditional currency and authority of old names. The geologist is often in despair in his attempt to express his ideas clearly and precisely, and at the same time to use terms which will be understandable by his readers and will not arouse needless controversy.

As illustrative of the above remarks reference may be made to a few terms commonly used in economic geology, such as *primary*, *secondary*, *syngenetic*, *epigenetic*, *supergene*, *hypogene*, *protore*, etc.

The most commonly used of these terms are *primary* and *secondary*. It is almost impossible to define them in a way which will cover all the conceptions for which they have been used, and yet in their context they have been very useful in conveying essential ideas. An ore formed by direct processes of sedimentation has sometimes been called primary, whereas an ore formed by later enrichment of these sediments has been called secondary. An ore formed directly by igneous processes has been called primary, while an ore formed by enrichment of such primary ore by later processes has been called secondary. It is clear, however, that these terms are merely relative, with application to a specific sequence, and that they do not fix the absolute position of the ore in a sequence applying to all ores. For instance, ores deposited directly as sediments or placers may be derived from the erosion of preëxisting ore bodies,—in which case it may sometimes be convenient to refer to the sedimentary ores or placers as secondary and the earlier ores as primary. Or a sulphide deposit originating through igneous agencies may undergo two or three successive enrichments, each successive one secondary to the preceding, but primary to the one following. In spite of these obvious difficulties, the terms primary and secondary may be entirely intelligible as indicating relative order of development under a given set of conditions.

The term *syngenetic* has been used for mineral deposits formed by processes similar to those which have formed the enclosing rocks and in general simultaneously with them, and *epigenetic* for those introduced into preëxisting rocks. In certain cases *syngenetic* may be roughly synonymous with *primary*, and *epigenetic* with *secondary*, and yet a primary ore may be epigenetic. For instance, zinc sulphides in the Mississippi valley limestones (pp. 54-55) are epigenetic, and yet are primary with reference to a later enrichment. The two sets of terms are meant to convey somewhat different ideas and are not interchangeable.

Ransome[4] has suggested, especially for vein and contact deposits, a series of names which has the considerable advantage of definiteness:—*hypogene ores*, formed in general by ascending non-oxidizing solutions, perhaps hot; *supergene ores*, formed in general by oxidizing and surface solutions, initially cold and downward moving; and *protores*, or metallized rock or vein substances which are too low in tenor to be classed as ores, but which would have been converted into ores had the enriching process been carried far enough. In this connection Ransome defines primary ore as unenriched material that can be profitably mined. In view of the general use of the terms primary and secondary as expressing a sequential relation of ore development, it is doubtful whether this more precise definition will supersede the older usage. Also it may be noted that commercial conditions might require, under these definitions, the designation of an ore as a protore at one time or place and as a primary ore at another. Hypogene ores are dominantly primary, and supergene ores are dominantly secondary, but either may include both primary and secondary ores.

The terms of these several classifications overlap, and seek to express different aspects of the same situation. While almost synonymous in certain applications they are not in others.

In this text the writer has certainly not escaped the difficulties in regard to names above referred to, nor in fact has he made any exceptional effort to do so. His chief purpose is to convey, in somewhat elementary terms, an understandable idea of the central features of economic geology. In the main, the most widely accepted terms are used. Almost at every turn it would be possible, in the interests of precision, to introduce qualifying discussions of names,—but at the expense of continuity and perspective in the presentation of the principal subject-matter. The writer does not wish to minimize the necessity for careful and precise nomenclature; but he regards it important that the student focus his attention on the central objective facts of the subject, and that he do not become misled by the sometimes over-strenuous advocacy of certain names or classifications in preference to others. If the facts are understood, he will ordinarily have no difficulty in judging the significance of the variety of names proposed to express these facts. If, on the other hand, the student approaches the subject with a ready-made set of names and definitions learned by rote, he is in danger of perceiving his facts from one angle only and through a distorted perspective.

MINERAL DEPOSITS AS MAGMATIC SEGREGATIONS IN IGNEOUS ROCKS

In this class are included deposits which crystallize within the body of igneous rock, almost, if not quite, simultaneously with the adjacent rock. These deposits form one of the main types of *syngenetic* deposits.

The titaniferous magnetites constitute a widely distributed but at present commercially unavailable class of iron ores. The magnetite crystals of these deposits interpenetrate with the other constituents of an igneous rock, commonly of a gabbro type, and the deposits themselves are essentially igneous rocks. Their shapes are for the most part irregular, their boundaries ill-defined, and their concentration varying. While their magmatic origin is clear, there is little agreement as to the precise conditions which determined their segregation in the molten rock. There is often a tendency for the ores to follow certain primary sheeted structures in the igneous mass, a fact for which the reason is not obvious.

The Sudbury nickel ores, of Ontario, Canada, the principal source of the world's nickel, lie mainly within and along the lower margin of a great intrusive igneous mass of a basic type called *norite*, and locally the ores project beyond the margin into adjacent rocks. Their textures and their intercrystallization with the primary minerals of the igneous rock have suggested that they are essentially a part of the norite mass, and that they crystallized during some segregative processes which were effective before the magma had solidified. Near the ores there are likely to be granitic rocks, which, like the ores, seem to be segregations from the norite magma. Locally both the ores and the associated granitic rocks replace the main norite body in such a fashion as to indicate their slightly later crystallization. However, the intimate association of the ores with the primary minerals in the magma, together with their absence from higher parts of the norite and from the extraneous rocks far from the contact, indicate to other investigators that they were not brought in from outside in vagrant solutions which followed the intrusion of the main magma, but that they were segregated within the magma essentially in place. The occurrence of these heavy ores near the base of the norite naturally suggests that they were segregated by sinking to the bottom of the molten magma, but this conclusion implies certain physical conditions of the magma which have not yet been proved. Again the precise nature of the process and the part played in it by aqueous and gaseous solutions are subject to some doubt and controversy. The settlement of this problem awaits the solution of the more general problem of the origin and crystallization of magmas.

In this general class of igneous deposits may be mentioned also diamonds, platinum, chromite, corundum, and other mineral products, although for the formation of commercial ores of many of these substances further concentration by weathering and sedimentation has been required.

Pegmatites are coarsely crystalline acid dike rocks which often accompany a large igneous intrusion and which have obviously crystallized somewhat later than the main igneous mass. They may constitute either sharply delimited dikes or more irregular bodies which grade into the surrounding igneous mass. They have a composition roughly similar to the associated igneous rock, but usually a different proportion of minerals. They are probably the result of the differentiation of the parent magma. The pegmatites are of especial interest to the economic geologist because of the frequency with which they carry commercial minerals, such as the precious stones, mica, feldspar, cassiterite (tin ore), and others. They show a complete gradation from dikes of definitely igneous characteristics to veins consisting largely of quartz in which evidence of igneous origin is not so clear. The pegmatites thus afford a connecting link between ores of direct igneous sources and ores formed as "igneous after-effects," which are discussed in the next paragraph. Aplites are fine-grained acid igneous rocks of somewhat the same composition as the pegmatites and often show the same general relations to ores.

MINERAL DEPOSITS WITHIN AND ADJACENT TO IGNEOUS ROCKS WHICH WERE FORMED IMMEDIATELY AFTER THE COOLING AND CRYSTALLIZATION OF THE MAGMAS THROUGH THE AGENCY OF HOT MAGMATIC SOLUTIONS.

These deposits are closely associated in place and age with igneous rocks, either intrusive or extrusive, and are usually considered to have come from approximately the same source; and yet they afford distinct evidence of having been deposited after the adjacent igneous rocks were completely crystallized and fractured. They are thus *epigenetic* deposits. They are not themselves igneous rocks and they do not constitute pegmatites, but they often grade into pegmatites and belong to the same general stage in the sequence of events. They include deposits formed by contact metamorphism. They are sometimes designated by the general term "igneous after-effects"—a term also applied in some cases to pegmatites. Some geologists discriminate between "deep vein" deposits (p. 43) and "contact-metamorphic" deposits, but the two are so closely related in place and origin that for our purposes they will be considered together.

The ores of this class are clearly deposited from vagrant solutions which wander through openings of all kinds in the igneous rock and outward into the adjacent country rocks. They also replace the wall rocks; limestone is especially susceptible. This is a phase of contact metamorphism. Some of the most important metalliferous deposits belong in this class, including most of the gold, silver, copper, iron, lead and zinc ores of the western United States and the copper deposits of Lake Superior.

In general, ores of this class are more abundant about intrusive igneous rocks, that is about igneous rocks which have stopped and cooled before reaching the surface,—than in association with extrusive igneous rocks which have poured over the surface as lava flows—but the latter are by no means insignificant, including as they do such deposits as the Lake Superior copper ores, the Kennecott copper ores of Alaska, some of the gold-silver deposits of Goldfield and other Nevada camps, and many others.

There is general similarity in the succession of events shown by study of ore bodies related to intrusives. First, the invasion of the magma, resulting in contact metamorphism of the adjacent rocks, sometimes with, and often without conspicuous crowding effects on the invaded rocks; second, the cooling, crystallization, and cracking of both the igneous rock and the adjacent rock; third, the introduction of ore-bearing solutions into these cracks,—sometimes as a single episode, sometimes as a long continued and complex process forming various types of minerals at successive stages. This order may in some cases be repeated in cycles, and overlapping of the successive events is a common feature.

One of the interesting facts is the way in which the igneous mass has invaded and extensively altered the country rocks in some mineral districts,—in some cases by crowding and crumpling them, and in others without greatly disturbing their structural attitudes. The latter is illustrated in the Bingham district of Utah and the Philipsburg district of Montana. In such cases there is so little evidence of crowding of the country rocks as to raise the question how such large masses of intrusives could be introduced without greater disturbing structural effect. This leads naturally to consideration of the general problem of the manner of progress of magmas through adjacent rocks,—a subject which is still largely in the realm of speculation, but which is not thereby eliminated from the field of controversy. Facts of this kind seem to favor the position of certain geologists that magmas may assimilate the rocks they invade.

EVIDENCE OF IGNEOUS SOURCE

No one ever saw one of these deposits in the process of formation; the conclusion, therefore, that they originated from hot solutions, either aqueous or gaseous, or both, which were essentially "after-effects" of igneous activity and came from the same primary source as the associated igneous rocks, is an inference based on circumstantial evidence of the kind below summarized:

(1) The close association both in place and age with igneous rocks. This applies not only to individual deposits, but to certain groups of deposits which have common characteristics, and which constitute a metallogenic province; also to groups of the same geologic age, which indicate a metallogenic epoch (pp. 308-309). The association with igneous rocks in one place might be a

coincidence but its frequent repetition can hardly be so explained. A zonal arrangement of minerals about intrusives is often noted. Geologic evidence often shows the processes of ore deposition to have been complete before the next succeeding geologic event,—as for instance in the Tonopah district of Nevada (p. 236), where the ores have been formed in relation to certain volcanic flows and have been covered by later flows not carrying ore, without any considerable erosion interval between the two events.

(2) The general contrast in mineralogical and chemical composition, texture, and mineral associations, between these ore minerals and the minerals known to be formed by ordinary surficial agencies under ordinary temperatures. The latter carry distinctive evidences of their origin. When, therefore, a mineral group is found which shows contrasting evidences, it is clear that some other agencies have been at work; and the natural assumption is that the solutions were hot rather than cold; that they came from below rather than above.

(3) The contrast between the character and composition of these ores (and their associated gangue) and the character and composition of the wall rocks, together with the absence of leaching of the wall rocks, favor the conclusion that the ore minerals are foreign substances introduced from extraneous sources. The source not being apparent above and the processes there observed not being of a kind to produce these results, it is concluded that the depositing solutions were hot and came from below.

(4) The fact that many of the ore minerals are never known to develop under ordinary temperatures at the surface. For some of them, experimental work has also indicated high temperature as a requisite to their formation.

Quartz, which is a common associate of the ores and often constitutes the principal gangue, serves as a geologic thermometer in that it possesses an inversion point or temperature above which it crystallizes in a certain form, below which in another. In deposits of this class it has often been found to crystallize at the higher temperatures.

The quartz sometimes shows bubbles containing liquid, gas, and small heavy crystals, probably of ferric oxide, as in the Clifton-

Morenci district of Arizona. It is clear that the ore-bearing solutions in these cavities, before the crystallization of the heavy mineral inclusions, held dissolved not only much larger quantities of mineral substances than can be taken up by water at ordinary temperatures, but also a substance like ferric oxide which is entirely insoluble under ordinary cool conditions.

(5) The association of the ores with minerals carrying fluorine and boron, with many silicate minerals, such as garnet, amphiboles, pyroxenes, mica (sericite) and others, and with other minerals which are known to be characteristic developments within or near igneous masses and which are not known to form under weathering agencies at the surface. Various characteristic groupings of these associated minerals are noted. In limestone much of the mass may be replaced by garnet and other silicates in a matrix of quartz. In igneous rock the ore-bearing solutions may have altered the wall rock to a dense mixture of quartz, sericite, and chlorite. Where sericite is dominant, the alteration is called sericitic alteration. Where chlorite is important, it is sometimes called chloritic or "propylitic" alteration. The chloritic phases are usually farther from the ore deposit than the sericitic phases, indicating less intense and probably cooler conditions of deposition. Locally other minerals are associated with the ores, as, for instance, in the Goldfield district of Nevada (p. 230), where alunite replaces the igneous rock. Alunite is a potassium-aluminum sulphate, which differs from sericite in that sulphur takes the place of silicon. In the quartzites of the lead-silver mines of the Coeur d'Alene district of Idaho (p. 212), siderite or iron carbonate is a characteristic gangue material replacing the wall rock.

Quartz in some cases, as noted above, gives evidence of high temperature origin and therefore of igneous association. Jasperoid quartz, as well illustrated in the Tintic district of Utah (p. 235), may show texture and crystallization suggestive of deposition from colloidal solution,—a process which can occur under both cold and hot conditions, but which is believed to be accelerated by heat.

Certain minerals, such as magnetite, ilmenite, spinel, corundum, etc., are often found as primary segregations within the mass of igneous rock. These and other minerals, including minerals of tin and tungsten, monazite, tourmaline, rutile, and various precious stones, are characteristically developed in pegmatites, which are known to be igneous rocks, crystallized in the later stages of igneous intrusion. When, therefore, such minerals are found in other ore deposits an igneous source is a plausible inference. For instance, in the copper veins of Butte, Montana (p. 201), are found cassiterite (tin oxide) and tungsten minerals. Their presence, therefore, adds another item to the evidence of a hot-water source from below.

(6) The occasional existence of hot springs in the vicinity of these ore deposits. Where hot springs are of recent age they may suggest by their heat, steady flow, and mineral content, that they are originating from emanations from the still cooling magmas. In the Tonopah camp (p. 236), cold and hot springs exist side by side, exhibiting such contrasts as to suggest that some are due to ordinary circulation from the surface and that others may have a deep source below in the cooling igneous rocks. This evidence is not conclusive. Hot springs in general fail to show evidence of ore deposition on any scale approximating that which must have been involved in the formation of this class of ore bodies. Much has been made of the slight amounts of metallic minerals found in a few hot springs, but the mineral content is small and the conclusion by no means certain that the waters are primary waters from the cooling of igneous rocks below.

In this connection the mercury deposits of California (p. 259), contribute a unique line of evidence. In areas of recent lavas, mercuric sulphide (cinnabar) is actually being deposited from hot springs of supposed magmatic origin, the waters of which carry sodium carbonate, sodium sulphide, and hydrogen sulphide,—a chemical combination known experimentally to dissolve mercury sulphide. The oxidation and neutralization of these hot-spring solutions near the surface throws out the mercury sulphide. At the same time the sulphuric acid thus formed extensively leaches and bleaches the surrounding rocks. Such bleaching is common about mercury deposits. When it is remembered that the mercury deposits contain minor amounts of gold and silver and sulphides of other metals; that they are closely associated with gold and silver deposits; and further that such gold, silver, and other sulphide deposits often contain minor amounts of mercury,—it is easy to assume the possibility that these minerals may likewise have had their origin in hot solutions from below. The presence of mercury in a deposit then becomes suggestive of hot-water conditions.

(7) Ores sometimes occur in inverted troughs indicating lodgment by solutions from below, as, for instance, in the saddle-reef gold ores of Nova Scotia and Australia, and in certain copper ores of the Jerome camp of Arizona (p. 204.) This occurrence does not indicate whether the solutions were hot or cold, magmatic or meteoric, but in connection with other evidences has sometimes been regarded as significant of a magmatic source beneath.

Perhaps no one of these lines of evidence is conclusive; but together they make a strong case for the conclusion that the solutions which deposited the ores of this class were hot, came from deep sources, and were probably primary solutions given off by cooling magmas.

The conclusion that some ores are derived from igneous sources, based on evidence of the kind above outlined, does not mean necessarily that the ore is derived from the immediately adjacent part of the cooling magma. In fact the evidence is decisive, in perhaps the majority of cases, that the source of the mineral solutions was somewhat below; that these solutions may have originated in the same melting-pot with the magma, but that they came up independently and a little later,—perhaps along the same channels, perhaps along others.

Possible Influence of Meteoric Waters in Deposition of Ores of This Class

It is hardly safe, with existing knowledge, to apply the above conclusion to all ore deposits with igneous associations, or in any case to eliminate entirely another agency,—namely, ground-waters of surface or meteoric origin, which are now present and may be presumed often to have been present in the rocks into which the ores were introduced. Such waters may have been heated and started in vigorous circulation by the introduction of igneous masses, and thereby may have been enabled to effectively search out and segregate minutely disseminated ore particles from wide areas. This has been suggested as a probability for the Kennecott copper ores of Alaska (p. 200) and for the copper ores of Ely, Nevada. In the Goldfield camp (p. 230) the ores are closely associated with alunite in such a manner as to suggest a common origin. It has been found difficult to explain the presence of the alunite except through the agency of surface oxidizing waters acting on hydrogen sulphide coming from below.

In the early days of economic geology there was relatively more emphasis on the possible effectiveness of ground-waters in concentrating ores of this type. With the recognition of evidence of a deeper source related to magmas, the emphasis has swung rapidly to the other extreme. While the evidence is sound that the magmatic process has been an important one, it is difficult to see how and to just what extent this process may have been related to the action of ground-waters,—which were probably present in a heated condition near the contact. It may never be possible to discriminate closely between these two agencies. It seems likely that at some stages the two were so intimately associated that the net result of deposition cannot be specifically assigned either to one or to the other.

Zonal Arrangement of Minerals Related to Igneous Rocks

Evidence is accumulating in many mining districts that ore deposits of these igneous associations were deposited with a rough zonal arrangement about the igneous rock. At Bingham, Tintic, and Butte (pp. 204, 208, 235), copper ores are on the whole closest to the igneous rock, and the lead, zinc, and silver ores are farther away. Furthermore, the quartz gangue near the igneous rock is likely to contain minerals characteristic of hot solutions, while farther away such minerals as dolomite and calcite appear in the gangue, suggestive of cooler conditions. In Cornwall (p. 262), tin ores occur close to the intrusives, and lead-silver ores farther away. The gradations are by no means uniform; shoots of one class of ore may locally cut abruptly across or through those of another class.

The existence of zones horizontally or areally arranged about intrusives suggests also the possibility of a vertical zonal arrangement with reference to the deep sources of the solutions. Of course when secondary concentration from the surface, described later, is taken into account, there may be a marked zonal distribution in a vertical direction, but this is not primary zoning. A few veins and districts show evidence of vertical zoning apparently related to primary deposition; for the most part, however, in any one mine or camp there is yet little evidence of primary vertical zoning. On the other hand, certain groups of minerals are characteristic of intense conditions of heat and pressure, as indicated by the coarse recrystallization and high degree of metamorphism of the rocks with which they are associated; and other groups have such associations as to indicate much less intense conditions of temperature and pressure. Depth is only one factor determining intensity of conditions, but it affords a convenient way to indicate them; so mineral deposits associated with igneous rocks are sometimes classified by economic geologists on the basis of deep, intermediate, and shallow depths of formation.

There are a considerable number of minerals which are formed in all three of these zones, although in differing proportions. There are comparatively few which are uniformly characteristic of a single zone. On the whole, it is possible to contrast satisfactorily mineral deposits representing very intense metamorphic conditions, usually associated with formation at great depth, with those formed at or near the surface; but there are many deposits with intermediate characteristics which it is difficult to place satisfactorily.

The accessible deposits of the deep zone are associated with plutonic igneous rocks which have been deeply eroded, and not with surface lavas. They are characterized by minerals of gold, tin, iron, titanium, zinc, and copper, and sometimes of tungsten and molybdenum, in a gangue of quartz, which contains also minerals such as garnet, corundum, amphibole, pyroxene, tourmaline, spinel, and mica. The deep-zone minerals are not unlike the pegmatite minerals in their grouping and associations.

Deposits formed at shallow depths are related to extrusive rocks and to intrusives near the surface. Erosion has not been deep. Mercury, silver and gold (tellurides, native metals, and silver sulphides), antimony, lead, and zinc minerals are characteristic, together with alunite, adularia, and barite. Metallic copper also is not infrequent. Very often the gangue material is more largely calcite than quartz, whereas calcite is not present in the deep zone.[5]

The trend of evidence in recent years has favored the conclusion that the principal ores associated with igneous rocks have not developed at very great depths. Even within our narrow range of observation there is a difference in favor of the shallower depths, and the greatest depths we can observe are after all but trivial on the scale of the earth.

A survey of the ore deposits of Utah has suggested the generalization that ores are more commonly related to intrusive stocks than to the forms known as laccoliths, and that within and about intrusive stocks the ores are much more abundant near the top or apex of a stock than lower down.[6] In parts of the region where erosion has removed all but the deeper portions of the stocks, ore bodies are less abundant. It will be of interest to follow the testing of this generalization in other parts of the world.

The scientist is constantly groping for underlying simple truth. Such glimpses of order and symmetry in the distribution of ore around igneous rocks as are afforded by the facts above stated, tempt the imagination to a conception of a simple type or pattern of ore distribution around intrusions. For this reason we should not lose sight of the fact that, in the present state of knowledge, the common and obvious case is one of irregular and heterogeneous distribution, and that there are many variations and contradictions even to the simplest generalization that can be made. The observer is repeatedly struck by the freakish distribution of ores about igneous masses, as compared with their regularity of arrangement under sedimentary processes to be discussed later. It is yet unexplained how an intrusive like the Butte granite can produce so many different types of ores at different places along its periphery or within its mass, and yet all apparently under much the same general conditions and range of time. It is difficult also to discern the laws under which successive migrations of magma, from what seems to be a single deep-seated source or melting-pot, may carry widely contrasting mineral solutions. Far below the surface, beyond our range of observation, it is clear that there is a wonderful laboratory for the compounding and refinement of ores, but as to its precise location and the nature of its processes we can only guess.

Other features of distribution of minerals associated with igneous rocks are indicated by their grouping in metallogenic provinces and epochs (see pp. 308-309).

THE RELATION OF CONTACT METAMORPHISM TO ORE BODIES OF THE FOREGOING CLASS.

The deposition of ores of igneous source in the country rock into which the igneous rocks are intruded is a phase of contact metamorphism. Ordinarily where this deposition occurs there are further extensive replacements and alterations of the country rock, resulting in the development of great masses of quartz, garnet, pyroxene, amphibole, and other silicates, and in some cases of calcite, dolomite, siderite, barite, alunite, and other minerals. Looked at broadly, the deposition of ores at igneous contacts under contact metamorphism is a mere incident in the much more widespread and extensive alterations of this kind. Hence it is that the subject of contact metamorphism is of interest to economic geologists. The minerals here formed which do not constitute ores throw much light on the nature of the ore-bearing solutions, the conditions of temperature and pressure, and the processes which locally and incidentally develop the ore bodies. The subject, however, is a complex one, the full discussion of which belongs in treatises on metamorphism.[7] We may note only a few salient features.

For many hundreds of yards the rocks adjacent to the intrusions may be metamorphosed almost beyond recognition. This is especially true of the limestone, which may be changed completely to solid masses of quartz and silicates. The shales and sandstones are ordinarily less vitally affected. The shales become dense, highly crystalline rocks of a "hornstone" type, with porphyritic developments of silicate minerals. The sands and sandstones become highly crystalline quartzites, spotted with porphyritic developments of silicates. Occasionally even these rocks may be extensively replaced by other minerals, as in the Coeur d'Alene district, where quartzites adjacent to the ore veins may be completely replaced by iron carbonate.

A question of special interest to economic geologists is the source of the materials for the new minerals in these extensively altered zones. In some cases the minerals are known to be the result of recrystallization of materials already in the rock, after the elimination of certain substances such as carbon dioxide and lime, under the pressures and temperatures of the contact conditions. In such cases there has obviously been large reduction in volume to close the voids created by the elimination of substances. In the majority of cases, the new substances or minerals are clearly introduced from the igneous source, replacing the wall rock volume for volume so precisely that such original textures and structures as bedding are not destroyed. In many cases the result is clearly due to a combination of recrystallization of materials already present and introduction of minerals by magmatic solutions from without. So obvious is the evidence of the introduction of materials from without, that there has been a tendency in some quarters to overlook the extensive recrystallization of substances already present; and the varying emphasis placed on these two processes by different observers has led to some controversy.

SECONDARY CONCENTRATION IN PLACE OF THE FOREGOING CLASSES OF MINERAL DEPOSITS THROUGH THE AGENCY OF SURFACE SOLUTIONS

Mineral deposits of direct magmatic segregation are seldom much affected by surficial alteration, perhaps because of their coarse crystallization and their intermingling with resistant crystalline rocks. Mineral deposits of the "igneous after-effect" type may be profoundly altered through surficial agencies. The more soluble constituents are taken away, leaving the less soluble. The parts that remain are likely to be converted into oxides, carbonates, and hydrates, through reaction with oxygen, carbon dioxide, and water, which are always present at the surface and at shallow depths. These processes are most effective at the surface and down to the level of permanent ground-water, though locally they may extend deeper. This altered upper part of the ore bodies is usually called the *oxide zone*. It may represent either an enrichment or a depletion of ore values, depending on whether the ore minerals are taken into solution less rapidly or more rapidly than the associated minerals and rocks; all are removed to some extent. In certain deposits, there is evidence that both zinc and copper have been taken out of the upper zone in great quantity; but they happen to be associated with limestone, which has dissolved still more rapidly, with the result that there is a residual accumulation of copper and zinc values. Manganese, iron, and quartz are usually more resistant than the other minerals and tend to remain concentrated above. The same is true to some extent of gold and silver. The abundance of iron oxide thus left explains the name "iron cap" or "gossan" so often applied to the upper part of the oxide zone. Not infrequently, and especially in copper ores, the upper part of the oxide zone is nearly or entirely barren of values and is called the *capping*.

The depth or thickness of the oxide zone depends on topography, depth of water table, climatic conditions, and speed of erosion. A fortunate combination of conditions may result in a deep oxide zone with important accumulations of values. In other cases erosion may follow oxidation so rapidly as to prevent the growth of a thick oxide zone.

It is clear from the study of many ore deposits that the process of oxidation has not proceeded uniformly to the present, but has depended upon a fortunate combination of factors which has not been often repeated during geologic time. As illustrative of this, the principal oxidation of the Bisbee copper ores of Arizona (p. 204) occurred before Tertiary time, with reference to a place that has since been covered by later sediments. The conditions in the Ray, Miami, and Jerome copper camps of Arizona (pp. 203-205) likewise indicate maximum oxidation at an early period. The Lake Superior iron ore deposits (pp. 167-170) were mainly concentrated before Cambrian time, during the base-leveling of a mountainous country in an arid or semi-arid climate. The oxide zone of these deposits has no close relation to the present topography or to the present ground-water level. In the Kennecott (Alaska) copper deposits all oxidation has been stopped since glacial time by the freezing of the aqueous solutions. At Butte and at Bingham the main concentration of the ores is believed to have occurred in an earlier physiographic cycle than the present one. The *cyclic* nature of the formation of oxide zones is of comparatively recent recognition, and much more will doubtless be found out about it in the comparatively near future. Its practical bearing on exploration is obvious (see p. 325).

It should be clearly recognized that oxidizing processes are not limited to the zone above the ground-water level. Locally oxidizing solutions may penetrate and do effective work to much greater depths, especially where the rocks traversed at higher elevations are of such composition or in such a stage of alteration as not to extract most of the oxygen. Consequently the presence of oxide ores below the water table is not necessarily proof that the water table has risen since their formation. On the other hand, the facts of observation do indicate generally a marked difference, in circulation and chemical effect, between waters above and below this horizon, and show that oxidation is dominantly accomplished above rather than below this datum surface.

During the formation of the oxide zone, erosion removes some of the ore materials entirely from the area, both mechanically and in solution. Part of the material in solution, however, is known to penetrate downward and to be redeposited in parts of the ore body below the oxide zone,—that is, usually below the water table. Evidence of this process is decisive in regard to several minerals. Copper is known to be taken into solution as copper sulphate at the surface, and to be redeposited as chalcocite where these sulphate solutions come in contact with chalcopyrite or pyrite below. Not only has the process been duplicated in the laboratory, but the common coating of chalcocite around grains of pyrite and chalcopyrite below the water level indicates that this process has been really effective. Sulphides of zinc, lead, silver, and other metals are similarly concentrated, in varying degrees. The zone of deposition of secondary sulphides thus formed is called the zone of *secondary sulphide enrichment*. Ores consisting mainly of secondary sulphides are also called *supergene* ores (p. 33). In some deposits, as in the copper deposits of Ray and Miami, there is found, below the secondary sulphide zone, a lean sulphide zone which is evidently of primary nature. The mineralized material of this zone, where too lean to mine, has been called a *protore*.

With the discovery of undoubted evidence of secondary sulphide enrichment, there was a natural tendency to magnify its importance as a cause of values. Continued study of sulphide deposits, while not disproving its existence and local importance, has in some districts shown clearly that the process has its limitations as a factor in ore concentration, and that it is not safe to assume its effectiveness in all camps or under all conditions. At Butte for instance, secondary chalcocite is clearly to be recognized. The natural inference was that as the veins were followed deeper the proportion of chalcocite would rapidly diminish, and that a leaner primary zone of chalcopyrite, enargite and other primary minerals would be met. However, the great abundance of chalcocite in solid masses which have now been proved to a depth of 3500 feet, far below the probable range of waters from the surface in any geologic period, seems to indicate that much of the chalcocite is primary. The present tendency at Butte is to consider as secondary chalcocite only certain sooty phases to be found in upper levels. The solid masses of chalcocite in the Kennecott copper mines seem hardly explainable as the result of secondary sulphide enrichment. No traces of other primary minerals are present and the chalcocite here is regarded as probably primary.

The possible magnification of the process of secondary enrichment above referred to has had for its logical consequence a tendency to over-emphasize the persistence of primary ores in depth. The very use of the terms "secondary" and "primary" has suggested antithesis between surficial and deep ores. Progress in investigation, as indicated on previous pages, seems to indicate that the primary ores are not uniformly deep and that in many cases they are distinctly limited to a given set of formations or conditions comparatively near the surface.

In general the processes of oxidation and secondary sulphide enrichment have been studied mainly by qualitative methods with the aid of the microscope and by considerations of possible chemical processes. These methods have disclosed the nature but not the quantitative range and relations of the different processes. Much remains to be done in the way of large scale quantitative analysis of ores at different depths, as a check to inferences drawn by other methods. One may know, for instance, that a mineral is soluble and is actually removed from the oxide zone and redeposited below. The natural inference, therefore, is that the mineral will be found to be depleted above and enriched below. In many cases its actual distribution is the reverse,—indicating that this process has been only one of the factors in the net result, the more rapid solution and deposition of other materials being another factor. If one were to approach the study of the concentration of iron ores with the fixed idea of insolubility of quartz from a chemical standpoint, and were to draw conclusions accordingly, he would fail to present a true picture of the situation. While quartz is insoluble as compared with most minerals, it is nevertheless more soluble than iron oxide, and therefore the net result of concentration at the surface is to accumulate the iron rather than the silica. Descriptions of enrichment processes as published in many reports are often misleading in this regard. They may be correct in indicating the actual existence of a process, but may lead the reader to assumptions as to net results which are incorrect.

RESIDUAL MINERAL DEPOSITS FORMED BY THE WEATHERING OF IGNEOUS ROCKS IN PLACE

Igneous rocks not containing mineral deposits may on weathering change to mineral deposits. The lateritic iron ores such as those of Cuba (p. 172), many bauxite deposits, many residual clays, and certain chromite and nickel deposits are conspicuous representatives of this class. The chemical and mineralogical changes involved in the formation of these deposits are pretty well understood. Certain constituents of the original rock are leached out and carried away, leaving other constituents, as oxides and hydrates, in sufficiently large percentage in the mass to be commercially available. The accumulation of large deposits depends on the existence of climatic and erosional conditions which determine that the residual deposit shall remain in place rather than be carried off by erosion as fast as made. In the glaciated parts of the world, deposits of this nature have usually been removed and dispersed in the glacial drift.

When the minerals of these deposits are eroded, transported, and redeposited in concentrated form, they come under the class of placer or sedimentary deposits described under the following heading. There are of course many intermediate stages, where the residual deposit is only locally moved and where the distinction between this class of deposits and that next described is an arbitrary one.

MINERAL DEPOSITS FORMED DIRECTLY AS PLACERS AND SEDIMENTS

Mineral deposits of this class are of large value, including as they do salt, gypsum, potash, sulphur, phosphates, nitrates, and important fractions of the ores of iron, manganese, gold, tin, tungsten, platinum, and precious stones; also many common rocks of commercial importance. The minerals of these deposits are derived from the weathering and erosion of land surfaces, either igneous or sedimentary. They are deposited both under air and under water, both mechanically and chemically (in part by the aid of organisms). These deposits form the principal type of *syngenetic* deposits (p. 32); the term *sedigenetic* deposits has also been applied to them.

MECHANICALLY DEPOSITED MINERALS

Mechanical erosion of preëxisting mineral deposits or rocks and their transportation, sorting, and deposition are responsible for the placers of gold, tin, tungsten, platinum, and various precious stones, and for certain iron sands and conglomerates. Sands, sandstones, shales, and certain clays and bauxites also belong in this group. These deposits may be formed under air or under water, and under various climatic and topographic conditions. During the process of formation the minerals of differing density are more or less sorted out and tend to become segregated in layers. The process is not unlike the artificial process of mechanical concentration where ores are crushed, shaken up, and treated with running water. The process is most effective for minerals which are resistant to abrasion and to solution, and of such density as to differentiate them from the other minerals of the parent rock.

The origin of deposits of this kind is fairly obvious where they are of recent age and have not been subsequently altered or buried. A considerable amount of experimental work has brought out clearly the main elements of the processes. Physiographic and climatic conditions play an important part, and cannot be safely overlooked by anyone studying such deposits.

Extensive copper deposits exist as sediments (pp. 205-206). It is not clear to what extent they are mechanically or to what extent chemically deposited. For the most part the concentration of copper in this manner has not been sufficient to yield deposits of large commercial value; the mineral is too much dispersed. Relatively small amounts are mined in the Mansfield shales of Germany and the Nonesuch shales and sandstones of the Lake Superior country.

The Clinton and similar iron ores of the United States and Newfoundland, the pre-Cambrian iron ores of Brazil, and the Jurassic iron ores of England and western Europe (pp. 166-167) are now commonly agreed to be direct sedimentary deposits in which mechanical agencies of sorting and deposition played a considerable part. How far chemical and bacterial agencies have also been effective is not clear. The climatic, topographic, and other physiographic and sedimentary conditions which cause the deposition of this great group of ores present one of the great unsolved problems of economic geology. The study of present-day conditions of deposition affords little clue as to the peculiar combination of conditions which was necessary to accomplish such remarkable results in the past.

On the whole, minerals of this mechanically deposited group are not greatly affected by later surficial alteration and concentration, because, having already been subjected to weathering, they are in a condition to resist such influences.

Chemically and Organically Deposited Minerals

The products of surface weathering and erosion are in part carried away in chemical solution and redeposited as sediments. Sediments thus formed include limestone and dolomite, siderite, salt, gypsum, potash, sulphur, phosphates, nitrates, and other minerals. Precipitation may be caused by chemical reactions, by organic secretion, or by evaporation of the solutions. The processes are qualitatively understood and it is usually possible to ascertain with reasonable accuracy the conditions of depth of water, relation to shore line, climate, nature of erosion, and other similar factors; yet the vast scale of some of these deposits, and their erratic areal and stratigraphic distribution, present unsolved problems as to the precise combinations of factors which have made such results possible.

Chemically and organically deposited minerals of this class are usually susceptible to further alteration by surface weathering, and some of them, for instance the phosphates and siderites, are thus secondarily concentrated. These processes are discussed under the next heading.

In general the great unsolved problem of the origin of the entire group of mineral deposits in placers and sediments relates to the scale of the results. Observation of present-day processes and conditions of deposition of these minerals affords satisfactory evidence of their nature, but fails to give us a clear idea of the precise combinations of agencies and conditions necessary to produce such vast results as are represented by the mineral deposits. For example, solution of iron on a land surface and redeposition in bogs and lagoons (as actually observed to be taking place today) show how some iron-ore sediments may be formed; but these processes are entirely inadequate to explain the deposition of iron ores in thick masses over broad areas without intermingling of other sediments—as represented by the Clinton iron ores of North America, the Jurassic ores of Europe and England, and the ancient iron ores of Brazil. The Paleozoic seas in northern and eastern United States encroached over land areas to the north and east and deposited ordinary sediments such as sandstone, shale, and limestone. Suddenly, without, so far as known, tapping any new sources of supply on the ancient land areas, and without any yet ascertainable change in topographic or climatic conditions, they deposited enormous masses of iron ore. There is clearly some cyclic factor in the situation which we do not yet understand.

The various deposits of salt, gypsum, potash, sulphur, and other minerals are known to be the result of evaporation, and the deposition of each of these minerals is known to be related to the degree of evaporation as well as to temperature, pressure, and factors such as mass action and crystallization of double salts. The nature of the processes is fairly well understood; but again, observation of the present-day operation of these processes fails to give us much clue to the enormous accumulations at certain times and places in the past. It is difficult to say just what conditions of climate, in combination with particular physiographic factors, could have preserved uniformity of conditions for the long periods necessary to account for some of the enormously thick salt deposits. Again some cyclic factor in the situation remains to be worked out.

SEDIMENTARY MINERAL DEPOSITS WHICH HAVE REQUIRED FURTHER CONCENTRATION TO MAKE THEM COMMERCIALLY AVAILABLE

The conditions for the direct deposition of sedimentary mineral deposits of the foregoing class are also responsible for the deposition of minerals in more dispersed or disseminated form, requiring further concentration through surface agencies to render them commercially available. Some of these deposits are discussed below.

The lead and zinc ores of the Mississippi Valley, Virginia, Tennessee, Silesia, Belgium, and Germany (pp. 211-212, 216-219) are in sedimentary rocks far removed from igneous sources. Lead and zinc were deposited in more or less dispersed form with the enclosing sediments. It is supposed that deposition was originally chemical and was favored by the presence of organic material, which is a rather common accompaniment of the sediments. It is supposed further that these organic participants were originally localized during sedimentation in so-called estuarine channels and shore-line embayments. When subsequently exposed to weathering, the lead and zinc minerals were dissolved and redeposited in more concentrated form in fissures and as replacements of limestone.

Agreement as to origin of these deposits, so far as it exists, does not go beyond these broad generalizations. There is controversy as to whether the original sources of the ore minerals were the sediments directly above, from which the mineral solutions have been transferred downward during weathering and erosion, or whether the original minerals were below and have been transferred upward by artesian circulation, or whether they were situated laterally and have been brought to their present position by movement along the beds, or whether there has been some combination of these processes. It is the writer's view that the evidence thus far gathered favors on the whole the conclusion of direct downward concentration from overlying sources which have been removed by erosion, although this conclusion fails to explain why certain sulphide deposits give so little evidence of important downward transfer from their present position. This matter is further discussed on pages 216-219. The choice of the various alternatives has some practical bearing on exploration.

Since these ores were brought into approximately their present position, they have undergone considerable oxidation near the surface and secondary sulphide enrichment below. The chemical and mineralogical changes are pretty well understood, but the quantitative range of these changes and their relative importance in determining the net result are far from known. Undoubted evidence of secondary sulphide enrichment has led in some quarters to an assumption of effectiveness in producing values which is apparently not borne out by quantitative tests.

A group of mineral deposits in sandstones in Utah is regarded as due to chemical concentration of material originally disseminated in the rock. They include silver, copper, manganese, uranium, and radium deposits. The Silver Reef deposits, including silver, copper, uranium, and vanadium, are commercially the most important of this type.[8] The ore minerals are commonly associated with carbonized material representing plant remains, and have replaced the calcareous and cementing material of the rock, and also some of the quartz grains. The deposits are regarded as having been formed by circulating waters which collected the minerals disseminated through the sedimentary rocks, and deposited them on contact with carbonaceous matter, earlier sulphides, or other precipitating agents. The circulation in some places is believed to have been of artesian character and to have been controlled to a large extent by structural features. The Silver Reef deposits are near the crest of a prominent anticline. Most of the minerals have been later altered by surface solutions.

Another great group of ores to be considered under this head are the iron ores of Lake Superior,—which were originally deposited as sediments, called jaspers or iron formations, with too low a percentage of iron to be of use, and which have required a secondary concentration by surficial agencies to render them valuable. The process of concentration has been a simple one. The iron minerals have been oxidized in place and the non-ferrous minerals have been leached out, leaving iron ores. This process contrasts with the concentration described above, in that there is little evidence of collection of iron minerals from disseminated sources. The Lake Superior iron ores are essentially residual concentrations in place. The outstanding problems of secondary concentration relate to the structural features which determined the channels through which the oxidizing and leaching waters worked, and to the topographic and climatic conditions which existed at the time the work was done. As with many other classes of ores, it was first assumed that these processes were related to the present erosion surface; but it is now known that concentration happened long ago under conditions far different from those now existing. These deposits contribute to the rapidly accumulating evidence of the *cyclic* nature of ore concentration.

Our least satisfactory knowledge of the Lake Superior ores relates to the peculiar conditions which determined the initial stage of sedimentation of the so-called iron formation. As in the case of the Clinton iron ores, no present-day sedimentation gives an adequate clue. Students of the problem have fallen back on the association of the iron formation with contemporaneous volcanic rocks, as affording a possible explanation of the wide departure from ordinary conditions of sedimentation evidenced by these formations.[9]

Coal deposits are direct results of sedimentation of organic material. They are mainly accumulations of vegetable matter in place. To make them available for use, however, they undergo a long period of condensation and distillation. Conditions of primary deposition may be inferred from modern swamps and bogs; but, as in the case of sediments described under the preceding heading, we are sometimes at a loss to explain the magnitude of the process, and especially to explain the maintenance of proper surface conditions of plant growth and accumulation for the long periods during which subsidence of land areas and encroachment of seas are believed to have been taking place. The processes of secondary concentration are also understood qualitatively, but much remains to be learned about the influences of pressure and heat, the effect of impervious capping rocks, and other factors.

Various oil shales and asphaltic deposits are essentially original sediments which have subsequently undergone more or less decay and distillation. The migration of the distillates to suitable underground reservoirs is responsible for the accumulation of oil and gas pools.

Oil and gas are distillates from these oil shales and asphaltic deposits, and also from other organic sediments such as carbonaceous limestones. The distillates have migrated to their present positions under pressure of ground-waters. The stratigraphic horizons favorable to their accumulation are generally recognized. The geologist is concerned in identifying these horizons and in ascertaining where they exist underground. He is further concerned in analysis of the various structural conditions which will give a clue to the existence of local reservoirs in which the oil or gas may have been accumulated. So capricious are the oil migrations that the most intensive study of these conditions still leaves vast undiscovered possibilities.

ANAMORPHISM OF MINERAL DEPOSITS

Mineral deposits formed in any one of the ways indicated above may undergo repeated vicissitudes, both at the surface and deep below the surface, with consequent modifications of character. They may be cemented or replaced by introduction of mineral solutions from without. They may be deformed by great earth pressures, undergoing what is called dynamic metamorphism (pp. 25-27), which tends to distort them and give them schistose and crystalline characters. They may be intruded by igneous rocks, causing considerable chemical, mineralogical, and structural changes. All these changes may take place near the surface, but on the whole they are more abundant and have more marked effects deep below the surface.

In general all these changes of the deeper zone tend to make the rocks more crystalline and dense and to make the minerals more complex. Cavities are closed. The process is in the main an integrating and constructive one which has been called *anamorphism*, to contrast it with the disintegrating and destructive processes near the surface, which have been called *katamorphism* (see also pp. 27-28). There is little in the process of anamorphism in the way of sorting and segregation which tends to enrich and concentrate the metallic ore bodies. On the contrary the process tends to lock up the valuable minerals in resistant combinations with other substances, making them more difficult to recover in mining. Later igneous intrusions or the ordinary ground-waters may bring in minerals which locally enrich ores under anamorphic conditions, but these are relatively minor effects. An illustration of the general effect is afforded by a comparison of the Cuban iron ores, which are soft and can be easily taken out, with the Cle Elum iron ores of Washington, which seem to be of much the same origin, but which have subsequently been buried by other rocks and rendered hard and crystalline. In the first case the ores can be mined easily and cheaply with steam shovels at the surface. In the second, underground methods of mining are required, which cost too much for the grade of ore recovered.

On the other hand, the same general kind of anamorphic processes, when applied to coal, result in concentration and improvement of grade. The same is true up to a certain point in the concentration of oil; but where the process goes too far, the oil may be lost (pp. 140-141).

CONCLUSION

Mineral deposits are formed and modified by practically all known geologic processes, but looked at broadly the main values are produced in three principal ways:

(1) As after effects of igneous intrusion, through the agency of aqueous and gaseous solutions given off from the cooling magma.

(2) Through the sorting processes of sedimentation,—the same processes which form sandstone, shale, and limestone. Organic agencies are important factors in these processes.

(3) Through weathering of the rock surface in place, which may develop values either by dissolving out the valuable minerals and redepositing them in concentrated form, or by dissolving out the non-valuable minerals and leaving the valuable minerals concentrated in place. The latter process is by far the more important.

The overwhelming preponderance of values of mineral deposits as a whole is found in the second of the classes named.

Under all these conditions it appears that the maximum results are obtained at and near the surface. On the scale of the earth even the so-called deep veins may be regarded as deposits from solutions reaching the more open and cooler outer portions of the earth. However, valuable mineral deposits are found in the deepest rocks which have been exposed by erosion, and the question of what would be found at still greater depths, closer to the center of the earth, is a matter of pure speculation.

Ultimately all minerals are derived from igneous sources within the earth. The direct contributions from these sources are only in small part of sufficient concentration to be of value; for the most part they need sorting and segregation under surface conditions.

We can only speculate as to causes of the occurrence of valuable minerals in certain igneous rocks and not in others. Many granites are intruded into the outer shell of the earth, but only a few carry "minerals"; also, of a series of intrusions in the same locality, only one may carry valuable minerals. It is clear that in some fashion these minerals are primarily segregated within the earth. Causes of this segregation are so involved with the problem of the origin of the earth as a whole that no adequate explanation can yet be offered. Our inductive reasoning from known facts is as yet limited to the segregation within a given mass of magma, and even here the conditions are only dimly perceived. A discussion of these ultimate problems is beyond the scope of this book.

FOOTNOTES:

[4] Ransome, Frederick Leslie, Copper deposits near Superior, Arizona: *Bull. 540, U. S. Geol. Survey*, 1914, pp. 152-153; The copper deposits of Ray and Miami, Arizona: *Prof. Paper 115, U. S. Geol. Survey*, 1919, p. 156; Discussion: *Econ. Geol.*, vol. 8, 1913, p. 721.

[5] For more specific definitions of vertical zones of ore deposition in association with igneous rocks see Spurr, J. E., Theory of ore deposition: *Econ. Geol.*, vol. 7, 1912, pp. 489-490; Lindgren, W., *Mineral deposits*, McGraw-Hill Book Co., 2d ed., 1919, Chapters XXIV-XXVI; and Emmons, W. H., *The principles of economic geology*, McGraw-Hill Book Co., 1918, Chapters VI-VIII.

An excellent discussion of a case of vertical and areal zoning of minerals is contained in *Ore deposits of the Boulder batholith of Montana*, by Paul Billingsley and J. A. Grimes, Bull. Am. Inst. Min. Engrs., vol. 58, 1918, pp. 284-368.

[6] Butler, B. S., Loughlin, G. F., Heikes, V. C., and others, The ore deposits of Utah: *Prof. Paper 111, U. S. Geol. Survey*, 1920, p. 201.

[7] Leith, C. K., and Mead, W. J., *Metamorphic Geology*, Pt. 2, Henry Holt and Company, New York, 1915.

[8] Butler, B. S., Loughlin, G. F., Heikes, V. C., and others, The ore deposits of Utah: *Prof. Paper 111, U. S. Geol. Survey*, 1920, pp. 152-158.

[9] Van Hise, C. R., and Leith, C. K., Geology of the Lake Superior region. *Mon. 52, U. S. Geol. Survey*, 1911, pp. 506-518; and references there given.

\

CHAPTER IV

MINERAL RESOURCES—SOME GENERAL QUANTITATIVE CONSIDERATIONS

Of the 1,500 known mineral species, perhaps 200 figure in commerce as mineral resources.

For the mineral substances used commercially, the term "mineral" is used in this chapter with a broad significance to cover any or all of the materials from which the needed elements are extracted,—whether these materials be single minerals or groups of minerals; whether they be rocks or ores; whether they be liquid or solid.

The following figures are generalizations based on the miscellaneous information available. The purpose is to indicate the general perspective rather than the detail which would be necessary for precise statement.

WORLD ANNUAL PRODUCTION OF MINERALS IN SHORT TONS

Exclusive of water, but inclusive of petroleum, the world's annual output of mineral resources amounts to two billions of tons. This figure refers to the crude mineral as it comes from the ground and not to the mineral in its concentrated form.

Of this total extraction, coal amounts to nearly 70 per cent, stone and clay 10 per cent, iron ore about 9 per cent, petroleum 4 per cent, copper ore 3 per cent, and all the remaining minerals constitute less than 6 per cent.

If spread out on the surface in a uniform mass with an estimated average density based on relative proportions of the crude minerals, this annual production would cover a square mile to a depth of 2,300 feet.

Of the total annual production 85 per cent comes from countries bordering the North Atlantic basin; 75 per cent is accounted for by the United States, England, and Germany; the United States has 39 per cent of the total, England 18 per cent, and Germany 18 per cent. By continents, Europe accounts for nearly 51 per cent, North America for nearly 42 per cent, Asia for nearly 4 per cent, and the remaining continents for nearly 4 per cent. The United States mineral production in recent years has been about 900,000,000 tons.

According to the United States census of 1920, nearly half of all the establishments or businesses engaged in quarrying or mining operations in this country are operating in oil and gas.

Of the crude materials extracted from the ground perhaps 10 per cent, including gold, silver, copper, lead, zinc, nickel, and other ores, are concentrated mainly at the mine, with the result that this fraction of the tonnage in large part does not travel beyond the mine. About 90 per cent of the total production, therefore, figures largely in the transportation of mineral resources.

It is estimated that roughly two-thirds of the annual world production is used or smelted within the countries of origin, the remaining one-third being exported. Of the minerals moving internationally, coal and iron constitute 90 per cent of the tonnage.

The metal smelting capacity of the world in terms of yearly production of crude metal is estimated at nearly 100,000,000 short tons. Of this amount about 80 per cent is located in the United States, England, and Germany. The United States alone has over half of the total. Of the oil-refining capacity the United States controls nearly 70 per cent.

One of the significant features of the situation above summarized is the concentration of production and smelting in a comparatively few places in the world. This statement applies with even more force to the individual mineral commodities.

Water may be regarded as a mineral resource in so far as it is utilized as a commodity for drinking, washing, power, irrigation, and other industrial uses. For purposes of navigation and drainage, or as a deterrent in excavation, it would probably not be so classed. While it is not easy to define the limits of water's use as a mineral resource, it is clear that even with a narrow interpretation the total tonnage extracted from the earth as a mineral resource exceeds in amount all other mineral resources combined.

WORLD ANNUAL PRODUCTION OF MINERALS IN TERMS OF VALUE

In terms of value, mineral resources appear in different perspective. The annual world value of mineral production, exclusive of water, is approximately $9,000,000,000. This figure is obtained by dividing the annual value of the United States output of each of the principal minerals by the percentage which the United States output constitutes in the world output, and adding the figures thus obtained. The values here used are mainly selling prices at the mines. It is impossible to reduce the figures absolutely to the value of the mineral as it comes from the ground; there are always some items of transportation included. This method of figuring is of course only the roughest approximation; the values as obtained in the United States cannot be accurately exterpolated for the rest of the world because of locally varying conditions. However, the figures will serve for rough comparative purposes.

Of this total value coal represents roughly 61 per cent, petroleum 12 per cent, iron 6 per cent, copper 5 per cent, and gold 3 per cent.

In terms of value, about 25 per cent of the world's mineral production is available for export beyond the countries of origin. Of this exportable surplus the United States has about 40 per cent, consisting principally of coal, copper, and formerly petroleum.

The value of the United States annual mineral production in recent years has been from about $3,500,000,000 to $5,500,000,000. Annual imports of mineral products into the United States have averaged recently in the general vicinity of $450,000,000, the larger items being copper, tin, fertilizers, petroleum, gems and precious stones, manganese, nickel, and tungsten.

Again the perspective is changed when the value of water resources is considered. As a physiologically indispensable resource, the value of water in one sense is infinite. There is no way of putting an accurate value on the total annual output used for drinking and domestic purposes,—although even here some notion of the magnitude of the figures involved may be obtained by considering the average per capita cost of water in cities where figures are kept, and multiplying this into the world population. This calculation would not imply that any such amount is actually paid for water, because the local use of springs, wells, and streams can hardly be figured on a cash basis; but, if human effort the world over in securing the necessary water is about as efficient as in the average American city, the figures would indicate the total money equivalent of this effort.

SIGNIFICANCE OF GEOGRAPHIC DISTRIBUTION OF MINERAL PRODUCTION

The remarkable concentration of the world's mining and smelting around the North Atlantic basin, indicated by the foregoing figures, does not mean that nature has concentrated the mineral deposits here to this extent. It is an expression rather of the localized application of energy to mineral resources by the people of this part of the world. The application of the same amount of energy in other parts of the world would essentially change the distribution of current mineral production. The controlling factor is not the amount of minerals present in the ground; this is known to be large in other parts of the world and more will be found when necessary. Controlling factors must be looked for in historical, ethnological, and environmental conditions. This subject is further discussed in the chapters on the several resources, and particularly in relation to iron and steel.

THE INCREASING RATE OF PRODUCTION

The extraction of mineral resources on the huge scale above indicated is of comparatively recent date.

From 1880 to the end of 1918 the value of the annual mineral production of the United States has increased from $367,000,000 to more than $5,500,000,000, or nearly fifteen times; measured in another way, it has increased from a little over $7 per capita to more than $52.[10]

More coal has been mined in the United States since 1905 than in all the preceding history of the country. More iron ore has been mined since 1906 than in all the preceding history. The gold production of the United States practically started with the California gold rush in 1849. The great South African gold production began in 1888. Production of diamonds in South Africa began about 1869. The large use of all fertilizer minerals is of comparatively recent date. The world's oil production is greater now each year than it was for any ten years preceding 1891, and more oil has come out of the ground since 1908 than in all the preceding history of the world. The use of bauxite on a large scale as aluminum ore dated practically from the introduction of patented electrolytic methods of reduction in 1889.

In one sense the world has just entered on a gigantic experiment in the use of earth materials.

The most striking feature of this experiment relates to the vast acquisition of power indicated by the accelerating rate of production and consumption of the energy resources—coal, oil, and gas (and water power). Since 1890 the per capita consumption of coal in the United States has trebled and the per capita consumption of oil has become five times as great as it was. If the power from these sources used annually in recent years be translated roughly into man power, it appears that every man, woman, and child in the United States has potential control of the equivalent of thirty laborers,—as against seven in 1890. Energy is being released on a scale never before approximated, with consequences which we can yet hardly ascertain and appraise. This consideration cannot but raise the question as to the ability of modern civilization to control and coödinate the dynamic factors in the situation.

CAPITAL VALUE OF WORLD MINERAL RESERVES

It is impossible to deduce accurately the capital value of mineral resources from values of annual output, but again some approximation may be made. The profit on the extraction of mineral resources on the whole, considering the cost of exploration, is probably no greater than in other industries (p. 330). If we assume a 6 per cent return, which perhaps is somewhere near the world-wide standard of interest rate for money, and capitalize the value of the world's annual output at this rate, we obtain a world capital value for mineral resources, exclusive of water, of 150 billions of dollars. This assumes an indefinitely long life for reserves. This assumption may need some qualifications, but it is the writer's view (Chapter XVII) that it is justified for a sufficiently long period to substantiate the above method of calculation.

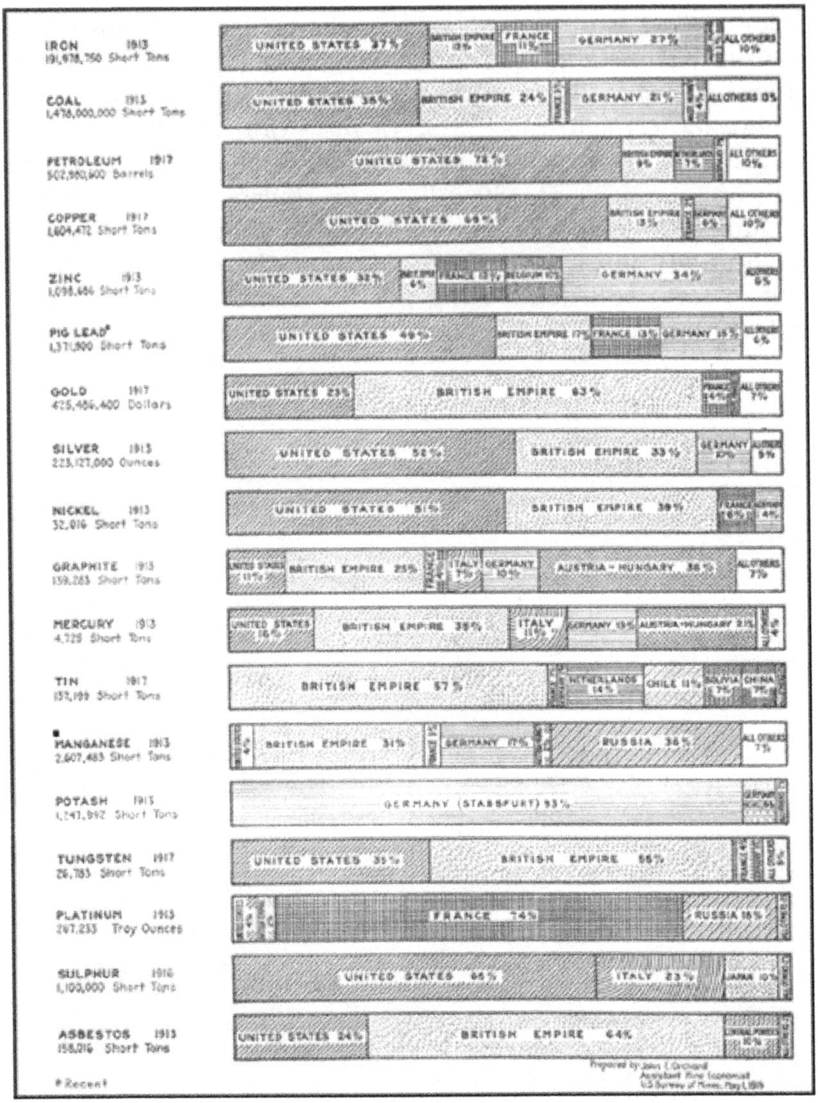

FIG. 2. Commercial (financial) control of the mineral resources of the world.ToList

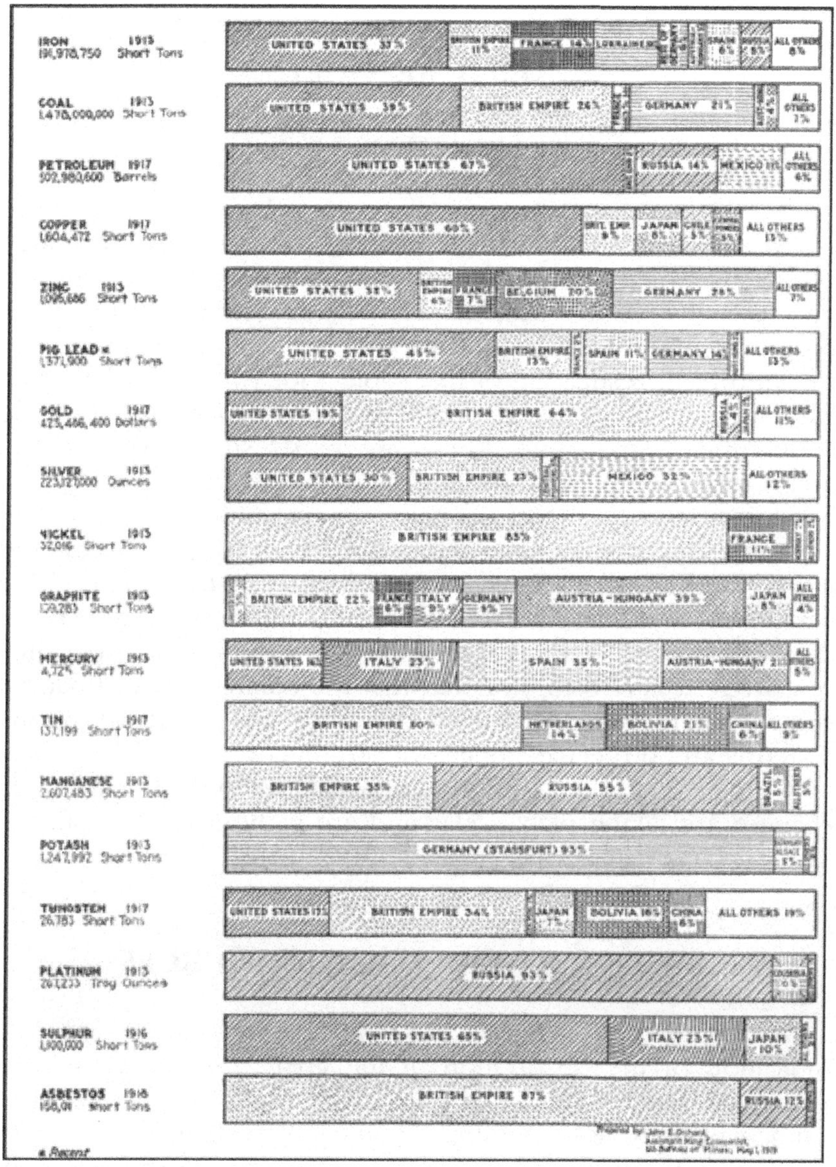

FIG. 3. Political (territorial) control of the mineral resources of the world. ToList

POLITICAL AND COMMERCIAL CONTROL OF MINERAL RESOURCES

The occurrence of a mineral resource within a country does not necessarily mean control by that particular political unit. A citizen of the United States may own a mineral resource in South America. Commercial control of this sort was demonstrated during the war to be of more far-reaching significance than had been supposed, and it became necessary to ascertain, not only the output of the different countries, but the commercial control of this output. Investigation of this subject for twenty-three leading commodities shows that the political and commercial control are by no means the same. These are partly summarized in the accompanying graphs from Spurr.[11] It is to be noted that the graphs show the control of many commodities as it existed in 1913, the last normal year before the war. Changes during and since the war have of course largely altered the situation for certain commodities, notably for iron, coal, and potash. These developments are summarized in the discussion of the individual resources. It is also to be noted that the commercial or financial control of the world's minerals, under the influence of the fostering and protective policies of certain governments discussed in Chapter XVIII, is at present in a state of flux. Considerable changes are taking place today and are to be looked for in the future.

RESERVES OF MINERAL RESOURCES

Annual production figures are only to a very partial extent an indication of the distribution of the great reserves of mineral resources. For instance, there are enormous reserves of coal in China which are not yet utilized to any large extent. The minerals of South America and Africa are in a very early stage of development. The total world reserves will of course not be known until exploration and development of the world's resources are complete—a time which will probably never come. Figures of reserves represent only our present partial state of knowledge and are likely to be considerably modified in the future. Furthermore, the quantitative accuracy of knowledge of reserves is so variable in different parts of the world that it is almost impossible to make up world figures which have any great validity. There are, however, certain broad facts ascertainable.

Every country in the globe is deficient in supplies of some minerals. The United States is better off than any other country, but still lacks many mineral commodities (see pp. 396-399.) No single continent has sufficient reserves of all mineral commodities.

For the world, however, it may be stated with reasonable certainty that the reserves of the principal minerals are now known to be ample with the exception of those of oil, tin, and perhaps gold and silver. By *ample* we mean sufficient to give no cause for worry for the next few decades. For many mineral commodities the amounts now actually in sight will not last long, but the possibilities of extension and discovery are so great that a long future availability of these commodities can be counted upon with reasonable safety.

The present shortages in oil, tin, and other minerals mentioned may be only temporary. There is a large part of the world still to be explored, and the present reserves merely mark a stage in this exploration. Nevertheless, the ratio of reserves and discovery on the one hand to accelerated use on the other gives cause for much concern. Looking forward to the future, the problem of mineral reserves in general is not one of the possible ultimate amount which the earth may contain—presumably in no case is this deficient—but of the success with which the resource may be found and developed to keep up with the rapid acceleration of demand. In the chapter on conservation the suggestion is made that future difficulties are more likely to arise from failure to coödinate the dynamic factors of supply and demand, than from absolute shortage of material in the earth.

FOOTNOTES:

[10] Bastin, Edson S., and McCaskey, H. D., The work on mineral resources done by the U. S. Geological Survey: *Min. Res. of the United States for 1918, U. S. Geol. Survey*, pt. 1, 1920, p. 3a.

[11] Spurr, J. E., Who owns the earth?: *Eng. and Min. Jour.*, vol. 109, 1920, pp. 389-390.

CHAPTER V

WATER AS A MINERAL RESOURCE

GENERAL GEOLOGIC RELATIONS

With the solid earth as the special care of geology, it may seem presumptuous for the geologist to claim the waters thereof, but he does not disclaim this inheritance. Water is so all-pervasive that it is more or less taken for granted; and so many and so intricate are its relations that it is not easy to make an objective survey of the water problem in its relation to geology.

The original source of water, as well as of air, is in molten magmas coming from below. These carry water and gases,—some of which are released and some of which are locked up in the rocks on cooling, to be later released during the alterations of the rocks. It is supposed, whatever theory of the origin of the earth we favor, that in its early stages the earth lacked both hydrosphere and atmosphere, and that during the growth of the earth these gradually accumulated on and near the surface in the manner stated.

During alterations at the surface water is added to the mineral constitution of the rocks, and by alterations deep below the surface it may be subtracted. Water is the agent through which most mineral and chemical changes of rocks are accomplished. It is the agent also which is mainly responsible for the segregation of mineral deposits. Water, both as running water and in the solid form of ice, plays an important part in determining the configuration of the earth's surface. Water is the medium in which most sedimentary rocks are formed. It is an important agent in the development of soil and in organic growth. These various influences of water on geological processes touch the economic field at many points, especially in relation to the concentration of ores and to the development of soils and surface forms.

Water comes even more directly into the field of economic geology as a mineral resource. Water supplies, for the greatest variety of purposes, involve geologic considerations at almost every turn.

Finally, water may be an aid or a hindrance to excavation and to a great variety of structural operations, both in war and in peace; and in this relation it again affords geologic problems.

The part played by water in geologic processes, such as that of mineral segregation, is more or less incidentally discussed in other chapters. We may consider more fully in this chapter the application of geology to the general subject of water supplies.

From the geological point of view, water is a mineral,—one of the most important of minerals,—as well as a constituent of other minerals. It becomes a mineral resource when directly used by man. It is ordinarily listed as a mineral resource when shipped and sold as "mineral water," but there is obviously no satisfactory line between waters so named and water supplies in general, for most of them are used for the same purposes and none of them are free from mineral matter. Water which is pumped and piped for municipal water supply is as much a mineral resource as water which is bottled and sold under a trade name. Likewise water which is used for irrigation, water power, and a wide variety of other purposes may logically be considered a mineral resource.

Notwithstanding the immense economic importance of water as a mineral resource its value is more or less taken for granted, and considerations of valuation and taxation are much less in evidence than in the case of other mineral resources. Water must be had, regardless of value, and market considerations are to a much less extent a limiting factor. Economic applications of geology to this resource are rather more confined to matters of exploration, development, total supply, and conservation, than to attempts to fix money value.

DISTRIBUTION OF UNDERGROUND WATER

Free water exists in the openings in rocks where it is sometimes called *hygroscopic* water. There is also a large amount of water combined molecularly with many of the minerals of rocks, in which form it is called *water of constitution*. This water is fixed in the rock so that it is not available for use, though some of the processes of rock alteration liberate it and contribute it to the free water. The immediate source of underground water, both free and combined, is mainly the surface or rain waters. A subordinate amount may come directly from igneous emanations or from destruction of certain hydrous minerals. Ultimately, as already indicated, even the surface water originates from such sources.

The openings in rocks consist of joints and many other fractures, small spaces between the grains of rocks (pore space), and amygdaloidal and other openings characteristic of surface volcanic rocks. Many of these openings are capillary and sub-capillary in size. Most rocks, even dense igneous rocks, are porous in some degree, and certain rocks are porous in a very high degree. The voids in some surface materials may amount to 84 per cent of the total volume. In general the largest and most continuous openings are near the surface,—where rocks on the whole are more largely of the sedimentary type and are more fractured, disintegrated, and decomposed, than they are deep within the earth. The largest supplies of water are in the unconsolidated sediments. The water in igneous and other dense rocks is ordinarily in more limited quantity.

APPROXIMATE QUANTITY OF WATER WHICH WILL BE ABSORBED BY SOILS AND ROCKS[1]

Material	*Volume of water absorbed per 100 of material*
Sandy soil[2]	45.4
Chalk soil[2]	49.5
Clay[2]	50-52.7
Loam[2]	45.1-60.1
Garden earth[2]	69.0
Coarse sand[2]	39.4
Peat subsoil[2]	84.0
Sand	30-40
Sandstone	5-20

Limestone and dolomite	1-8
Chalk	6-27
Granite	03.-.8

[1] Mead, Daniel W., *Hydrology*: McGraw-Hill Book Co., New York, 1919, p. 393.

[2] Woodward, H. B., *Geology of soils and substrata*: Edward Arnold, London, 1912.

Immediately at the surface, the openings of rocks may not be filled with water; but below the surface, at distances varying with climatic and topographic conditions, the water saturates the openings of the rocks and forms what is sometimes called the *zone of saturation* or the *sea of underground water*. The top surface of this zone is called the *water table*, or the *ground-water level*. The space between the water table and the earth's surface is sometimes referred to as the *vadose zone* or the *zone of weathering*, since it is the belt in which weathering processes are most active. The zone of weathering is not necessarily dry. Water from the surface enters and sinks through it and water also rises through it from below; it may contain suspended pockets of water surrounded by dry rocks; it is not continuously and fully saturated.

The water table or ground-water level may be near or at the surface in low and humid areas, and it may be two thousand feet or more below the surface in arid regions of high topographic relief. Because of the influence of capillarity, the water table is not a horizontal surface. It shows irregularities more or less following the surface contours, though not nearly so sharply accentuated.

The lower limit of the ground-water is more irregular than the upper surface and is less definitely known. In general, openings in rocks tend to diminish with depth, due to cementation and to closing of cavities by pressures which are too great for the rock to withstand. But rocks differ so widely in their original character, and in their response to physical and chemical environment, that it is not unusual to find dense and impervious rocks above, and open and porous rocks below. The lower limit of the zone of abundant underground water varies accordingly. A well may encounter nearly dry rock at a comparatively shallow depth, or it may reach a porous water-bearing stratum at considerable depth. At the greater depths pockets of water are sometimes found which have a composition different from that of the surface water, and which evidently are isolated from the surface water by zones of non-pervious rock.

Attempts have been made to calculate the total volume of underground water by measuring the openings of rocks and making assumptions as to the depth to which such openings may extend. In this manner it has been estimated that, if all the ground-water were assembled in a single body, it would make a shell between eighty and two hundred feet thick (depending on the assumptions) over all the continental areas.

MOVEMENT OF UNDERGROUND WATER

Availability of water supplies is determined by the movement or flow of water as well as by its distribution and amount. The natural flow of water underground is caused by gravity in the larger openings, but in the smaller openings adhesion and capillarity are also important forces.

Of all the water falling on the surface, some may not go below the surface at all but may immediately evaporate or join the runoff—that is, the surface streams. Another part may penetrate a little distance into the zone of weathering and then join the runoff. Of the water which reaches the zone of saturation, a part may soon come to the surface in low areas and join the runoff, and a part may penetrate deeply.

Above the zone of saturation gravity carries the water downward in devious courses until it reaches the water table. Thereafter its course is determined largely by the lowest point of escape from the water table. In other words, the water table is an irregular surface; and under the influence of gravity the water tends to move from the high to the low points of this surface. Between the point of entrance and the point of escape from the water table, the water follows various courses, depending upon the porosity and the openings in the rocks. In general it fills all of the available openings, and uses the entire available cross section in making its progress from one point to another. The difference in height or the "head" between the point of entrance and the point of escape, together with the porosity of the rock and other factors, determine the general speed of its movement (see p. 73). With equal porosity the flow is at a maximum along a line directly connecting the two points, and on more devious courses the flow is less.

The surface water first enters the ground through innumerable small openings. Soon, however, it tends to be concentrated into channels of easiest flow, with the result that in the later part of its underground course it may be much concentrated in large trunk channels. These channels may consist of joints, or frequently of very coarse and pervious beds. The sedimentary rocks as a whole contain the most voids, and therefore the largest flow and largest supply of water is often localized in them. Of the sedimentary rocks, sandstones and limestones usually contain the largest and most continuous openings, and thus afford the freest circulation for water. The voids in fine-grained shales may equal in volume those in sandstones and limestones, but the openings are so small and discontinuous that the water does not flow freely. Regardless of total amount of water, unless there are continuous openings of some size the flow may be small.

The relations of more porous rocks to containing impervious strata also profoundly affect the flow of underground water. Between impervious strata the circulation may be concentrated and vigorous within the porous bed. Where the porous bed is not so contained, the movement may be more dispersed and less vigorous locally. The inclination of the beds, of course, also affects the direction and amount of the flow.

The influence of gravity upon underground water may locally tend toward a state of equilibrium in which there is little movement. In such a case the water is substantially ponded, and moves only when tapped by artificial openings.

WELLS AND SPRINGS

Underground water becomes available for use by means of springs and through wells or bore holes. Water rises to the surface in natural springs at points where the pressure or *head*, due to its entrance into the ground at a higher level, is sufficient to force it to the surface after a longer or shorter underground course. The movement may be all downward and lateral to the point of escape, or it may be downward, lateral, and upward. Ordinarily, the course of spring waters does not carry them far below the surface. Heat and gases may be added beneath the surface by contact with or contributions from cooling igneous rocks. These may accelerate the upward movement of spring waters, and yield thermal and gas-charged waters, as in the springs and geysers of Yellowstone Park.

When a well is sunk to tap the underground water supply, the water may not rise in the artificial opening but may have to be lifted to the surface.

If, however, the water is confined beneath an impervious stratum and is under pressure from the water of higher areas, a well opening may simply allow it to move upward under its own pressure or head. This pressure may carry it upward only a few feet or quite to the surface or beyond, in which latter case the well is called an *artesian* well. The essential condition for an artesian circulation is a porous zone, inclining downward from the surface beneath an impervious stratum which tends to confine and pond the water. The water at any point in the water-bearing rock is under pressure which is more or less equivalent to the weight of the column of water determined by the difference in height between this point and the point of entrance or feeding area of the water. If the feeding area is higher than the collar of the well, the water will rise quite to the surface; if not, it will rise only part way. Capillary resistance, however, may and usually does lessen the theoretical pressure so figured.

The flow in deep artesian circulations is ordinarily a slow one. For the artesian wells of southern Wisconsin, it has been calculated that waters entering the outcrop of the southward dipping sandstone and limestone layers in the northern part of the state have required two or three hundred years to reach a point in the southern part of the state where they are tapped.

Because of this slow movement, a large number of wells in any one spot may exhaust the local supply faster than it is replenished from the remainder of the formation. The drilling of additional wells near at hand in such cases does not increase the total yield, but merely divides it among a larger number of wells.

The porosity of the rocks, and therefore the flow of an artesian circulation, may in some cases be artificially increased by blasting and shattering.

COMPOSITION OF UNDERGROUND WATERS

Underground waters are never entirely free from dissolved mineral substances, and seldom are they free from suspended particles. Some waters are desired because they contain very small quantities of dissolved mineral matter. Others are prized because they have an unusually high content of certain mineral substances. In determining the deleterious or beneficial effect of dissolved substances, much depends on the purpose for which the water is to be used,—whether for drinking, washing, steam boilers, or irrigation. Near the surface underground waters may carry bacteria, as well as animal and vegetable refuse, which from a sanitary standpoint are usually objectionable. Deeper waters are more likely to lack this contamination because of filtration through rocks and soils.

The dissolved mineral substances of underground water are derived for the most part from the solution of rocks with which the waters come in contact, particularly at or near the surface. Through the agency of underground water most of the mineral and chemical changes of rocks are produced. The dissolved substances in solution at any time and place may therefore be regarded as by-products of rock alterations. Locally they may to some extent be derived from direct emanations from cooling igneous masses.

The most common mineral substances contained in waters are lime and magnesia. Less common, but abundant locally, are soda, potash, iron, and silica. Waters contain also certain acid and gaseous substances, the most common of which is carbon dioxide; and less widespread, but locally abundant, are chlorine and sulphur dioxide. Where lime and magnesia are abundant the water is ordinarily classed as a hard water. Where absent, or subordinate to soda and potash, the water is ordinarily classed as a soft water. Large amounts of the acid substances like chlorine and sulphur are detrimental for most purposes. Where there are unusual amounts of carbon dioxide or other gases present, they may by expansion cause the water to bubble.

If we were to attempt to describe and define the characteristics, with reference to dissolved mineral content and temperature, which make a given water more desirable than another, we should enter a field of the most amazing complexity and one with many surprising contradictions. For the most widespread use, the most desirable water is a cold water as free from mineral content as possible, and especially one lacking an excess of lime and magnesia which make it hard; also lacking an excess of acid constituents like sulphur dioxide, carbon dioxide, or chlorine, which give the water a taste, or which make impossible its use in boilers. Locally and for special reasons, waters of other qualities are in demand. Waters so excessively carbonated as to bubble, sulphureted waters, chlorine waters, waters high in iron, high in silica, high in potash, high in soda, or high in magnesia, or waters of high temperature, may come to be regarded as desirable. It is an interesting fact that any water with unusual taste, or unusual mineral content, or unusual temperature, is likely to be regarded as having medicinal value. Sometimes this view is based on scientific knowledge; sometimes it is an empirical conclusion based on experience; and again it may be merely superstition. In one case the desirable feature may be the presence of a large amount of carbon dioxide; in another case it may be its absence. In one case the desirable feature may be high temperature; in another case low temperature. The same combination of qualities which in a certain locality may be regarded as highly desirable, may be regarded as highly detrimental somewhere else where certain other types of waters are in vogue.

Proprietary rights and advertising have brought certain waters into use for drinking purposes which are not essentially different from more widely available waters which are not regarded as having special value. Two springs located side by side, or a spring and a deep well, whose waters have exactly the same chemical characteristics, may be used and valued on entirely different scales. Any attempt to classify mineral waters sold to the public in any scientific way discloses a most intricate and confused situation. One can only conclude that the popularity of certain waters is not based alone on objective qualities of composition, but rather on causes which lie in the fields of psychology and commerce.

The part played by sentiment in putting value on water is well illustrated by the general preference for spring waters as compared with well waters. In the public mind, "spring water" denotes water of unusual purity and of more desirable mineral content than well water. Illustrations could be cited of districts in which the surface or spring waters have a composition not different from that of the deeper well waters, and are much more likely to be contaminated because of proximity to the surface; and yet people will pay considerable sums for the spring water in preference to the cheaply available well water.

RELATION OF GEOLOGY TO UNDERGROUND WATER SUPPLY

It is obvious that a knowledge of geology is helpful in locating an underground water supply. Locally the facts may become so well known empirically that the well driller is able to get satisfactory results without using anything but the crudest geologic knowledge; but in general, attention to geologic considerations tends to eliminate failures in well drilling and to insure a more certain and satisfactory water supply.

In drilling for water, it is essential to know the nature, succession, and structure of the rocks beneath the surface in order to be able to identify and correlate them from drill samples. The mere identification of samples is often sufficient to determine whether a well has been drilled far enough or too far to secure the maximum results. In order to arrive at any advance approximation of results for a given locality, a knowledge of the general geology of the entire region may be necessary. Especially for expensive deep artesian wells it is necessary to work out the geologic possibilities well in advance. It is useless, for instance, to look for artesian water in a granite; but in an area of gently inclined strata, with alternations of porous and impervious layers, the expert may often figure with a considerable degree of certainty the depth at which a given porous stratum will be found, and the pressure under which the water will be in this particular stratum at a given point. Even the mineral content of the water may in some cases be predicted from geologic study.

One of the most obvious and immediately useful services of the geologist in most localities is the collection and preservation of well samples for purposes of identification and correlation of rock formations, and as a guide to further drilling. Failure to preserve samples has often led to useless and expensive duplication of work.

The problem of water supply in some localities is comparatively simple and easy. In other areas there is an infinite variety of geologic conditions which affect the problem, and the geologist finds it necessary to bring to bear all the scientific knowledge of any sort which can be used,—particularly knowledge in relation to the type of rock, the stratigraphy and the structure.

SURFACE WATER SUPPLIES

Where underground water is not abundant or not cheaply available, or where larger amounts of water are needed, as in large cities or for irrigation purposes, surface water is used. In general, surface waters are more likely to be contaminated by vegetable and animal matter and to require purification for drinking purposes.

Surface waters are also used for irrigation, water power, drainage, the carrying of sewage, etc. This great variety of uses brings the consideration of surface waters into many fields other than geology, but an understanding and interpretation of the geological conditions is none the less fundamental. This is evidenced by the inclusion of geologic discussions in most textbooks of hydrology, and in the reports of the Hydrographic Branch of the U. S. Geological Survey. The very fact that this important branch of governmental investigation is in a charge of the U. S. Geological Survey indicates its close relation to geology.

The principles of geology used in the study of surface waters relate chiefly to physiography (see Chapter I). It is usually necessary to know the total quantity of flow, its annual and seasonal variation, and the possible methods of equalization or concentration; the maximum quantity of flow, the variation during periods of flood, and the possibilities of reduction or control; the minimum flow and its possible modification by storage or an auxiliary supply. These questions are obviously related to the size and shape of the catchment area, the topography, the rock structure, the relation between underground flow or absorption and the runoff, and other physiographic factors. Quoting from D. W. Mead:[12]

Geological conditions are frequently of great importance in their influence on the quantity and regularity of runoff. If the geological deposits of the drainage area are highly impervious, the surface flow will receive and transmit the water into the mass only through the cracks and fissures in the rock. Pervious materials, such as sandstones, sands, gravels, and cracked or fissured rocks, induce seepage, retard runoff, and, if such deposits are underlaid with an impervious bed, provide underground storage which impounds water away from the conditions which permit evaporation, and hence tends to increase runoff and equalize flow. On the other hand, if such pervious deposits possess other outlets outside of the stream channel and drainage area, they may result in the withdrawal of more or less of the seepage waters entirely from the ultimate flow of the stream. Coarse sands and gravels will rapidly imbibe the rainfall into their structure. Fine and loose beds of sand also rapidly receive and transmit the rainfall unless the precipitation is exceedingly heavy under which conditions some of it may flow away on the surface.

Many of the highly pervious indurated formations receive water slowly and require a considerable time of contact in order to receive and remove the maximum amount.

In flat, pervious areas, rainfalls of a certain intensity are frequently essential to the production of any resulting stream flow. In a certain Colorado drainage area, the drainage channel is normally dry except after a rainfall of one-half inch or more. A less rainfall, except under the condition of a previously saturated area, evaporates and sinks through the soil and into the deep lying pervious sand rock under the surface which transmits it beyond the drainage area. Such results are frequently greatly obscured by the interference of other factors, such as temperature, vegetation, etc.

The natural storage of any drainage area and the possibilities of artificial storage depend principally upon its topography and geology. Storage equalizes flow, although the withdrawal of precipitation by snow or ice storage in northern areas often reduces winter flow to the minimum for the year. Both surface and sub-surface storage sometimes hold the water from the streams at times when it might be advantageously used. Storage, while essential to regulation, is not always an advantage to immediate flow conditions.

UNDERGROUND AND SURFACE WATERS IN RELATION TO EXCAVATION AND CONSTRUCTION

Scarcely more than a mention of this subject is necessary. In mining, the pumping charge is one of the great factors of cost. A forecast of the amount and flow of water to be encountered in mining is based on the geologic conditions. The same is true in excavating tunnels, canals, and deep foundations. Detailed study of the amount and nature of water in the rock and soil of the Panama Canal has been vital to a knowledge of the cause and possibilities of prevention of slides. Rock slides in general are closely related to the amount and distribution of the water content.

The importance of ground-water as a detriment in military operations was shown during the recent war in trenching and other field works. At the outset, with the possible exception of the German army, a lack of scientific study of ground-water conditions led to much unnecessary difficulty. It soon became necessary to study and map the water conditions in great detail in advance of operations. Much of this work was done by geologists (see Chapter XIX).

Geological considerations are involved in a great variety of engineering undertakings related to river and harbor improvements, dam sites, etc., mentioned in Chapter XX.

FOOTNOTES:

[12] Mead, Daniel W., *Hydrology*: McGraw-Hill Book Co., New York, 1919, pp. 447-448, 456.

CHAPTER VI

THE COMMON ROCKS AND SOILS AS MINERAL RESOURCES

ECONOMIC FEATURES OF THE COMMON ROCKS

Under the general heading of common rocks are included the ordinary igneous, sedimentary, and "metamorphic" rocks, and the unconsolidated clays, sands, and gravels characteristic of surface conditions, which are mined and quarried for commercial use. Soils are closely related to this group; but since they present special problems of their own, they are discussed under a separate heading at the end of the chapter. Names of the common rocks will be used with the general commercial significance given them by the United States Geological Survey in its mineral resource reports.

Because of their inexhaustible quantity and ready availability, the value of the common rock products is not large per unit of weight; but in the aggregate it ranks high among mineral products. In respect to tonnage, common rocks constitute perhaps 10 per cent of the world annual output of all mineral commodities (exclusive of water).

The greater tonnage of the common rocks is used commercially in crushed or comminuted forms for road material, for railroad ballast, and for cement, brick, concrete, and flux. In blocks and structural shapes, of less aggregate tonnage, they are used as building stone, monumental stone, paving blocks, curbing, flagging, roofing, refractory stone, and for many other building and manufacturing purposes.

The common rocks are commodities in which most countries of the globe are self-sufficing. International trade in these commodities is insignificant, being confined to small quantities of materials for special purposes, or to local movements of short distances, allowed by good transportation facilities.

The common rocks are so abundant and widespread that the conservation of raw materials is not ordinarily a vital problem.

Conservational principles do apply, however, to the human energy factor required for their efficient use. In the valuation of common rocks, also, the more important factors are not the intrinsic qualities of the stones, but rather the conditions of their availability for use.

Because of bulk and comparatively low intrinsic value, the principal commercial factors in the availability of the common rocks are transportation and ease of quarrying, but these are by no means the only factors determining availability. Their mineral and chemical composition, their texture and structure, their durability, their behavior under pressure and temperature changes, and other factors enter in to important degrees. The weighting and integration of these factors, for the purpose of reaching conclusions as to the availability of particular rock materials, depend also on the purposes for which these materials are to be used. The problem is anything but simple. The search for a particular rock to meet a certain demand within certain limits of cost is often a long and arduous one. On account of the abundance and widespread distribution of common rocks and their variety of uses, there is a good deal of popular misapprehension as to their availability. Many building and manufacturing enterprises have met disastrous checks, because of a tendency to assume availability of stone without making the fullest technical investigation. Many quarrying ventures have come to grief for the same reason. It is easy to assume that, because a granite in a certain locality is profitably quarried and used, some other granite in the same locality has equal chances. However, minor differences in structure, texture, and composition, or in costs of quarrying and transportation, may make all the difference between profit and loss. Even though all these conditions are satisfactorily met, builders and users are often so conservative that a new product finds difficulty in breaking into the market. A well-established building or ornamental stone, or a limestone used for flux, may hold the market for years in the face of competition from equally good and cheaper supplies. The very size of a quarry undertaking may determine its success or failure.

GRANITE

The term granite, as used commercially, includes true granite and such allied rocks as syenite and gneiss. In fact even quartzite is sometimes called granite in commerce, as in the case of the

Baraboo quartzites of Wisconsin, but this is going too far. For statistical purposes, the United States Geological Survey has also included small quantities of diorite and gabbro. The principal uses of granite are, roughly in order of importance, for monumental stone, building stone, crushed stone, paving, curbing, riprap and rubble. Thirty states in the United States produce granite, the leaders being Vermont, Massachusetts, North Carolina, Maine, Wisconsin, Minnesota, and California.

Basalt and Related Types

Basalt and related rocks are sometimes included under the name "trap rock," which comprises,—besides typical basalt and diabase,—fine-grained diorite, gabbro, and other basic rocks, which are less common in occurrence and are similar in chemical and physical properties. The principal use of these rocks is as crushed stone for road and ballast purposes and for concrete. They are produced in some fifteen states, the leaders being New Jersey, Pennsylvania, California, and Connecticut.

Limestone, Marl, Chalk

In the United States limestone is used principally as crushed stone for road material, railroad ballast, concrete, and cement, as fluxing stone for metallurgical purposes, and in the manufacture of lime. Minor uses are as building stones, paving blocks, curbing, flagging, rubble, and riprap; in alkali works, sugar factories, paper mills, and glass works; and for agricultural purposes. For the making of cement, in metallurgical fluxes, and in most of the manufacturing and agricultural uses, both limestone and lime (limestone with the CO_2 driven out by heating) are used. Lime is also extensively used in the making of mortar for building operations, in tanning leather, and in a great variety of chemical industries. The total quantity of limestone used for all purposes in the United States nearly equals that of iron ore. Nearly every state in the union produces limestone, but the more important producers are Pennsylvania (where a large amount is used for fluxing), Ohio, Indiana, New York, Michigan, and Illinois.

Closely associated with limestone in commercial uses, as well as in chemical composition, is calcareous marl, which is used extensively in the manufacture of Portland cement.

Chalk is a soft amorphous substance of the same composition as limestone. The main uses of chalk are as a filler in rubber, and as a component of paint and putty. It is also used for polishing. The principal producers of this commodity are England, Denmark, and France, and the chief consumer is the United States. The United States depends upon imports for its supply of chalk for the manufacture of whiting. Before the war two-thirds came from England and a third from France. During the war importation was confined to England, with a small tonnage from Denmark. No deposits of domestic chalk have been exploited commercially. A somewhat inferior whiting, but one capable of being substituted for chalk in most cases, is manufactured from the waste fine material of limestone and marble quarries.

MARBLE

Marble is limestone which has been coarsely recrystallized by metamorphism. The marble of commerce includes a small quantity of serpentine as quarried and sold in Massachusetts, California, Maryland, Pennsylvania, and Vermont, and also a small amount of so-called onyx marble or travertine obtained from caves and other deposits in Kentucky and other states. The principal uses of marble are for building and monumental stones. Of the twenty-two states producing marble, the leaders are Vermont, Georgia, and Tennessee.

A small amount of marble of special beauty, adapted to ornamental purposes, is imported from European countries, especially from Italy. Marble imports from Italy constitute about two-thirds, both in tonnage and value, of all stone imported into the United States.

SAND, SANDSTONE, QUARTZITE (AND QUARTZ)

Sand is composed mainly of particles of quartz or silica, though sometimes feldspar and other minerals are present. Sandstones are partially cemented sands. Quartzites are completely cemented sands. To some extent these substances are used interchangeably for the same purposes.

The principal uses of sand in order of commercial totals are for building purposes—for mortar, concrete, sand-lime brick, etc.,—as molding sand in foundries, as a constituent of glass, in grinding and polishing, in paving, as engine sand, as fire or furnace sand, in the manufacture of ferrosilicon (a steel alloy), and in filters. Reference is made to sand as an abrasive and in the manufacture of steel in Chapters XIII and IX. Almost every state produces some sand, but for some of the more specialized uses, such as glass sand, molding sand, and fire or furnace sand, the distribution is more or less limited. The United States Geological Survey has collected information concerning the distribution of various kinds of sand and gravel, and serves a very useful function in furnishing data as to supplies of material for particular purposes. Fine molding sands have been imported from France, but during the war domestic sources in New York and Ohio were developed sufficiently to meet any requirements.

The sandstone of commerce includes the quartzites of Minnesota, South Dakota, and Wisconsin, and the fine-grained sandstones of New York, Pennsylvania, and elsewhere, known to the trade as "bluestone." In Kentucky most of the sandstone quarried is known locally as "freestone." The principal uses of sandstone are for building stone, crushed stone, and ganister (for silica brick and furnace-linings). Other uses are for paving blocks, curbing, flagging, riprap, rubble, grindstones, whetstones, and pulpstones (see also Chapter XIII). Sandstone is sometimes crushed into sand and is used in the manufacture of glass and as molding-sand. Most of the states of the union produce sandstone, the principal producers being Pennsylvania, Ohio, and New York.

"SAND AND GRAVEL"

Where sand is coarse and impure and mixed with pebbles, it is Ordinarily referred to as "sand and gravel." For sand and gravel the principal uses are for railroad ballast, for road building, and for concrete. Sand and gravel are produced in almost every state in the union, the largest producers being Pennsylvania, Ohio, Illinois, New Jersey, and North Carolina.

CLAY, SHALE, SLATE

Shale is consolidated clay, usually with a fine lamination due to bedding. Slate is a more dense and crystalline rock, produced usually by the anamorphism of clay or shale under pressure, and characterized by a fine cleavage which is usually inclined to the sedimentary bedding.

Clays are used principally for building and paving brick and tile, sewer-pipe, railroad ballast, road material, puddle, Portland cement, and pottery. Clay is mined in almost every state. Ohio, Pennsylvania, New Jersey, and Illinois have the largest production. There has been a considerable importation of high-grade clays, principally from England, for special purposes—such as the filling and coating of paper; the manufacture of china, of porcelain for electrical purposes, and of crucibles; and for use in ultramarine pigments, in sanitary ware, in oilcloth, and as fillers in cotton bleacheries. War experience showed the possibility of substitution of domestic clays for most of these uses; but results were not in all cases satisfactory, and the United States will doubtless continue to use imported clays for some of these special purposes.

Shales, because of their thinly bedded character and softness, are of no value as building stones, but are used in the manufacture of brick, tile, pottery, and Portland cement.

Slates owe their commercial value primarily to their cleavage, which gives well-defined planes of splitting. The principal uses are for roofing and, in the form of so-called mill stock for sanitary, structural, and electrical purposes. Small amounts are used for tombstones, roads, slate granules for patent roofing, school slates, blackboard material, billiard table material, etc. The color, fineness of the cleavage, and size of the flakes are the principal features determining the use of any particular slate. Ten states produce slate, the principal production coming from Pennsylvania and Vermont.

The Feldspars

Feldspars are minerals, not rocks, but mention of them is made here because, with quartz, they make up such an overwhelming percentage of earth materials. It is estimated that the feldspars make up 50 per cent of all the igneous rocks and 16 per cent of the sedimentary rocks. As the igneous rocks are so much more abundant than the sedimentary rocks, the percentage of feldspars in the earth approaches the former rather than the latter figure. In most rocks feldspar is in too small grains and is too intimately associated with other minerals to be of commercial importance; in only one type of rock, pegmatite, which is an igneous rock of extremely coarse and irregular texture, are the feldspar crystals sufficiently large and concentrated to be commercially available.

Feldspar is used principally in the manufacture of pottery, china ware, porcelain, enamel ware, and enamel brick and tile. In the body of these products it is used to lower the fusing point of the other ingredients and to form a firm bond between their particles. Its use in forming the glaze of ceramic products is also due to its low melting point. A less widespread use of feldspar is as an abrasive (Chapter XIII). One of the varieties of feldspar carries about 15 per cent of potash, and because of the abundance of the mineral there has been much experimental work to ascertain the possibility of separating potash for fertilizer purposes; but, because of cost, this source of potash is not likely for a long time to compete with the potash salts already concentrated by nature.

Feldspar is mined in eleven states, but the important production comes from North Carolina and Maine. The United States also imports some feldspar from Canada.

HYDRAULIC CEMENT (including Portland, natural, and Puzzolan cements)

Cement is a manufactured product made from limestone (or marl) and clay (or shale). Sometimes these two kinds of substances are so combined in nature (as in certain clayey limestones) that they are available for cement manufacture without artificial mixing. It is not our purpose in this volume to discuss manufactured products; but the cement industry involves such a simple transformation of raw materials, and is so closely localized by the distribution of the raw materials, that a mention of some of its outstanding features seems desirable.

Hydraulic cement is used almost exclusively as a structural material. It is an essential ingredient of concrete. Originally used chiefly for the bonding of brick and stone masonry and for foundation work, its uses have grown rapidly, especially with the introduction of reinforced concrete. It is being used in the construction of roads, and its latest use is in ship construction.

With the exception of satisfactory fuels, the raw materials required for the manufacture of cement are found quite generally throughout the world. While practically all countries produce some cement, much of it of natural grade, only the largest producers make enough for their own requirements and as a result there is a large world movement of this commodity. The world trade is chiefly in Portland cement.

Next to the United States, the producing countries having the largest exportable surplus of cement in normal times are Germany and Great Britain. France and Belgium were both large producers and exporters before the war, but the war greatly reduced their capacity to produce for the time being. Sweden, Denmark, Austria, Japan, and Switzerland all produce less extensively but have considerable surplus available for export. Italy and Spain have large productions, which are about sufficient for their own requirements. Holland and Russia import large amounts from the other European countries. The far eastern trade absorbs the excess production of Japan. In South Africa and Australasia, production nearly equals demand. In Canada, although the industry has been growing very rapidly, the demand still exceeds production. In South and Central America, Mexico and the West Indies, the demand is considerable and will probably increase; production has thus far been insufficient. Several modern mills are either recently completed or under construction in these countries, and concessions have been granted for several others. These new mills are largely financed by American capital.

The United States is the largest single producer of cement in the world, its annual production being about 45 per cent of the world's total. Domestic consumption has always been nearly as great as the production, and exports have usually not exceeded 4 per cent of the total shipments from the mills. South and Central America offer fields for exportation of cement from the United States.

GEOLOGIC FEATURES OF THE COMMON ROCKS

To describe the geologic features of the common rocks used in commerce would require a full treatise on the subject of geology. These are the bulk materials of the earth and in them we read the geologic history of the earth. In preceding chapters a brief outline has been given of the relative abundance of the common earth materials and of the processes producing them. In comparison, the metalliferous deposits are the merest incidents in the development of this great group of mineral resources.

In this section reference will be made only to a few of the rock qualities and other geologic features which require first attention in determining the availability of a common rock for commercial use. The list is very fragmentary, for the reason that the uses are so many and so varied that to describe all the geologic features which are important from the standpoint of all uses would very soon bring the discussion far beyond the confines of a book of this scope.[13]

BUILDING STONE

For building stones, the principal geologic features requiring attention are structure, durability, beauty, and coloring.

The structures of a rock include jointing, sedimentary stratification, and secondary cleavage. Nearly all rocks are jointed. The joints may be open and conspicuous, or closed and almost imperceptible. The closed joints or incipient joints cause planes of weakness, known variously as rift, grain, etc., which largely determine the shapes of the blocks which may be extracted from a quarry. Where properly distributed, they may facilitate the quarrying of the stone. In other cases they may be injurious, in that they limit the size of the blocks which can be extracted and afford channels for weathering agents. Some rocks of otherwise good qualities are so cut by joints that they are useless for anything but crushed stone. The bedding planes or stratification of sedimentary rocks exercise influences similar to joints, and like joints may be useful or disadvantageous, depending on their spacing. The secondary cleavage of some rocks, notably slates, enables them to be split into flat slabs and thus makes them useful for certain purposes.

Proper methods of extraction and use of a rock may minimize the disadvantageous effects of its structural features. The use of channelling machines instead of explosives means less shattering of the rock. By proper dressing of the surface the opening of small crevices may be avoided. Stratified rocks set on bed, so that the bedding planes are horizontal, last longer than if set on edge.

The durability of a rock may depend on its perviousness to water which may enter along planes of bedding or incipient fracture planes, or along the minute pore spaces between the mineral particles. The water may cause disastrous chemical changes in the minerals and by its freezing and thawing may cause splitting. For this reason, the less pervious rocks have in general greater durability than the more pervious. Highly pervious rocks used in a dry position or in a dry climate will last longer than elsewhere.

Durability is determined also by the different coefficients of expansion of the constituent minerals of the rock. Where the minerals are heterogeneous in this regard, differential stresses are more likely to be set up than where the minerals are homogeneous. Likewise a coarse-textured rock is in general less durable than a fine-textured one. Expansion and contraction of a stone under ordinary temperature changes, and also under fire and freezing, must necessarily be known for many kinds of construction.

Minerals resist weathering to different degrees, therefore the mineral composition of a rock is another considerable factor in determining its durability. Where pyrite is present in abundance it easily weathers out, leaving iron-stained pits and releasing sulphuric acid which decomposes the rock. Abundance of mica, especially where segregated along the stratification planes, permits easy splitting of the rock under weathering. Likewise the mica often weathers more quickly than the surrounding minerals, giving a pitted appearance; in marbles and limestones its irregular occurrence may spoil the appearance. Flint or chert in abundance is deleterious to limestones and marbles, because, being more resistant, it stands out in relief on the weathered surface, interferes with smooth cutting and polishing, and often causes the rock to split along the lines of the flint concretions. Abundance of tremolite may also be disadvantageous to limestones and marbles, because it weathers to a greenish-yellow clay and leaves a pitted surface.

The crushing strength of a rock has an obvious relation to its structural uses. The rock must be strong enough for the specified load. Most hard rocks ordinarily considered for building purposes are strong enough for the loads to which subjected, and this factor is perhaps ordinarily less important than the structural and mineral features already mentioned.

It is often necessary to know the modulus of elasticity and other mechanical constants of a rock, as in cases where it is to be combined with metal or other masonry or to be subjected to exceptional shock.

The beauty and coloring of a rock are its esthetic rather than its utilitarian features. They are particularly important in the construction of buildings and monuments for public or ornamental purposes.

Crushed Stone

The largest use of rock or stone is in the crushed form for road building, railway embankments, and concrete, and the prospect is for largely increased demands for such uses in the future. For the purpose of road building, it is necessary to consider a stone's resistance to abrasion, hardness, toughness, cementing value, absorption, and specific gravity. Limestone cements well, but in other qualities it is not desirable for heavy traffic. Shales are soft and clayey, and grind down to a mass which is dry and powdery, and muddy in wet weather. Basalt and related rocks resist abrasion, and cement well. Granites and other coarse-grained igneous rocks do not cement well and are not resistant to abrasion. Many sandstones are very hard and brittle and resist abrasion, but do not cement.

The application of geology on a large scale to the study of sources and qualities of crushed stone is now being required in connection with the great state and national projects of highway building. This work is by no means confined to a mere testing of the physical qualities of road-building materials found along the proposed route, but includes a careful study of their geologic occurrence, distribution, and probable amounts. In certain of the northern states specialists in glacial geology are preferred for this purpose.

Stone for Metallurgical Purposes

The use of limestone and other rock for metallurgical fluxes is dependent very largely on chemical composition. Comparatively few limestones are sufficiently pure for this purpose. For furnace linings, the quartzite or ganister must be exceptionally pure. The field search for rocks of the necessary composition has required geologic service.

Clay

For a variety of uses to which clay is put, it is necessary to know its degree of plasticity, tensile strength, shrinkage (both under air and fire), fusibility, color, specific gravity, and chemical properties. The testing of clay for its various possible uses is a highly specialized job, usually beyond the range of a geologist, although certain geologists have been leaders in this type of investigation. More commonly within the range of a geologist are questions concerning origin, field classification, distribution, quantities, and other geologic conditions affecting quality and production.

Clay originates from the weathering of common rocks containing silicates, by pretty well understood weathering processes (see Chapter II). It may remain in place above the parent rock, or may be transported and redeposited, either on land or under water, by the agencies of air, water, and ice. The kind of parent rock, the climatic conditions and nature of the weathering, and the degree of sorting during transportation, all determine the composition and texture of the resulting clay,—with the result that a classification on the basis of origin may indicate the broad group characteristics which it is desirable to know for commercial purposes. For instance, residual clays from the weathering of granite may be broadly contrasted with residual clays formed by the weathering of limestone, and both differ in group characteristics from clays in glacial deposits. Classification according to origin also may be useful in indicating general features of depth, quantity, and distribution. However, a genetic classification of clays is often not sufficient to indicate the precise characteristics which it is necessary to know in determining their availability for narrow and special technical requirements. Furthermore, clays suitable for certain commercial requirements may be formed in several different ways, and classification based on specific qualities may therefore not correspond at all to geologic classification based on origin.

Geologists have been especially interested in the causes of plasticity of clay and in its manner of hardening when dried. In general these phenomena have been found to be due to content of colloidal substances of a clayey nature, which serve not only to hold the substance together during plastic flow but to bind it during drying. The part played by colloids in the formation of clays, as well as of many other mineral products, is now a question which is receiving intensive study.

The same processes which produce clay also produce, under special conditions, iron ores, bauxites, the oxide zones of many sulphide ore bodies, and soils, all of which are referred to on other pages.

Limitations of Geologic Field in Commercial Investigation Of Common Rocks

In general the qualities of the earth materials which determine their availability for use are only to a minor extent the qualities which the geologist ordinarily considers for mapping and descriptive purposes. The usual geological map and report on a district indicate the distribution and general nature of the common rocks, and also the extent to which they are being used as mineral resources. Seldom, however, is there added a sufficiently precise description, for instance of a clay, to enable the reader to determine which, if any, of the many different uses the material might be put to. The variety of uses is so great, and the technical requirements for different purposes are so varied and so variable, that it is almost impossible to make a description which is sufficiently comprehensive, and at the same time sufficiently exact, to give all the information desired for economic purposes. If the geologist is interested in disclosing the commercial possibilities in the raw materials of an area, he may select some of the more promising features and subject them to the technical analysis necessary to determine their availability for special uses. In this phase of his work he may find it necessary to enlist the coöperation of skilled technicians and laboratories in the various special fields. The problem is simplified if the geologist is hunting for a particular material for a specific purpose, for then he fortifies himself with a knowledge of the particular qualities needed and directs his field and laboratory study accordingly.

Too often the geologist fails to recognize the complexity and definiteness of the qualities required, and makes statements and recommendations on the use of raw materials based on somewhat general geologic observations. On the other hand, the engineer, or the manufacturer, or the builder often goes wrong and spends money needlessly, by failing to take into consideration general geologic features which may be very helpful in determining the distribution, amount, and general characters of the raw materials needed.

It is difficult to draw the line between the proper fields of the geologist and those of the engineer, the metallurgist, and other technicians. It is highly desirable that the specialist in any one of these fields know at least of the existence of the other fields and something of their general nature. Too often his actions indicate he is not acutely conscious even of the existence of these related branches of knowledge. The extent and detail to which the geologist will familiarize himself with these other fields will of course vary with his training and the circumstances of his work. Whatever his limit is, it should be definitely recognized; his work should be thorough up to this limit and his efforts should not be wasted in fields which he is not best qualified to investigate.

These remarks apply rather generally to mineral resources, but they are particularly pertinent in relation to the common rock materials which the geologist is daily handling,—for he is likely to assume that he knows all about them and that he is qualified to give professional advice to industries using them. In connection with metallic resources, the metallurgical and other technical requirements are likely to be more definitely recognized and the lines more sharply drawn, with the result that the geologist is perhaps not so likely to venture into problems which he is not qualified to handle.

The limits to geologic work here discussed are not necessarily limits separating scientific from non-scientific work. The study and determination of the qualities of rocks necessary for commercial purposes is fully as scientific as a study of the qualities commonly considered in purely geologic work, and the results of technical commercial investigations may be highly illuminating from a purely geological standpoint. When a field of scientific endeavor has been established by custom, any excursion beyond traditional limits is almost sure to be regarded by conservatives in the field as non-scientific, and to be lightly regarded. The writer is fully conscious of the existence of limits and the necessity for their recognition; but he would explain his caution in exceeding these limits on the ground of training and effectiveness, rather than on fear that he is becoming tainted with non-scientific matters the moment he steps beyond the boundaries of his traditional field.

SOILS AS A MINERAL RESOURCE

Soils are not ordinarily listed as mineral resources; but as weathered and altered rock of great economic value, they belong nearly at the head of the list of mineral products.

ORIGIN OF SOILS

Soil originate from rocks, igneous, sedimentary, and "metamorphic" by processes of weathering, and by the mixing of the altered mineral products with decayed plant remains or *humus*. The humus averages perhaps 3 or 4 per cent of the soil mass and sometimes constitutes as much as 75 per cent. Not all weathered rock is soil in the agricultural sense. For this purpose the term is mainly restricted to the upper few inches or feet penetrated by plant roots.

The general process of soil formation constitutes one of the most important phases of katamorphism—the destructive side of the metamorphic cycle, described in Chapter II. Processes of katamorphism or weathering, usually accompanied by the formation of soils, affect the surface rocks over practically all the continental areas.

The weathering of a highly acid igneous rock with much quartz produces a residual soil with much quartz. The weathering of a basic igneous rock without quartz produces a clay soil without quartz, which may be high in iron. Where disintegration has been important the soil contains an abundance of the original silicates of the rock, and less of the altered minerals.

The production of soil from sedimentary rocks involves the same processes as alter igneous rocks; but, starting from rocks of different composition, the result is of course different in some respects. Sandstones by weathering yield only a sandy soil. Limestones lose their calcium carbonate by solution, leaving only clay with fragments of quartz or chert as impurities. A foot of soil may represent the weathering of a hundred feet of limestone. Shales may weather into products more nearly like those of the weathering of igneous rocks. Silicates in the shales are broken down to form clay, which is mixed with the iron oxide and quartz.

In some localities the soil may accumulate to a considerable depth, allowing the processes of weathering to go to an extreme;

in others the processes may be interrupted by erosion, which sweeps off the weathered products at intermediate stages of decomposition and may leave a very thin and little decomposed soil.

Soils formed by weathering may remain in place as residual soils, or they may be transported, sorted, and redeposited, either on land or under water. It is estimated by the United States Bureau of Soils[14] that upward of 90 per cent of the soils of the United States which have been thus far mapped owe their occurrence and distribution to transportation by moving water, air, and ice (glaciers), and that less than 10 per cent have remained in place above their parent rock. Glaciers may move the weathered rock products, or they may grind the fresh rocks into a powder called *rock flour*, and thus form soils having more nearly the chemical composition of the unaltered rocks. Glacial soils are ordinarily rather poorly sorted, while wind and water-borne soils are more likely to show a high degree of sorting.

The character of a transported soil is less closely related to the parent rock than is that of a residual soil, because the processes of sorting and mixture of materials from different sources intervene to develop deposits of a nature quite different from residual soils; but even transported soil may sometimes be traced to a known rock parentage.

Where deposited under water, soil materials may be brought above the water by physiographic changes, and exposed at the surface in condition for immediate use. Or, they may become buried by other sediments and not be exposed again until after they have been pretty well hardened and cemented,—in which case they must again undergo the softening processes of weathering before they become available for use. Where soils become buried under other rocks and become hardened, they are classed as sedimentary rocks and form a part of the geologic record. Many residual and transported soils are to be recognized in the geologic column; in fact a large number of the sedimentary rocks ordinarily dealt with in stratigraphic geology are really transported soils.

The development of soils by weathering should not be regarded as a special process of rock alteration, unrelated to processes producing other mineral products. Exactly the same processes that produce soils may yield important deposits of iron ore, bauxite, and clay, and they cause also secondary enrichment of many metallic mineral deposits. For instance the weathering of a syenite rock containing no quartz, under certain conditions, as in Arkansas, results in great bauxite deposits which are truly soils and are useful as such,—but which happen to be more valuable because of their content of bauxite. The weathering of a basic igneous rock, as in Cuba, may produce important residual iron ore deposits, which are also used as soils. Weathering of ferruginous limestone may produce residual iron and manganese ores in clay soils.

Composition of Soils and Plant Growth

The mineral ingredients in soils which are essential for plant growth include water, potash, lime, magnesia, nitrates, sulphur, and phosphoric acid—all of which are subordinate in amount to the common products of weathering (pp. 20-22, 23-24). Of these constituents magnesia is almost invariably present in sufficient quantity; while potash, nitrates, lime, sulphur, and phosphoric acid, although often sufficiently abundant in virgin soil, when extracted from the soils by plant growth are liable to exhaustion under ordinary methods of cultivation, and may need to be replenished by fertilizers (Chapter VII). Some soils may be so excessively high in silica, iron, or other constituents, that the remaining constituents are in too small amounts for successful plant growth.

Even where soils originally have enough of all the necessary chemical elements, one soil may support plant growth and another may not, for the reason that the necessary constituents are soluble and hence available to the plant roots in one case and are not soluble in the other. Plainly the mineral combinations in which the various elements occur are important factors in making them available for plant use. Similarly a soil of a certain chemical and mineralogical composition may be fruitful under one set of climatic conditions and a soil of like composition may be barren at another locality—indicating that availability of constituents is also determined by climatic and other conditions of weathering. Even with the same chemical composition and the same climatic conditions, there may be such differences in texture between various soils as to make them widely different in yield.

The unit of soil classification is the *soil type*, which is a soil having agricultural unity, as determined by texture, chemical character, topography, and climate. The types commonly named are clay, clay loam, silt loam, loam, fine sandy loam, sandy loam, fine sand, and sand. In general the soil materials are so heterogeneous and so remote from specific rock origin, that in such classification the geologic factor of origin is not taken into account. More broadly, soils may be classified into provinces on the basis of geography, similar physiographic conditions, and similarity of parent rocks; for instance, the soils of the Piedmont plateau province, of the arid southwest region, of the glacial and loessal province, etc. In such classification the geologic factors are more important. Soils within a province may be subdivided into "soil series" on the basis of common types of sub-soils, relief, drainage, and origin.

USE OF GEOLOGY IN SOIL STUDY

While the desirability of particular soils is related in a broad way to the character of the parent rocks, and while by geologic knowledge certain territories can be predicated in advance as being more favorable than others to the development of good soils, so many other factors enter into the question that the geologic factor may be a subordinate one. A soil expert finds a knowledge of geology useful as a basis for a broad study of his subject; but in following up its intricacies he gives attention mainly to other factors, such as the availability of common constituents for plant use, the existence and availability of minute quantities of materials not ordinarily regarded as important by the geologist, the climatic conditions, and the texture. As the geologic factors are many of them comparatively simple, much of the expert work on soils requires only elementary and empirical knowledge of geology. The geologist, although he may understand fully the origin of soils and may indicate certain broad features, must acquire a vast technique not closely related to geology before he becomes effective in soil survey work and diagnosis.

For these reasons the mapping and classification of soils, while often started by geologists of state or federal surveys, have in their technical development and application now passed largely into the hands of soil experts in the special soil surveys affiliated with the U. S. Department of Agriculture and with agricultural colleges.

FOOTNOTES:

[13] A good summary of this subject may be found in *Engineering Geology*, by H. Ries and T. L. Watson, Wiley and Sons, 2d ed., 1915.

[14] Marbut, Curtis F., Soils of the United States: *Bull. 96, Bureau of Soils*, 1913, p. 10.

CHAPTER VII

THE FERTILIZER GROUP OF MINERALS

GENERAL COMMENTS

Soils are weathered rock more or less mixed with organic material. The weathering processes forming soils are in the field of geologic investigation, but the study of soils in relation to agriculture requires attention to texture and to several of their very minor constituents which have little geologic significance. Soil study has therefore become a highly specialized and technicalized subject,—for which a geological background is essential, but which is usually beyond the range of the geologist. To supply substances which are deficient in soils, however, requires the mining, quarrying, or extraction of important mineral resources, and in this part of the soil problem the geologist is especially interested.

Soils may be originally deficient in nitrates, phosphates, or potash; or the continued cropping of soils may take out these materials faster than the natural processes of nature supply them. In some soils there are sufficient phosphates and potash to supply all plant needs indefinitely; but the weathering and alteration processes, through which these materials are rendered soluble and available for plant life, in most cases are unable to keep up with the depletion caused by cropping. A ton of wheat takes out of the soil on an average 47 pounds of nitrogen, 18 pounds of phosphoric acid, 12 pounds of potash. On older soils in Europe it has been found necessary to use on an average 200 pounds of mixed mineral fertilizers annually per acre. On the newer soils of the United States the average thus far used has been less than one-seventh of this amount. The United States has thus far been using up the original materials stored in the soil by nature, but these have not been sufficient to yield anything like the crop output per acre of the more highly fertilized soils of Europe.

In addition to the nitrates, phosphates, and potassium salts, important amounts of lime and sulphuric acid, and some gypsum, are used in connection with soils. Lime is derived from crushed limestone (pp. 82-83), and is used primarily to counteract acidity or sourness of the soil; it is, therefore, only indirectly related to fertilizers. Sulphuric acid is used to treat rock phosphates to make them more soluble and available to plant life. It requires the mining of pyrite and sulphur. Gypsum, under the name of "land-plaster," is applied to soils which are deficient in the sulphur required for plant life; increase in its use in the future seems probable. There are also considerable amounts of inert mineral substances which are used as fillers in fertilizers to give bulk to the product, but which have no agricultural value. The proportions of the fertilizer substances used in the United States are roughly summarized in Figure 4.

The United States possesses abundant supplies of two of the chief mineral substances entering into commercial fertilizers,—phosphate rock and the sulphur-bearing materials necessary to treat it. For potash the United States is dependent on Europe, unless the domestic industry is very greatly fostered under protective tariff. For the mineral nitrates the United States has been dependent on Chile, and because of the cheapness of the supply will doubtless continue to draw heavily from this source. However, because of the domestic development of plants for the fixation of nitrogen from the air, the recovery of nitrogen from coal in the by-product processes, and the use of nitrogenous plants, the United States is likely to require progressively less of the mineral nitrates from Chile.

The fertilizer industry of the United States is yet in its infancy and is likely to have a large growth. Furthermore much remains to be learned about the mixing of fertilizers and the amounts and kinds of materials to be used. The importance of sulphur as a plant food has been realized comparatively recently. The use of fertilizers in the United States has come partly through education and the activity of agricultural schools and partly through advertising by fertilizer companies. The increased use of potash has been due largely to the propaganda of the German sales agents. An examination of a map showing distribution of the use of fertilizers over the country indicates very clearly the erratic distribution of the effects of these various activities. One locality may use large amounts, while adjacent territory of similar physical conditions uses little. The sudden withdrawal of fertilizers for a period of three or four years during the war had very deleterious effects in some localities, but was not so disastrous as expected in others,—emphasizing the fact that the use of fertilizers has been partly fortuitous and not nicely adjusted to specific needs.

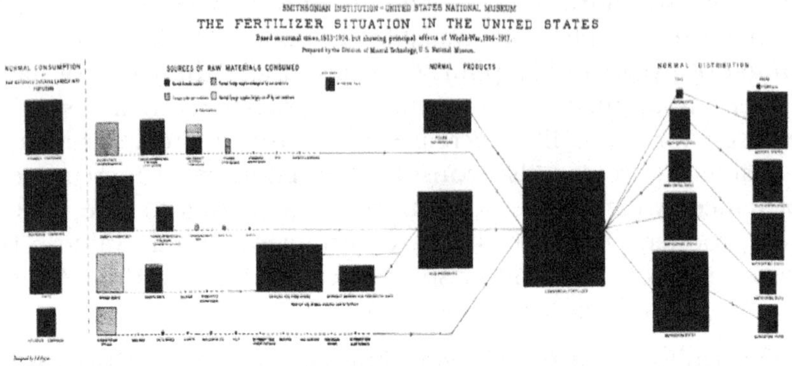

FIG. 4. FERTILIZER SITUATION IN THE UNITED STATES. SMITHSONIAN INSTITUTION—UNITED STATES NATIONAL MUSEUMToList

NITRATES

ECONOMIC FEATURES

There are several sources of nitrogen for fertilizer purposes: mineral nitrates, nitrogen taken from the air by certain plants with the aid of bacteria and plowed into the soil, nitrogen taken directly from the air by combining nitrogen and oxygen atoms in an electric arc, or by combining nitrogen and hydrogen to form ammonia, nitrogen taken from the air to make a compound of calcium, carbon, and nitrogen (cyanamid), nitrogen saved from coal in the form of ammonia as a by-product of coke-manufacture, and nitrogen from various organic wastes. Nitrogen in the form of ammonia is also one of the potential products of oil-shales (p. 150). While the principal use of nitrogenous materials is as fertilizers, additional important quantities are used in ammonia for refrigerating plants, and in the form of nitric acid in a large number of chemical industries. During the war the use of nitrates was largely diverted to explosives manufacture. The geologist is interested principally in the mineral nitrates as a mineral resource, but the other sources of nitrogen, particularly its recovery from coal, also touch his field.

Almost the single source of mineral nitrates for the world at present is Chile, where there are deposits of sodium nitrate or Chile saltpeter, containing minor amounts of potassium nitrate. About two-thirds of the Chilean material normally goes to Europe and about one-fourth to the United States. The supply has been commercially controlled chiefly by Great Britain and by Chilean companies backed by British and German capital.

The dependence of the world on Chile became painfully apparent during the war. Germany was the only nation which had developed other sources of nitrogenous material to any great extent. The other nations were dependent in a very large degree on the mineral nitrates, both for fertilizer and munition purposes. Total demands far exceeded the total output from Chile, requiring international agreement as to the division of the output among the nations. The stream of several hundred ships carrying nitrates from Chile was one of the vital war arteries. This situation led to strenuous efforts in the belligerent countries toward the development of other sources of nitrogen. The United States, under governmental appropriation, began the building of extensive plants for the fixation of nitrogen from the air, and the building of by-product coke ovens in the place of the old wasteful beehive ovens was accelerated. Germany before the war had already gone far in both of these directions, not only within her own boundaries, but in the building of fixation plants in Scandinavia and Switzerland. War conditions required further development of these processes in Germany, with the result that this country was soon entirely self-supporting in this regard. One of the effects was the almost complete elimination in Germany of anything but the by-product process of coking coal.

War-time development of the nitrogen industry in the United States for munition purposes brought the domestic production almost up to the pre-war requirements for fertilizers alone. With the increasing demand for fertilizers and with the cheapness of the Chilean supply of natural nitrates, it is likely that the United States will continue for a good many years to import considerable amounts of Chilean nitrates. It may be noted that, although this country normally consumes about one-fourth of the Chilean product, American interests commercially control less than one-twentieth of the output. Presumably, if for no other purpose than future protection, effort will be made to develop the domestic industry to a point where in a crisis the United States could be independent of Chile. Particularly may an increase in the output of by-product ammonia from coke manufacture be looked for (see also pp. 118-119), since nitrogenous material thus produced need bear no fixed part of the cost of production, and requires no protective tariff.

The reserves of Chilean nitrate are known to be sufficient for world requirements for an indefinitely long future.

GEOLOGIC FEATURES

Mineral nitrates in general, and particularly those of soda and potash, are readily soluble at ordinary temperatures. Mineral nitrate deposits are therefore very rare, and are found only in arid regions or other places where they are protected from rain and ground-water. The only large deposits known are those of northern Chile and some extensions in adjacent parts of Peru and Bolivia. These are located on high desert plateaus, where there is almost a total absence of rain, and form blankets of one to six feet in thickness near the surface. The most important mineral, the sodium nitrate or Chile saltpeter, is mingled with various other soluble salts, including common salt, borax minerals, and potassium nitrate, and with loose clay, sand, and gravel. The nitrate deposits occur largely around and just above slight basin-like depressions in the desert which contain an abundance of common salt. The highest grade material contains 40 to 50 per cent of sodium nitrate, and material to be of shipping grade must run at least 12 to 15 per cent.

The origin of the nitrate beds is commonly believed to be similar to that of beds of rock salt (pp. 295-298), borax, and other saline residues. The source of the nitrogen was probably organic matter in the soil, such as former deposits of bird guano, bones (which are actually found in the same desert basin), and ancient vegetable matter. By the action of nitrifying bacteria on this organic matter, nitrate salts are believed to have formed which were leached out by surface and ground waters, and probably carried in solution to enclosed bodies of water. Here they became mingled with various other salts, and all were precipitated out as the waters of the basins evaporated. Deliquescence and later migration of the more soluble nitrates resulted in their accumulation around the edges of the basins. The nitrate beds are thus essentially a product of desiccation.

While the origin just set forth is rather generally accepted, several other theories have been advanced. It has been suggested that the deposits were not formed in water basins, but that ground water carrying nitrates in solution has been and is rising to the surface,—where, under the extremely arid conditions, it evaporates rapidly, leaving the nitrates mixed with the surface clays. One group of writers accounts for the deposits by the fixation of atmospheric nitrogen through electrical phenomena. Still others note the frequent presence of nitrogen in volcanic exhalations and the association of the Chilean nitrate beds with surface volcanic rocks; they suggest that these rocks were the source of the nitrogen, which under unusual climatic conditions was leached out and then deposited by evaporation.

PHOSPHATES

Economic Features

The principal use of natural phosphates is in the manufacture of fertilizers. They are also used in the manufacture of phosphorus, phosphoric acid, and other phosphorus compounds, for matches, for certain metallurgical operations, and for gases used in military operations.

The material mined is mainly a phosphate of lime (tricalcium phosphate). To make it available for plant use, it is treated with sulphuric acid to form a soluble superphosphate; hence the importance of sulphuric acid, and its mineral sources pyrite and sulphur, in the fertilizer industry. A small percentage of the phosphate is also ground up and applied directly to the soil in the raw form. Other phosphatic materials are the basic slag from phosphatic iron ores made into Thomas-process steel, guano from the Pacific islands, and bone and refuse (tankage) from the cattle raising and packing countries. These materials are used for the same purposes as the natural phosphates.

The United States is the largest factor in the world's phosphate industry, with reference both to production and reserves.

The largest and most available of the European sources are in Tunis and Algeria, under French control, and in Egypt, under English control. Belgium and northern France have been considerable producers of phosphates, but, with the development of higher grade deposits in other countries, their production has fallen to a very small fraction of the world's total. There also has been very small and insignificant production in Spain and Great Britain. Russia has large reserves which are practically unmined.

While there is comparatively little phosphate rock in western Europe, a considerable amount of the phosphate supply is obtained as a by-product from Thomas slag, derived from phosphatic iron ores. These ores are chiefly from Lorraine and Sweden, but English and Russian ores can be similarly used.

Outside of Europe and the United States, there are smaller phosphate supplies in Canada, the Dutch West Indies, Venezuela, Chile, South Australia, New Zealand, and several islands of the Indian and South Pacific Oceans. None of these has yet contributed largely to world production, and their distance from the principal consuming countries bordering the North Atlantic basin is so great that there is not likely to be any great movement to this part of the world. On the other hand, some of the South Sea islands have large reserves of exceptionally high grade guano and bone phosphates, which will doubtless be used in increasing amounts for export to Japan, New Zealand, and other nearby countries. The most important of these islands are now controlled by Great Britain, Japan, and France.

A striking feature of the situation is that the central European countries, which have been large consumers of phosphate material, have lost not only the Pacific island phosphates but the Lorraine phosphatic iron ores, and are now almost completely dependent on British, French, and United States phosphate.

In the United States, reserves of phosphate are very large. They are mined principally in Florida, Tennessee, and South Carolina; but great reserves, though of lower grade, are known in Arkansas, Montana, Idaho, Wyoming, and Utah. There are possibilities for the development of local phosphate industries in the west, in connection with the manufacture of sulphuric acid from waste smelting gases at nearby mining centers. The Anaconda Copper Mining Company has taken up the manufacture of superphosphate as a means of using sulphuric acid made in relation to its smelting operations. The United States is independent in phosphate supplies and has a surplus for export. This country, England, and France exercise control of the greater part of the world's supply of phosphatic material. In competition for world trade, the Florida and Carolina phosphates are favorably situated for export, but there is strong competition in Europe from the immense fields in French North Africa, which are about equally well situated.

GEOLOGIC FEATURES

Small amounts of phosphorus are common in igneous rocks, in the form of the mineral apatite (calcium phosphate with calcium chloride or fluoride). Apatite is especially abundant in some pegmatites. In a few places, as in the Adirondacks where magnetic concentration of iron ores leaves a residue containing much apatite, and in Canada and Spain where veins of apatite have been mined, this material is used as a source of phosphate fertilizer. The great bulk of the world's phosphate, however, is obtained from other sources—sedimentary and residual beds described below.

Phosphorus in the rocks is dissolved in one form or another by the ground-waters; a part of it is taken up by land plants and animals for the building of their tissues, and another part goes in solution to the sea to be taken up by sea plants and animals. In places where the bones and excrements of land animals or the shells and droppings of sea animals accumulate, deposits of phosphatic material may be built up.

In certain places where great numbers of sea birds congregate, as on desert coasts and oceanic islands, guano deposits have been formed. Some of them, like the worked-out deposits of Peru and Chile, are in arid climates and have been well preserved. Others, like those of the West Indies and Oceania, are subjected to the action of occasional rains; and to a large extent the phosphates have been leached out, carried down, and reprecipitated, permeating and partially replacing the underlying limestones. In this way deposits have been formed containing as high as 85 per cent calcium phosphate.

Even more important bodies of phosphates have been produced by the accumulation of marine animal remains, probably with the aid of joint chemical, bacterial, and mechanical precipitation. These processes have formed the chief productive deposits of the world, including those of the United States, northern Africa, and Russia, and also the phosphatic iron ores of

England and central Europe. The sedimentary features of many phosphate rocks, particularly their oölitic textures, show a marked similarity to the features of the Clinton type of iron ores (pp. 166-167).

The marine phosphate beds originally consist principally of calcium phosphate and calcium carbonate in varying proportions. Depending on the amount of secondary enrichment, they form two main types of deposits. The extensive beds of the western United States (in the upper Carboniferous) are hard, and very little enrichment by weathering has taken place; they carry in their richer portions 70 to 80 per cent calcium phosphate, and large sections range only from about 30 to 50 per cent. In the southeastern deposits (Silurian and Devonian in Tennessee and Tertiary in the Carolinas and Florida), there has been considerable enrichment, the rock is softer, and the general grade ranges from 65 to 80 per cent. Both calcium carbonate and calcium phosphate are soluble in ordinary ground waters, but the carbonate is the more soluble of the two. Thus the carbonate has been dissolved out more rapidly, and in addition descending waters carrying the phosphate have frequently deposited it to pick up the carbonate. These enriching processes, sometimes aided by mechanical concentration, have formed high-grade deposits both in the originally phosphatic beds and in various underlying strata. Concretionary and nodular textures are common. The "pebble" deposits of Florida consist of the phosphatic materials broken up and worked over by river waters and advancing shallow seas.

PYRITE

ECONOMIC FEATURES

The principal use of pyrite is in the manufacture of sulphuric acid. Large quantities of acid are used in the manufacture of fertilizers from phosphate rock, and during war times in the manufacture of munitions. Sulphuric acid converts the phosphate rock into superphosphate, which is soluble and available for plant use. Other uses of the acid are referred to in connection with sulphur. Pyrite is also used in Europe for the manufacture of paper from wood-pulp, but in the United States native sulphur has thus far been exclusively used for this purpose. The residue from the roasting of pyrite is a high-grade iron ore material frequently very low in phosphorus, which is desirable in making up mixtures for iron blast furnaces.

Most of the countries of Europe are producers of pyrite, and important amounts are also produced in the United States and Canada. The European production is marketed mainly on that continent, but considerable amounts come to the United States from Spain.

Before the war domestic sources supplied a fourth to a third of the domestic demand for pyrite. Imports came mainly from Spain and Portugal to consuming centers on the Atlantic seaboard. The curtailment of overseas imports of pyrite during the war increased domestic production by about a third and resulted also in drawing more heavily on Canadian supplies, but the total was not sufficient to meet the demand. The demand was met by the increased use of sulphur from domestic deposits (p. 109). At the close of the war supplies of pyrite had been accumulated to such an extent that, with the prospect of reopening of Spanish importation, pyrite production in the United States practically ceased. War experience has demonstrated the possibility of substitution of sulphur, which the United States has in large and cheaply mined quantities. The future of the pyrite industry in the United States therefore looks cloudy, except for supplies used locally, as in the territory tributary to the Great Lakes, and except for small amounts locally recovered as by-products in the mining of coal or from ores of zinc, lead, and copper. Pyrite production in the past has been chiefly in the Appalachian region, particularly in Virginia and New York, and in California.

Geologic Features

Pyrite, the yellow iron sulphide, is the commonest and most abundant of the metallic sulphides. It is formed under a large variety of conditions and associations. Marcasite and pyrrhotite, other iron sulphide minerals, are frequently found with pyrite and are used for the same purposes.

The great deposits of Rio Tinto, Spain, which produce about half of the world's pyrite, were formed by replacement of slates by heated solutions from nearby igneous rocks. The ores are in lenticular bodies, and consist of almost massive pyrite with a small amount of quartz and scattered grains and threads of chalcopyrite (copper-iron sulphide). They carry about 50 per cent of sulphur, and the larger part carries about 2 per cent of copper which is also recovered.

Similar occurrences of pyrite on a smaller scale are known in many places. Pyrite is very commonly found in vein and replacement deposits of gold, silver, copper, lead, and zinc. In the Mississippi valley it is extracted as a by-product from the lead and zinc ores, and in the Cordilleran region large quantities of by-product pyrite could easily be produced if there were a local demand. The pyrite deposits of the Appalachian region are chiefly lenses in schists; they are of uncertain origin though some are believed to have been formed by replacement of metamorphosed limestones and schists.

Under weathering conditions pyrite oxidizes, the sulphur forming sulphuric acid,—an important agent in the secondary enrichment of copper and other sulphides,—and the iron forming the minerals hematite and limonite in the shape of a "gossan" or "iron-cap."

Pyrite is likewise frequently found in sediments, apparently being formed mainly by the reducing action of organic matter on iron salts in solution. In Illinois and adjacent states it is obtained as a by-product of coal mining.

SULPHUR

ECONOMIC FEATURES

Sulphur is used for many of the same purposes as pyrite. Under pre-war conditions, the largest use in the United States was in the manufacture of paper pulp by the sulphite process. Minor uses were in agriculture as a fungicide and insecticide, in vulcanizing rubber, and in the manufacture of gunpowder. About 5 per cent of the sulphur of the United States was used in the manufacture of sulphuric acid. During the war this use was greatly increased because of the shortage of pyrite and the large quantities of sulphuric acid necessary for the manufacture of explosives. The replacement of pyrite by sulphur in the manufacture of sulphuric acid has continued since the war, and in the future is likely to continue to play an important part. Sulphuric acid is an essential material for a great range of manufacturing processes. Some of its more important applications are: in the manufacture of superphosphate fertilizer from phosphate rock; in the refining of petroleum products; in the iron, steel, and coke industries; in the manufacture of nitroglycerin and other explosives; and in general metallurgical and chemical practice.

The United States is the world's largest sulphur producer. The principal foreign countries producing important amounts of sulphur are Italy, Japan, Spain, and Chile. Europe is the chief market for the Italian sulphur. In spite of increased demands in Europe the Italian production has decreased as the result of unfavorable labor, mining, and transportation conditions, and the deficit has had to be met from the United States. Japan's sulphur production has been increasing. Normally about half of the material exported comes to the United States to supply the needs of the paper industry in the Pacific states, and half goes to Australia and other British colonies. Spain's production is relatively small and has been increasing slowly; most of it is consumed locally. Chile's small production is mainly consumed at home and large additional amounts are imported.

The sulphur output of the United States, which in 1913-14 was second to Italy, now amounts to three-fourths of the entire output of the world, and the United States has become a large exporter of sulphur. Supplies are ample and production increasing, with the result that the United States can not only meet its own demands, but can use this commodity extensively in world trade. Small amounts of sulphur are mined in some of the western states, but over 98 per cent of the production comes from Louisiana and Texas.

GEOLOGIC FEATURES

Native sulphur is found principally in sedimentary beds, where it is associated with gypsum and usually with organic matter. Deposits of this type are known in many places, the most important being those of Sicily and of the Gulf Coast in the United States. In the latter region beds of limestone carry lenses of sulphur and gypsum which are apparently localized in dome-like upbowings of the strata. The deposits are overlain by several hundred feet of loose, water-bearing sands, through which it is difficult to sink a shaft. An ingenious and efficient process of mining is used whereby superheated water is pumped down to melt the sulphur, which is then forced to the surface by compressed air and allowed to consolidate in large bins. The Sicilian deposits are similar lenses in clayey limestones containing 20 to 25 per cent of sulphur, associated with gypsum and bituminous marl; they are mined by shafts.

Concerning the origin of these deposits several theories have been advanced. It has been thought that the materials for the deposits were precipitated at the same time as the enclosing sediments; and that the sulphur may have been formed by the oxidation of hydrogen sulphide in the precipitating waters through the agency of air or of sulphur-secreting bacteria, or that it may have been produced by the reduction of gypsum by organic matter or bacteria. Others have suggested that hot waters rising from igneous rocks may have brought in both the sulphur and the gypsum, which in crystallizing caused the upbowing of the strata which is seen in the Gulf fields (see also p. 298).

Native sulphur is also found in mineral springs from which hydrogen sulphide issues, where it is produced by the oxidation of the hydrogen sulphide. It likewise occurs in fissures of lava and around volcanic vents, where it has probably been formed by reactions between the volcanic gases and the air. The Japanese and Chilean deposits are of the volcanic type.

POTASH

Economic Features

Potash is used principally as a component of fertilizers in agriculture. It is also used in the manufacture of soap, certain kinds of glass, matches, certain explosives, and chemical reagents.

For a long time potash production was essentially a German monopoly. The principal deposits are in the vicinity of Stassfurt in north central Germany (about the Harz Mountains). Stassfurt salts are undoubtedly ample to supply the world's needs of potash for an indefinite future. However, other deposits, discovered in the Rhine Valley in Alsace in 1904, have been proved to be of great extent; and though the production has hitherto been limited by restrictions imposed by the German Government, it has nevertheless become considerable.[15] The grade (18 per cent K_2O) is superior to the general run of material taken from the main German deposits, and the deposits have a regularity of structure and uniformity of material favorable to cheaper mining and refining than obtains in the Stassfurt deposits.

Other countries have also developed supplies of potash, some of which will probably continue to produce even in competition with the deposits of recognized importance referred to above. Noteworthy among the newer developments are those in Spain.[16] These have not yet produced on any large scale, but their future production may be considerable. Less important deposits are known in Galicia, Tunis, Russia, and eastern Abyssinia, and the nitrate deposits of Chile contain a small percentage of potash which is being recovered in some of the operations.

Prior to the war the United States obtained its potash from Germany. The German potash industry was well organized and protected by the German Government, which made every effort to maintain a world monopoly. During the war the potash exports from Germany were cut off, excepting exports to the neutrals immediately adjoining German territory. The result in the United States was that the price of potash rose so far as to greatly diminish its use as fertilizer.

The consequent efforts to increase potash production in the United States met with considerable success, but the maximum production attained was only about one-fourth of the ordinary pre-war requirements. The principal American sources are alkaline beds and brines in Nebraska, Utah, and California, and especially at Searles Lake, California. These furnished 75 per cent of the total output. Minor amounts have been extracted in Utah from the mineral alunite (a sulphate of potassium and aluminum), in Wyoming from leucite (a potassium-aluminum silicate), in California from kelp or seaweed, and in various localities from cement-mill and blast-furnace dusts, from wood ashes, from wool washings, from the waste residues of distilleries and beet-sugar refineries, and from miscellaneous industrial wastes. At the close of the war, sufficient progress had been made in the potash industry to indicate that the United States might become self-supporting in the future, though at high cost. The renewal of importation of cheap potash from Germany, with probable further offerings from Alsace and Spain, makes it impossible for the United States potash production to continue; except, perhaps, for the recovery of by-products which will go on in connection with other industries. Demand for a protective tariff has been the inevitable result (see Chapters XVII and XVIII).

GEOLOGIC FEATURES

Potassium is one of the eight most abundant elements in the earth. It occurs as a primary constituent of most igneous rocks, some of which carry percentages as high as those in commercial potash salts used for fertilizers. It is present in some sediments and likewise occurs in many schists and gneisses. Various potassium silicates—leucite, feldspar, sericite, and glauconite—and the potassium sulphate, alunite, have received attention and certain of them have been utilized to a small extent, but none of them are normally able to compete on the market. Potential supplies are thus practically unlimited in amount and distribution. Deposits from which the potash can be extracted at a reasonable cost, however, are known in only a few places, where they have been formed as saline sediments.

In the decomposition of rocks the potash, like the soda, is readily soluble, but in large part it is absorbed and held by clayey materials and is not carried off. Potash is therefore more sparingly present in river and ocean waters than is soda, and deposits of potash salts are much rarer than those of rock salt and other sodium compounds. The large deposits in the Permian beds of Stassfurt, as well as those in the Tertiary of Alsace and Spain, have been formed by the evaporation of very large quantities of salt water, presumably sea water. They consist of potassium salts, principally the chloride, mixed and intercrystallized with chlorides and sulphates of magnesium, sodium, and calcium. In the Stassfurt deposits the potassium-magnesium salts occupy a relatively thin horizon at the top of about 500 feet of rock salt beds, the whole underlying an area about 200 miles long and 140 miles wide. The principal minerals in the potash horizon are carnallite (hydrous potassium-magnesium chloride), kieserite (hydrous magnesium sulphate), sylvite (potassium chloride), kainite (a hydrous double salt of potassium chloride and magnesium sulphate), and common salt (sodium chloride). The potash beds represent the last stage in the evaporation of the waters of a great closed basin, and the peculiar climatic and topographic conditions which caused their formation have been the subject of much speculation. This subject is further treated in the discussion of common salt beds (pp. 295-298).

In the United States the deposits at Searles Lake, California, have been produced by the same processes on a smaller scale. In this case evaporation has not been carried to completion, but the crystallization and separation out of other salts has concentrated the potassium (with the magnesium) in the residual brine or "mother liquor." The deposits of this lake or marsh also contain borax (see p. 276), and differ in proportions of salts from the Stassfurt deposits. This is due to the fact that they were probably derived, not from ocean waters, but from the leaching of materials from the rocks of surrounding uplands, transportation of these materials in solution by rivers and ground waters, and concentration in the desert basin by evaporation.

The alkali lakes of Nebraska are believed to be of very recent geologic origin. They lie in depressions in a former sand dune area, and contain large quantities of potash supposedly accumulated by leaching of the ashes resulting from repeated burnings of the grass in the adjacent country.

Of other natural mineral sources, alunite is the most important. The principal deposits worked are at Marysville, Utah, but the mineral is a rather common one in the western part of the United States, associated with gold deposits, as at Goldfield, Nevada. Alunite occurs as veins and replacement deposits, often in igneous associations, and is supposed to be of igneous source. Its origin is referred to in connection with the Goldfield ores (p. 230).

FOOTNOTES:

[15] Gale, Hoyt S., The potash deposits of Alsace: *Bull. 715-B, U. S. Geol. Survey*, 1920, pp. 17-55.

[16] Gale, Hoyt S., Potash deposits in Spain: *Bull. 715-A, U. S. Geol. Survey*, 1920, pp. 1-16.

CHAPTER VIII

THE ENERGY RESOURCES—COAL, OIL, GAS (AND ASPHALT)

COAL

Economic Features

Coal overshadows all other mineral resources, except water, in production, value, and demand. It is the greatest of the energy sources—coal, petroleum, gas, and water power. Roughly two-thirds of the world's coal is used for power, one-sixth for smelting and metallurgical industries, and one-sixth for heating purposes. Coal constitutes over one-third of the railroad tonnage of the United States and is the largest single tonnage factor in international trade; 70 per cent of the pre-war tonnage of outgoing cargoes from England was coal.

World production and trade. The great coal-producing countries of the world border the North Atlantic basin. The United States produces about 40 per cent of the world's total, Great Britain about 20 per cent, and Germany about 20 per cent. Other countries producing coal stand in about the following order: Austria-Hungary, France, Russia, Belgium, Japan, China, India, Canada, and New South Wales. There is similarity in the major features of the distribution of coal production and of iron ore production. The great centers of coal production—the Pennsylvania and Illinois fields of the United States, the Midlands district of England, and the lower Rhine or Westphalian fields of Germany—are also the great centers of the iron and steel industries of these countries. As in the case of iron ore, there is rather a striking absence of important coal production in the southern hemisphere and in Asia. A significant item in the world's distribution of coal supplies is England's world-wide system of coaling stations for shipping.

The principal coal-producing countries all have large reserves of coal. Outside of these countries the world's most important reserves are in China, which may be looked to for great future development. For the most part, except for the probable Chinese development, it is likely that countries now producing most of the coal will continue to do so in the future, and that outlying parts of the world will continue to be supplied mainly from these countries.

The quantity and distribution of the coal reserves of the world have been estimated with perhaps a greater degree of accuracy than those of any other mineral resource. From these estimates it appears that the North American continent contains about half of the world reserves (principally in the United States, with lesser amounts in Canada) and Asia about one-fourth (principally in China, with some in India). Europe contains only one-sixth of the world total, chiefly in the area of the former German Empire and in Great Britain, with smaller quantities in Russia, Austria-Hungary, France, and Belgium. Australasia (New South Wales), Africa (British South Africa), and South America (Chile, Brazil, Peru, and Colombia), together contain less than a tenth of the total reserves. Coal being one of the great bases for modern industrialism, the large reserves of high grade-coals in China have led to the belief that China may some day develop into a great manufacturing nation. Similarly, the deficiency in coal of most of the South American and African countries seems to preclude their developing any very large manufacturing industries, except where water power is available. Coal reserves and the conservation of coal are further discussed in Chapter XVII.

The war resulted in considerable disturbances in coal production and distribution. There has not yet been a return to normal conditions, and some of the changes are probably permanent. The great overseas movement of coal from Germany was stopped and that from England curtailed. To some extent the deficiency was supplied by coal exports from the United States, particularly to South America. The shutting off of the normal German export to France and Mediterranean countries, the occupation of the French and Belgian coal fields by the Germans, and the partial restriction of German exports to Scandinavian countries, resulted in Europe's absorbing most of the British coal available for export, and in addition requiring coal from the United States. The stress in the world's coal industry to meet the energy requirements of war is too recent and vivid to require more than mention. The world was made to realize almost for the first time the utterly vital and essential nature of this industry.

Since the war, there has been a gradual resumption of England's export of coal along old lines of international trade. The German overseas export trade has not been reëstablished, and cannot be for a long time to come if Germany fulfills the terms of the Peace Treaty. Indeed, because of slow recovery in output of German coal, there is yet considerable lag in the supply available for European countries. The terms of the Peace Treaty lessened the territory of German coal reserves and required considerable additional contributions of coal to be delivered to France, Belgium, Luxemburg, and Italy.

The increased export of coal from the United States during the war is likely to be in part continued in the future, although the great bulk of the United States production will in the future, as in the past, be absorbed locally. Most of the coal in the United States available for export is higher in volatile matter than the British and German export coal. This quality will in some degree be a limiting factor in exportation. On the other hand, it may result in wider introduction of briquetting, coking, and other processes, which will tend to improve the local industry and be conservational in their effect.

Japan will doubtless hold some of the Asiatic coal market gained during the war.

International coal relations are further discussed in Chapter XVIII.[17]

Production in the United States. The main features of the distribution of coal supplies in the United States are:

(1) Localization of the anthracite production and reserves in a limited area in the Lawton region of Pennsylvania. Low-grade anthracite coal also occurs in Rhode Island, North Carolina, Colorado, and Idaho.

(2) Localization of the bituminous production in the eastern and interior states of Pennsylvania, West Virginia, Ohio, Indiana, Illinois, and Kentucky. The principal reserves of bituminous coal occur in the same provinces, but important additional reserves are known in Texas, in North and South Carolina, and in the Rocky Mountain and Pacific Coast provinces.

(3) The existence of large tonnages of subbituminous coal in the west, which have not been mined to any extent.

(4) The existence of large fields of lignite in the Gulf Coast region, and in the Northern Plains region, which have not been mined.

Coke. About one-sixth of the bituminous coal mined in the United States is made into *coke*, that is, it is subjected to heat in ovens from which oxygen is excluded in order to drive off the volatile gases (chiefly hydrocarbons and water) which constitute about 40 per cent of the weight of the coal. The residual product, the coke, is a light, porous mass with a considerably higher percentage of fixed carbon than bituminous coal. In regard to composition, coking accomplishes artificially somewhat the same result reached by nature in its slow development of high-grade coals, but the texture of coke is far different from that of coal. Not all bituminous coals are suitable for coke manufacture; and such coals are frequently divided into two classes, known as *coking* and *non-coking* coals. Coke is used principally for smelting purposes. Because of its spongy, porous texture, it burns more rapidly and intensely than coal.

The gases eliminated in coking are wasted in the old-fashioned "beehive" ovens, but in modern "by-product" coke ovens these gases by proper treatment yield valuable coal tar products and ammonia. It is estimated that the sum of the value of the products thus recovered from a ton of coal multiplies the value of the ton of coal at the mine by at least thirteen times. The importance of this fact from the conservational standpoint cannot be too much emphasized. At present over half of the total coke produced in the United States comes from by-product ovens, and this proportion will doubtless increase in the future.

BALANCE SHEET SHOWING CONTRAST BETWEEN VALUE OF 1 TON OF BITUMINOUS COAL AT MINE AND VALUE OF PRODUCTS WHICH IT CONTAINS, BASED ON CONDITIONS PREVAILING IN 1915.[1]

	Value of mine, 1915	Quantity	Value at point of production, 1915
1 ton (2,000 pounds)		1,500 pounds smokeless fuel	$5.00[2]
bituminous coal		10,000 cubic feet gas,	9.00[3]
contains	$1.13 =	at 90c. per 1,000	
		22 pounds ammonium	.61
		sulphate at 2.8c.	
		2-½ gallons benzol, at 30c.	.75[4]
		9 gallons tar, at 2.6c.	.23[4]
Total	$1.13[5]		$15.59

[1]Gilbert, Chester G., and Pogue, Joseph E., The energy resources of the United States—A field for reconstruction: *Bull. 102, U. S. National Museum*, vol. 1, 1919, p. 11.

[2]Figure based upon approximate selling price of anthracite.

[3]Figure based upon average price of city gas.

[4]These figures would be much higher if an adequate coal products industry were in existence.

This figure shows clearly that lowering the cost of production cannot be expected to lower the price of coal. Even if the cost of production were eliminated, the price of coal would merely be a dollar less.

Classification of coals. The accurate naming and classification of different varieties of coal is not an easy matter. The three main classes,—anthracite, bituminous, and lignite,—have group characteristics determined by their composition, color, texture, origin, and uses, and for general purposes these names have reasonably definite significance. However, there is complete gradation in coal materials from peat through lignite to bituminous and anthracite coals; many varieties fall near the border lines of the main groups, and their specific naming then becomes difficult. In addition, coal is made up of several substances which vary unequally in their proportions. It is difficult to arrange all of these variables in a graded series in such a fashion as to permit of precise naming of the coal. Furthermore, the scientific naming of a coal may not serve the purpose of discriminating coals used for different commercial purposes. Even the commercial names vary among themselves, depending on the use for which the coal is being considered.

Thus it is that the naming and classification of coals is a perennial source of difficulty and controversy. The earliest and most widely used classification is based on the ratio between fixed (or non-volatile) carbon and volatile constituents, called the "fuel ratio." For this purpose "proximate" analyses of coal are made, in terms of fixed carbon, volatile matter, moisture, ash, and sulphur. Anthracite has a higher fuel ratio than bituminous coal; that is, it has more fixed carbon in relation to volatile matter. Similarly bituminous coal has a higher fuel ratio than lignite. The fuel ratio measures roughly the heat or calorific power of the coal, in other words, its fuel value. However, some bituminous coals have a higher calorific power than some anthracites, because a large part of their volatile matter is combustible and yields more heat than the corresponding weight of fixed carbon in the anthracite. The fuel ratio pretty well discriminates coals of the higher ranks, and gives a classification corresponding roughly with their commercial uses. For the lower ranks of coal it is not so satisfactory, because the volatile constituents of such coals contain large and varying percentages of non-combustible hydrogen, oxygen, and nitrogen. Also such coals contain larger and more variable amounts of moisture, which is inert to combustion and requires heat for its evaporation. Two coals of the lower ranks with the same fuel ratio may have very different fuel qualities and different commercial uses, because of their different amounts of inert volatile matter and of water. For these coals it is sometimes desirable to supplement the chemical classification by physical criteria. For instance, subbituminous coal may be distinguished from lignite, not by its fuel ratio alone, but by its shiny, black appearance as contrasted with the dull, woody appearance of lignite. Bituminous may be distinguished from subbituminous by the manner of weathering. Other classifications have attempted to recognize these difficulties and still maintain a purely chemical basis by considering separately the combustible and non-combustible volatile constituents. For this purpose, it is necessary to have not merely approximate analyses, but the ultimate analyses in terms of elements.

Definitions of the principal kinds of coal by Campbell,[18] of the United States Geological Survey, are as follows:

Anthracite. Anthracite is generally well known and may be defined as a hard coal having a fuel ratio (fixed carbon divided by the volatile matter) of not more than 50 or 60 and not less than 10.

Semianthracite. Semianthracite is also a hard coal, but it is not so hard as true anthracite. It is high in fixed carbon, but not so high as anthracite. It may be defined as a hard coal having a fuel ratio ranging from 6 to 10. The lower limit is uncertain, as it is difficult to say where the line should be drawn to separate "hard" from "soft" coal and at the same time to divide the two ranks according to their fuel ratio.

Semibituminous. The name "semibituminous" is exceedingly unfortunate, as literally it implies that this coal is half the rank of bituminous, whereas it is applied to a kind of coal that is of higher rank than bituminous—really superbituminous. Semibituminous coal may be defined as coal having a fuel ratio ranging from 3 to 7. Its relatively high percentage of fixed carbon makes it nearly smokeless when it is burned properly, and consequently most of these coals go into the market as "smokeless coals."

Bituminous. The term "bituminous," as generally understood, is applied to a group of coals having a maximum fuel ratio of about 3, and hence it is a kind of coal in which the volatile matter and the fixed carbon are nearly equal; but this criterion cannot be used without qualification, for the same statement might be made of subbituminous coal and lignite. As noted before, the distinguishing feature which serves to separate bituminous coal from coals of lower rank is the manner in which it is affected by weathering.

Subbituminous. The term "subbituminous" is adopted by the Geological Survey for what has generally been called "black lignite," a term that is objectionable because the coal is not lignitic in the sense of being distinctly woody, and because the use of the term seems to imply that this coal is little better than the brown, woody lignite of North Dakota, whereas many coals of this rank approach in excellence the lowest grade of bituminous coal. Subbituminous coal is generally distinguishable from lignite by its black color and its apparent freedom from distinctly woody texture and structure, and from bituminous coal by its loss of moisture and the consequent breaking down of "slacking" that it undergoes when subjected to alternate wetting and drying.

Lignite. The term "lignite," as used by the Geological Survey, is restricted to those coals which are distinctly brown and either markedly woody or claylike in their appearance. They are intermediate in quality and in development between peat and subbituminous coal.

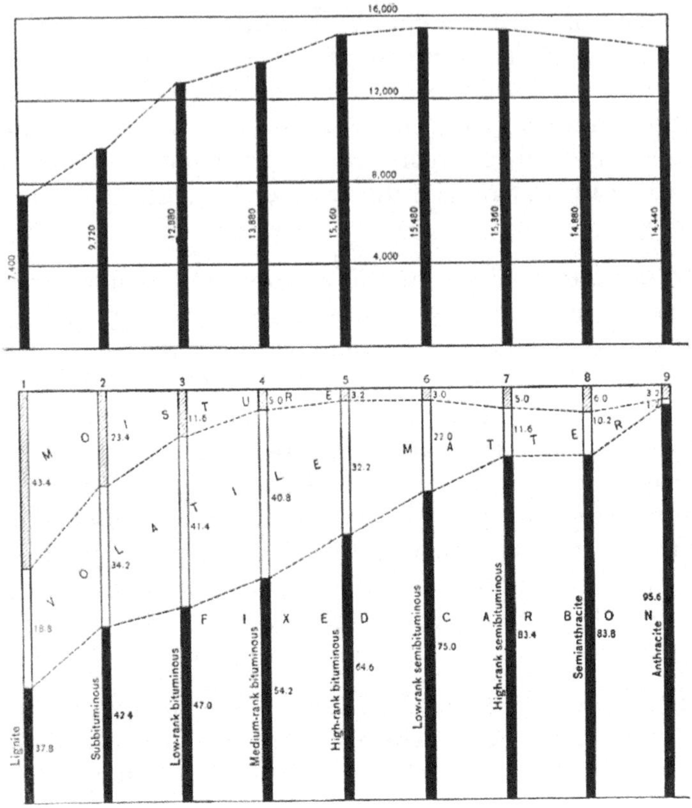

FIG. 5. Diagrams showing the chemical composition and heat efficiency of the several ranks of coal. Upper diagram: Comparative heat value of the samples of coal represented in the lower diagram, computed on the ash-free basis. Lower diagram: Variation in the fixed carbon, volatile matter, and moisture of coals of different ranks, from lignite to anthracite, computed on samples as received, on the ash-free basis. After Campbell.ToList

GEOLOGIC FEATURES

Geologic features of coal may be conveniently described in terms of origin or genesis. Coal has essential features in common with asphalt, oil, and gas. They are all composed of carbon, hydrogen, and oxygen, with minor quantities of other materials, combined in various proportions. They are all "organic" products which owe their origin to the decay of the tissues of plants and perhaps animals. They have all been buried with other rocks beneath the surface. The common geologic processes affecting all rocks have in the main determined the evolution of these organic products and the forms in which we now find them. Originating at the surface, they have participated in the constructive or anamorphic changes of the metamorphic cycle, which occur beneath the surface, and under these influences have undergone various stages of condensation, refinement, distillation, and hardening.

All stages in the development of coal have been traced. In brief, the story is this:

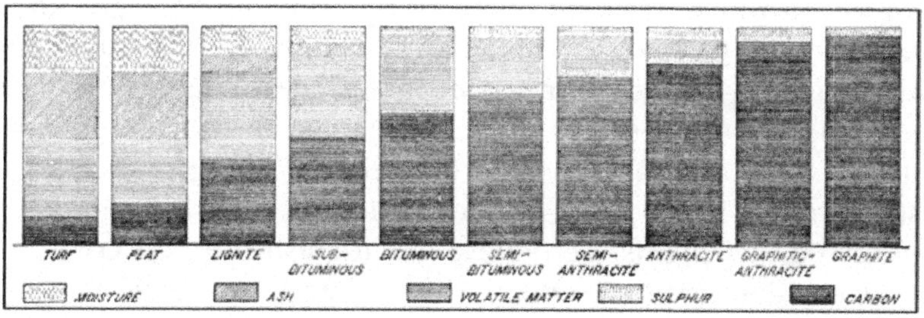

FIG. 6. Origin and development of coal. After Gilbert.ToList

This exhibit shows the successive chemical stages in the evolution of coal. The striking qualities of the original are lost in the reproduction through the use of designs in the place of realistic coloring, but the effect is retained sufficiently to indicate the nature of the sequence and the directness with which it leads back to an origin in vegetal accumulations. The evolutionary process is seen to take the form of increasing density through the progressive expulsion of volatilizable matters in the course of geologic time. This inference is substantiated beyond reasonable question by the actual presence of organic remains in coal beds.

Grasses, trees, and other plants growing in swamps and bogs decay and form a vegetable mold in the nature of *peat*. A peat bog from the top downward consists of (1) living plants, (2) dead plants, and (3) a dense brownish-black mass, of decayed and condensed vegetable material, in which the vegetable structure is more or less indistinct. Peat consists chiefly of fixed carbon and volatile matter, also of sulphur, moisture, and ash. The volatile matter consists mainly of various combinations of hydrogen and carbon, called hydrocarbons; it goes off in gas or smoke when the peat is heated to a red heat. The fixed carbon is the carbon left after the volatile matter has been driven off. The ash represents the more incombustible mineral matter, usually of the nature of clay or slate. The moisture in peat may be as high as 90 per cent.

The essential condition for thick accumulation of peat seems to be abundance of moisture, which favors luxuriant growth and protects the plant remains from complete oxidation or decay. Without moisture the vegetable material would completely oxidize, leaving practically no residue, as it does in dry climates. For the formation of thick peat beds, there seems to be implied some sort of a balance between the slow building up of organic accumulations and the settling of the area to keep it near the elevation of the water table. Present day bog deposits are known in some cases to have a thickness of forty feet. This thickness is not enough to account for some of the great coal seams within the earth; but there seems to be no escape from the conclusion that the same sort of deposits, formed on a larger scale in the past, were the first step in the formation of the coal seams. Flat, swampy coastal plains are believed to furnish the best conditions for thick accumulation of peat. There is good evidence that most of the deposits accumulate essentially in place, without appreciable transportation.

In time these surface accumulations of vegetable material may subside and be buried under clay, sand, or other rock materials. The processes of condensation begun in the peat bog are then carried further. They result in the second stage of coal formation, that of *lignite* or *brown coal*. This is brown, woody in texture, and has a brown streak. It has a higher percentage of fixed carbon, and less volatile matter and water, than peat.

Continuation of the processes of induration produces *subbituminous coal*, or *black lignite*, which is usually black and sometimes has a fairly bright luster. It is sometimes distinguished from bituminous coal, where weathered or dried, by the manner in which it checks irregularly or splits parallel to the bedding,—the characteristic feature of bituminous coal being columnar fracture.

The next stage in coal formation is *bituminous coal*. It has greater density than the lignites or subbituminous coals, is black, more brittle, and breaks with a cubical or conchoidal fracture. It is higher in fixed carbon, lower in volatile matter and water. A variety of bituminous coal, called *cannel coal*, is characterized by an unusually high percentage of volatile matter, which causes it to ignite easily. This material has a dull luster and a conchoidal fracture. It is composed almost entirely of the spores and spore cases, which are resinous or waxy products, of such plants as lived in the parent coal swamp.

There are gradations from bituminous coal into *anthracite coal*. *Semibituminous* and *semianthracite* are names used to some extent for these intermediate varieties. The final stage of coal formation is anthracite,—hard, brittle, black, with high luster and conchoidal fracture. It has a higher percentage of fixed carbon and correspondingly less of the volatile constituents, than any of the other coals.

The coals form a completely graded series from peat to the hard anthracite. Comparison of the compositions of the coal materials at different stages shows clearly what has happened. Moisture has diminished, certain volatile hydrocarbons have been eliminated as gases, and oxygen has decreased. On the other hand, the residual fixed carbon, sulphur, and usually ash, have remained in higher percentage. This change in composition is graphically represented in Figure 6.

During this process volume has been progressively reduced and density increased. Five feet of wood or plant may produce about one foot of bituminous coal, or six-tenths of a foot of anthracite.

The exact physical conditions in the earth which determine the progressive changes in coals, above outlined, cannot be fully specified. Time is one of the factors—the longer the time, the greater the opportunity for accomplishing these results. Another factor is undoubtedly pressure, due to the weight of overlying sediments, or to earth movements. In peat condensational changes of this nature are accomplished artificially by the pressure of briquetting machines. Another factor is believed to be the heat developed by earth movements and vulcanism, which presumably facilitates the elimination of volatile materials, and thus accelerates the gradational changes above described. This is suggested by the fact that in places where hot volcanic lavas have gone through coal beds they have locally produced coals of anthracitic and coke-like varieties. In general, however, it has not been possible to determine the degree to which heat has been responsible for the changes. Coals which have been developed in different localities, under what seem to be much the same heat conditions, may show quite different degrees of progress toward the anthracite stage. Another factor that has been suggested as possibly contributing to the change, is the degree of permeability of the rocks overlying the coal to the volatile materials which escape from the coal during its refinement. It is argued that in areas of folding or of brittle rock where the cover is cracked, volatile gases have a better chance to escape, and that the change toward anthracite is likely to advance further here than elsewhere.

Bacterial action is an important factor in the earlier stages, in the partial decay of vegetable matter to form peat; accumulation of waste products from this action, however, appears to inhibit further bacterial activity.

Coal deposits have the primary shapes of sedimentary beds. They are ordinarily thin and tabular, and broadly lenticular,—on true scale being like sheets of thin paper. At a maximum they seldom run over 100 feet in thickness, and they average less than 10 feet. Seldom is a workable coal bed entirely alone; there are likely to be several superposed and overlapping seams of coal, separated by sandstones, shales, or other rocks. In Illinois and Indiana there are nine workable coal seams, in Pennsylvania in some places about twenty, and in Wales there are over one hundred, many of which are worked. Some of the seams are of very limited extent; others are remarkably persistent, one seam in Pennsylvania having an average thickness of 6 to 10 feet over about 6,000 square miles of its area. Only 2 per cent of the coal-bearing measures of the eastern United States is actually coal.

Even where not subsequently disturbed by deformation, coal beds are not free from structural irregularity. They are originally deposited in variable thicknesses on irregular surfaces. During their consolidation there is a great reduction of volume, resulting in minor faults and folds. Subsequent deformation by earth forces may develop further faults and folds, with the result that the convolutions of a coal bed may be very complex. The beds of a coal-bearing series are usually of differing thickness and competency, and as a consequence they do not take the same forms under folding. Shearing between the beds may result in an intricate outline for one bed, while the beds above and below may have much more simple outlines. In short, the following of a coal seam requires at almost every stage the application of principles of structural geology. It is obvious, also, that the identification and location of sedimentary geologic horizons are essential, and hence the application of principles of stratigraphy.

The folios of the United States Geological Survey on coal-bearing areas present highly developed methods of mapping and representing the geologic features of coal beds. On the surface map are indicated the topography, the geologic horizons, and the lines of outcrop of the coal seams. In addition, there are indicated the sub-surface contours of one or more of the coal seams which are selected as datum horizons. The sub-surface structure, even though complex, can be readily read from one of these surface maps. With the addition of suitable cross sections and comparative columnar sections, the story is made complete. In the study of the occurrence of coal seams, the reader cannot do better than familiarize himself with one or more of the Geological Survey folios.

The high-grade coals of the eastern and central United States are found in rocks of Carboniferous age. The very name Carboniferous originated in the fact that the rocks of this geologic period contain productive coal beds in so many parts of the world. The coal measures of Great Britain, of Germany, Belgium, and northern France, of Russia, and the largest coal beds of China are all of Carboniferous age. Deposits of this period include the bulk of the world's anthracite and high-grade bituminous coal. Coal deposits of more recent age are numerous, but in general they have had less time in which to undergo the processes of condensation and refinement, and hence their general grade is lower. In the western United States there are great quantities of subbituminous coal of Cretaceous age, and of Tertiary lignites which have locally been converted by mountain upbuilding into bituminous and semibituminous coals. Jurassic coals are known in many parts of the world outside of North America, and lignites of Tertiary age are widely distributed through Asia and Europe.

PETROLEUM

ECONOMIC FEATURES

Petroleum is second only to coal as an energy resource. The rapid acceleration in demand from the automobile industry and in the use of fuel oil for power seems to be limited only by the amounts of raw material available.

Production and reserves. The distribution by countries of the present annual production of petroleum, the past total production, and the estimated reserves, is indicated in terms of percentages of the world's total in the table[19] on the opposite page.

This table indicates the great dominance of the United States both in present and past production of petroleum, as well as the concentration of the industry in a few countries. In addition the United States controls much of the Mexican production as well as production in other parts of the world, making its total control of production at least 70 per cent. of the world's total. Notwithstanding its large domestic production, the United States has recently consumed more oil than it produces. Imports of crude oil are about balanced by exports of kerosene, fuel oils, lubricants, etc. The per capita consumption of petroleum in the United States is said to be twenty times greater than in England. On the other hand, the remaining principal producers consume far less than they produce, the excess being exported.

The oil from the United States, Russia, the Dutch East Indies, India, Roumania, and Galicia is for the most part treated at refineries near the source of supply or at tidewater, and exports consist of refined products. The Mexican oil is largely exported in crude form to the United States though increasing quantities are being refined within Mexico.

The figures shown in the table for oil reserves are of course the roughest approximations, particularly for some of the less explored countries. However, they are compiled from the best available sources and may serve at least to show the apparent relative positions of the different countries at this time. Further exploration is likely to change the percentages and add very greatly to the totals. The significant feature of these figures is the contrast which they indicate between distribution of reserves and distribution of past production. Particularly do they show that the reserves of the United States, which are more closely estimated than those of any other country, are in a far lower ratio to past production than are the reserves in other countries. It was estimated in 1920 that about 40 per cent of the United States reserves are exhausted.[20]

PRESENT AND PAST PRODUCTION AND RESERVES OF OIL, BY COUNTRIES, IN TERMS OF PERCENTAGE OF WORLD'S TOTAL

Country	*Per cent of production, 1918*	*Per cent of total production, 1857-1918*	*Per cent of total oil resources*
United States and Alaska	69.15	61.41	16.26
Mexico	12.40	3.80	10.51
Russia (southeastern Russia, southwestern	7.86	24.96	15.69
Siberia, region of the Caucasus, northern			
Russia, and Saghalien			
East Indies	2.58	2.51	7.00
Roumania, Galicia, and western Europe	2.79	4.07	2.64
India	1.55	1.41	2.31
Persia and Mesopotamia	1.40	.19	13.51

Japan and Formosa	.48	.51	2.87
Egypt and Algeria	.40	.07	2.15
Germany	.14	.22	—
Canada	.06	.33	2.31
Northern South America, including Peru,	.93	.43	13.31
Trinidad and Venezuela			
China	—	—	3.19
Italy	}		
Cuba	}	.02	
Other countries	}		
World total	100.00	100.00	100.00

Looking forward to the future, it is clear that there will be considerable shifts in the centers of principal production of petroleum in the directions indicated by the reserve figures. In particular, conspicuous development of production may be expected in the immediate future in the countries bordering the Caribbean Sea and the Gulf of Mexico. In the eastern hemisphere production is rapidly increasing in Persia and Mesopotamia; and Russia, with the stabilization of political conditions, may become ultimately the world's leading oil producer. At the now indicated rate of production, world reserves now estimated would be exhausted in eighty-six years and the peak of production would be passed earlier. With continuing acceleration of production, total reserves would be exhausted in considerably less time,—providing physical conditions would allow the oil to be pumped from the ground at the necessary speed, which they probably will not. These figures taken at face value are alarming; but the earth offers such huge possibilities for further discoveries that the life of oil reserves above indicated is likely to be considerably extended. At many times in the history of the mineral industry the end has apparently been in sight for certain products; but with the increased demand for these products has come increased activity in exploration, with the result that as yet no definite end has been approached for any one of them. The more immediate problems of the petroleum industry seem to the writer to be of rather different nature: first, whether the discovery and winning of the oil can be made to keep pace with the enormous acceleration of demand; and second, the adjustment of political and financial control of oil resources, the possession of which is becoming so increasingly vital to national prosperity.

In regard to the first question, it is a much more difficult problem today to locate and develop a supply of oil to replace the annual world production (recently half a billion barrels), than it was twenty years ago, when it was necessary for this purpose to find only one-fifth this amount; and if the demand is unchecked, it will be still more difficult to replace the three-quarters of a billion barrels of oil which will doubtless be required in a very few years. Regardless of the amount of oil actually in the ground, it is entirely possible that physical limitations on its rate of discovery and recovery will prevent its being made available as fast as necessary to meet the increasing demand. This fact is likely to make itself felt through increase of price. Other natural results should be the development of substitutes, such as alcohol or benzol for gasoline; the larger recovery of oil from oil shales; and the general speeding up of conservational measures of various kinds. These are all palliatives and not essential remedies. To make enough alcohol to substitute for the gasoline now coming from oil would use a very considerable fraction of the world's food supply. To make enough benzol (a by-product of coke) to replace gasoline would necessitate the manufacture of many times the amount of coke now required by the world's industries. To develop the oil shale industry to a point where it could supply anything like the amount of oil now derived from oil pools would mean the building of great plants, including towns, railroads, and other equipment, equivalent to the plants of the coal mining industry. To apply any one of the various conservational measures discussed on later pages would only temporarily alleviate the situation.

The question of political and financial control of oil supplies may be illustrated by particular reference to the United States. On present figures it appears that within three to five years the peak of production in this country will be passed; and at the present rate of production the life of the reserves may not be over seventeen to twenty years. Of course production could not continue to the end at this rate, and the actual life will necessarily be longer. Again the doubtful factor is the possibility for further discoveries. Many favorable structures have been mapped which have not yet been drilled, and there are considerable unexplored areas where the outcrops are so few that there is no clue at the surface to the location of favorable structures. The future is likely to see a considerable amount of shallow drilling for the sole purpose of geological reconnaissance. For upwards of ten years important parts of the public domain have not been available for exploration, but Congress has now enacted legislation which opens up vast territories for this purpose.

Even with large allowance for these possibilities, it seems unlikely that production in the United States can increase very long at the accelerating rate of the domestic demand, which is already in excess of domestic production. The supplies of Mexico are in a large part controlled by American capital and are thus made available to the United States (subject, of course, to political conditions); but even with these added, the United States is in a somewhat unfavorable situation as compared with certain other countries. This situation is directing attention to the possibility of curtailment of oil exports, and to the possibility of acquiring additional oil supplies in foreign countries. In this quest the United States is peculiarly handicapped in that most foreign countries, in recognition of the vital national importance of the oil resource, have imposed severe restrictions on exploration by outsiders. Nationals of the United States are excluded from acquiring oil concessions, or permitted to do so only under conditions which invalidate control, in the British Empire, France, Japan, Netherlands, and elsewhere, and the current is still moving strong in the direction of further exclusion. As the United States fields are yet open to all comers, it has been suggested that some restriction by the United States might be necessary for purposes of self-protection, or as an aid in securing access to foreign fields. The activity of England during and since the war has increased the amount of oil controlled by that country from an insignificant quantity to potentially over half of the world's oil reserves. The problem of future oil supplies for the United States presents an acute phase of the general question of government coöperation or participation in mineral industries, which is further discussed in Chapter XVIII.

The following table summarizes the distribution of the oil production in the United States, together with the salient features of its geologic distribution and character.

This table, in conjunction with Fig. 8 below, shows clearly that the bulk of the United States production of oil comes from two great sources—the Pennsylvanian sandstones of the Mid-Continent field in Kansas and Oklahoma, and the Cretaceous and Tertiary sediments of the southern half of California. Phenomenal development of the Central and North Texas field in 1919 increased its yield to about one-sixth of the country's total. The older Appalachian oil field, extending from New York to West Virginia and Tennessee, was the earliest area discovered; it is still one of the more productive fields, though it has long since passed its maximum production. The other principal sources of oil are the Gulf Coast field in Louisiana and Texas, the North Louisiana field, the southern Illinois field, and the Rocky Mountain region. This last region, containing large amounts of government land recently opened to exploration, bids fair to produce increasing quantities of oil for some time.

PAST PRODUCTION OF PETROLEUM IN THE UNITED STATES.
(FIGURES FROM U. S. GEOLOGICAL SURVEY)

State	*Age of containing rocks*	*Base*	*Production for 1919 (barrels)*	*Total production including 1918 (barrels)*
Alaska	East-Low. Tertiary West-Jurassic	Paraffi n	(a)	(a)
California	Cretaceous:	Ashpal	97,531,997	1,110,226,57

	Tertiary	t		6
Colorado	Pierre-Cretaceous	Paraffi n	143,286	11,319,370
Illinois	Mississippian-Pennsylvanian	Paraffi n	13,365,974	298,225,380
Indiana	East-Ordovician (Trenton)	Paraffi n	877,558	106,105,584
	West-Pennsylvanian			
Kansas	Pennsylvanian	Par.-Asph.	45,451,017	148,450,298
Kentucky,	Mississippian	Paraffi n	4,376,342	18,213,188
Tennessee				
Louisiana	Cretaceous-Quat.	Paraffi n	16,042,600	150,769,911
	Cretaceous-Eocene			
Michigan,	Carboniferous	Paraffi n	(a)	(a)
Missouri				
Montana	—	—	69,323	213,639
New Mexico	Carboniferouos -Cretaceous	—	(a)	(a)
New York,	Devonian-Carboniferous	Paraffi n	8,216,655	788,202,717
Pennsylvani a				
Ohio, East	Ordovician-Carboniferous	Paraffi n	7,285,005	463,367,386

and West				
Oklahoma	Pennsylvanian	Paraffin	103,347,070	851,320,457
Texas	Pennsylvanian, Cretaceous-Quat.	Asph.-Par.	38,750,031	327,550,005
Utah	—	—	(b)	(b)
West Virginia	Devonian-Carboniferous	—	7,866,628	294,474,710
Wyoming	Carboniferous-Cretaceous	Asph.-Par.	12,596,287	40,019,573
Other	—	—	7,943	112,925
			355,927,716	4,608,571,719

(a) Included in "Other." (b)Included in Wyoming.

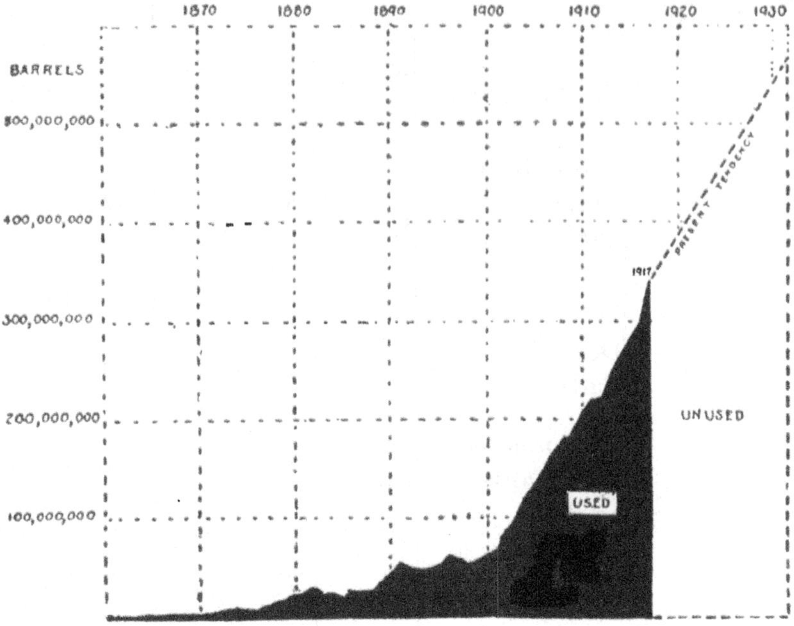

FIG. 7. Chart showing the present tendency of the United States in respect to its unmined reserve of petroleum. Data from U.S. Geological Survey. After Gilbert and Pogue.ToList

Methods of estimating reserves. It may be of interest to inquire into the basis on which predictions are made of the life of an oil pool. The process is essentially a matter of platting curves of production, and of projecting them into the future with the approximate slopes exhibited in districts which are already approaching exhaustion.[21] While no two wells or two districts act exactly alike, these curves have group characteristics which are used as a rough basis for interpreting the future.

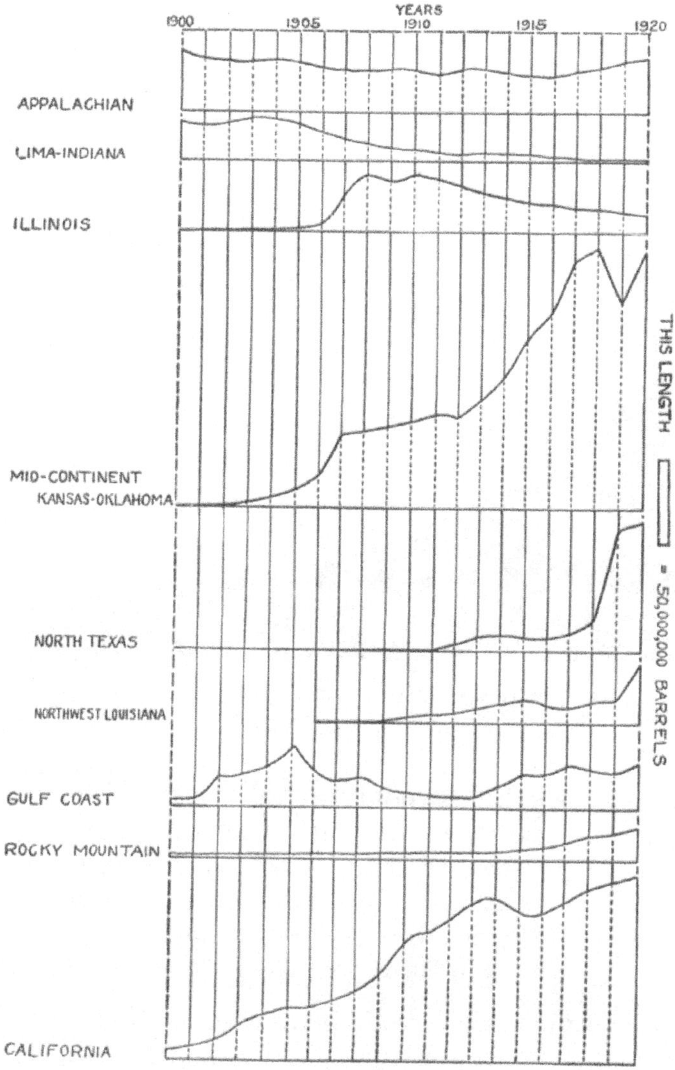

FIG. 8. The annual output of the principal oil fields of the United States for the last twenty years. Data from U.S. Geological Survey.ToList

A less reliable method is to calculate from geologic data the volume and porosity of the oil-bearing reservoirs, and to estimate the percentage of recovery on the basis of current practices and conditions. Complete data for this method are often not available;

but in the early years of a field, before production curves are established, this method may serve for a rough approximation.

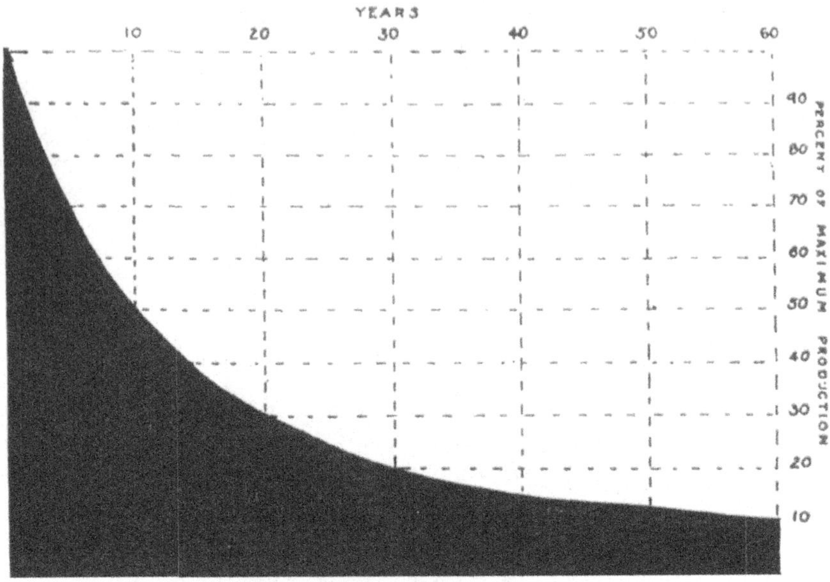

FIG. 9. Curve showing the usual decline in oil field production after the period of maximum output is reached. After Ralph Arnold. The Petroleum Resources of the United States, Smithsonian report for 1916, p. 283. Compare this theoretical curve of final decrease with the production curve shown in Fig. 8.ToList

Classes of oils. When crude petroleum is distilled, it gives off in succession various substances and gradually thickens until it leaves a solid residue, which may be largely either paraffin wax or asphalt. The two main classes of oils are determined by the nature of this solid residual. The products given off are natural gas and then liquid hydrocarbons of various kinds, which evaporate in the order of their lightness. Petroleum is thus a mixture or mutual solution of different liquids, gases, and solids. Nearly one-fifth of the domestic consumption of crude petroleum is burned directly as fuel, and four-fifths are refined. The several principal primary products of refinement are gasoline, kerosene, fuel oil, and lubricating oil; but these may be broken up into other substances, each the starting point of further refinements, with the result that present commercial practice yields several hundred substances of commercial value. With increasing chemical and technical knowledge these products are being multiplied. The rapidly increasing demand for gasoline has led to the use of processes which extract a large proportion of this substance from the raw material, by "cracking" or breaking up other substances; but while, under the stress of necessity, there is possibility of slight modification of the proportions of principal substances extracted from the crude oil, it is not possible to change these proportions essentially. It is, therefore, a problem to adjust relative demands to supplies of the different products. The domestic demand for gasoline is greater than the supply. On the other hand, the demand for kerosene, which must be produced at the same time, is much less than the domestic supply. Hence the importance of maintaining export markets for kerosene.

The nature or grade of the oil of various fields is an important matter in considering reserves for the future. Perhaps half of the United States reserves consist of the asphalt-base oils of the California and certain of the Gulf fields, which yield comparatively small amounts of gasoline and other valuable light products, though they are very satisfactory for fuel purposes. Similarly the large reserve tonnages of oil in Mexico and the Caribbean countries, in Peru, and probably in Russia, are essentially of the heavier, lower grade oils. The oils of the Mid-Continental and eastern fields of the United States, of Ontario, of the Dutch East Indies, of Burma, and of Persia and Mesopotamia are reported to be largely of the paraffin base type, which, because of its larger yield of gasoline and light oils, is at present considerably more valuable. These generalizations are of course subject to qualifications, in that the oils of a given region may vary considerably, and that some oils are intermediate in character, containing both asphalt and paraffin wax.

Conservation of oil. The rapid increase in demand for oil as compared with discovery of new sources is leading naturally to a more intensive study of the conservational aspects of the industry. This is a complex and difficult subject which we shall not take up in detail, but we may point out some of the phases of the problem which are receiving especial attention.

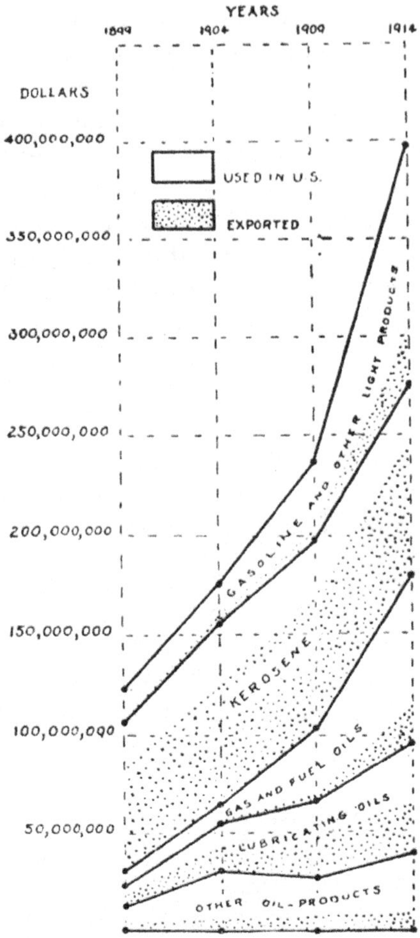

FIG. 10. Chart showing the relative values of the principal petroleum products manufactured in the United States from 1899 to 1914. After Gilbert and Pogue. Note the decreasing importance of kerosene in sustaining the cost of refining, and the necessity of exports for maintaining a balanced outlet of products. Data from Story B. Ladd, Petroleum Refining. Census of Manufactures: 1914, Bureau of Census, Washington, 1917, p. 10.ToList

About 50 per cent of the oil in the porous strata, of oil pools is ordinarily not recovered, because it clings to the rock. Efforts are being made along various lines to increase the percentage of recovery,—as, for instance, in preventing infiltration of water to the oil beds and in the use of artificial pressures and better pumping. "Casing-head gasoline" is being recovered to an increasing extent from the natural gas which was formerly allowed to dissipate in the air.

Minute division of the ownership of a pool, with consequent multiplication of wells and unrestricted competition, tends to gross over-production and highly wasteful methods. The more rapid exhaustion of one well than the others may result in the flooding of the oil sands by salt waters coming in from below. Various efforts have been made toward a more systematic and coördinated development of oil fields.

In general, the organization and technique involved in the development of an oil field are improving in the direction of extracting a greater percentage of the total available oil.

Better methods of refining the oil, and the refining of a larger percentage of the crude oil, make the oil more available for a greater variety of purposes and therefore more valuable. Great advances have been made along these lines, particularly in the application of the "cracking" method for a greater recovery of the more valuable light oils at the expense of the less valuable heavy oils. Similarly, modifications of internal combustion engines will probably permit the use, in an increasing number of cases, of products of lower volatility than gasoline.

One of the conservational advances in coming years will probably be a restriction in the amount of crude oil used directly for fuel and road purposes without refining. These crude uses cut down the output of much desired products from the distillation of the oil. Various other restrictions in the use of oil have been proposed, such as the curtailment of the use of gasoline in pleasure cars. The gasless Sundays during the war represented an attempt of this kind. In general, it seems likely that such restrictions will come mainly through increase in the price of oil products.

The substitution of oil from oil shales, and of alcohol for gasolene, already mentioned, will be conservational so far as the oil is concerned, though perhaps not so in regard to other elements of the problem.

GEOLOGIC FEATURES

Organic theory of origin. According to this theory, accumulations of organic materials in sedimentary beds, usually muds or marls, have been slowly altered and distilled during geologic ages; the products of distillation have migrated chiefly upward to porous strata like sandstones or cavernous limestones, where, under suitable conditions, they have become trapped.

The original organic material is believed to have been plants of low order and animal organisms (such as foraminifera) which were deposited as organic detritus with mud and marl in the bottoms of ponds, lakes, estuaries, and on the sea bottom,—in both salt and fresh waters. Bacteria are supposed to have played a part in the early stage of alteration, sometimes called the biochemical stage. When the organic matter was buried under later sediments and subjected to pressure, physical conditions were responsible for further volatilization or distillation. This stage is called the geochemical stage. There is much in common as to origin between coal, oil shales, and petroleum. According to White,[22]

whether the ingredient organic matter, be it plant or animal, will be in part transformed to coal of the ordinary type, to cannel, to oil shale, to the organic residues in so-called bituminous shales and carbonaceous shales, or to petroleum and natural gas, is dependent upon the composition of the ingredient organic débris, the conditions of its accumulation or deposition, and the extent of the microbian action.

White has further developed the important principle that, in the geochemical stage of development, both coal and oil react to physical influences in much the same way; and that therefore when both are found in the same geologic series, the degree of concentration of the coal, measured by its percentage of carbon, may be an indication of the stage of development of the oil. More specifically where the coal contains more than 65 to 70 per cent of fixed carbon, chances for finding oil in the vicinity are not good (though commercial gaspools may be found), probably for the reason that the geochemical processes of distillation have gone so far as to volatilize the oils, leaving the solid residues in the rock.

White also finds that the lowest rank oils, with considerable asphalt, are found in regions and formations where the coal deposits are the least altered, and the lighter, higher rank oils, on the whole, where the coal has been brought to the correspondingly higher ranks; in other words, up to the point of complete elimination of the oil, improvement in quality of the oil accompanies increased carbonization of the coal. The principle, therefore, becomes useful in exploration in geologic series where oil is associated with coal. Where the coal is in one series and the oil in another, separated by unconformity (indicating different conditions of development), the principle may not hold, even though there is close geographical association.

The oil and gas distillates migrate upward under gas pressure and under pressure of the ground-water. If there are no overlying impervious beds to furnish suitable trapping conditions, or conditions to retard the flow, the oil may be lost. The conditions favorable for trapping are overlying impervious beds bowed into anticlines, or other structural irregularities, due either to secondary deformation or to original deposition, which may arrest the oil in its upward course. A dome-like structure or anticline may be due to stresses which have buckled up the beds, or to unequal settling of sediments varying in character or thickness; thus some of the anticlinal structures of the Mid-Continent field may be due to settling of shaley sediments around less compressible lenses of sandstone which may act as oil reservoirs, or around islands which stood above the seas in which the oil-bearing sediments were deposited and on the shores of which sands capable of acting as oil reservoirs were laid down. Favorable conditions for trapping the oil may be furnished by impervious clay "gouges" along fault planes, or by dikes of igneous rock. Favorable conditions may also be merely differences in porosity of beds in irregular zones, determined by differences either in original deposition or in later cementation.

The thickness of oil-producing strata may vary between 2 or 3 feet and 200 feet. The porosity varies between 5 and 50 per cent. In sandstones the average is from 5 to 15 per cent. In shales and clays, which are commonly the impervious "cap-rocks," porosity may be equally high, but the pores are too small and discontinuous to permit movement.

When the impervious capping is punctured by a drill hole, gas is likely to be first encountered, then oil, and then water, which is usually salty. The gas pressure is often released with almost explosive violence, which has suggested that this is an important cause of the underground pressures. It has been supposed also that the pressures are partly those of artesian flows. The vertical arrangement of oil, gas, and water under the impervious capping is the result of the lighter materials rising to the top. In certain fields, oil and gas have been found in the tops of anticlines in water-saturated rocks, and farther down the flanks of folds or in synclines in unsaturated rocks.

The localization of oil pools is evidently determined partly by original organic deposition, often in alignment with old shore lines, and partly by the structural, textural, and other conditions which trap the oil in its migration from the source.

Effect of differential pressures and folding on oil genesis and migration. Another organic hypothesis proposed somewhat recently[23] is that oil is formed by differential movement or shearing in bituminous shales, which are often in close relationship with the producing sand of an oil field, and that the movement of oil to the adjacent sands is accomplished by capillary pressure of water and not by ordinary free circulation of water under gravity. The capillary forces have been shown to be strong enough to hold the oil in the larger pores against the influence of gravity and circulation. The accumulation of the oil into commercial pools is supposed to take place in local areas where the oil-soaked shale, due to jointing or faulting, is in direct contact with the water of the reservoir rock. This suggests lack of wide migration. This hypothesis is based on experimental work with bituminous shales. The general association of oil pools with anticlinal areas is explained on the assumption that anticlines on the whole are areas of maximum differential movement, resulting in oil distillation, and that they are ordinarily accompanied by tension joints or faults, affording the conditions for oil migration. Data are insufficient, however, to indicate the extent to which the anticlinal areas are really areas of maximum shearing. As regards the exact nature of the process, it is not clear to what extent differential movement may involve increase in temperature which may be the controlling factor in distillation,—although in McCoy's experiment oil was formed when no appreciable amount of heat was generated.

The development of petroleum by pressure alone acting on unaltered shale, as shown by these experiments, has been taken by White[24] to have a significant bearing on the geochemical processes of oil formation. Under differential stresses acting on fine-grained carbonaceous strata under sufficient load, there is considerable molecular rearrangement, as well as actual movement of the rock grains,—thus promoting the distillation of oil and gas from the organic matter in the rocks, and the squeezing out of the oil, gas, and water into adjacent rocks, such as coarse round-grained sandstones and porous limestones, which are more resistant to change of volume under pressure. Migration, concentration and segregation of the oil, gas, and water is supposed to be brought about, partly through the effect of capillary forces—the water, by reason of its greater capillary tension, tending to seize and hold the smaller voids, and thus driving the oil and gas into the larger ones—and partly through the action of gravity.

White also suggests that the process may go further where the parent carbonaceous strata are of such thickness and under such load of overlying rocks that they undergo considerable interior adjustment and volume change before yielding to stress by anticlinal buckling,—than where the strata yield quickly. It is not clear to the writer that the interior adjustment assumed under this hypothesis is necessarily slowed up or stopped by anticlinal buckling. Interior stresses are inherent in any sedimentary formation, when settling and consolidating and recrystallizing under gravity, and these may be independent of regional thrusts from without.

The first oils evolved by pressure from the organic mother substance are probably heavy, the later oils lighter, and the oils from formations and regions where the alteration is approaching the carbonization limit are characteristically of the highest grade. This is the reverse of the order of products obtained by heat distillation. Whether there is also a natural fractionation and improvement of the first heavy oils as they undergo repeated migrations is not known.

Inorganic theory of origin. Another theory of the source of oil has had some supporters, although they are much in the minority. This is the so-called "inorganic" theory, that oil comes from magmas and volcanic exhalations. In support of this theory attention is called to the fact that igneous rocks and the gases associated with them frequently carry carbides or hydrocarbons; that many oil fields have a suggestive geographic relationship with volcanic rocks; and that certain of the oil domes, as for instance in Mexico, are caused by plugs of igneous rocks from below. It has been suggested that deep within the earth carbon is combined with iron in the form of an iron carbide, and that from the iron carbide are generated the hydrocarbons of the oil, either by or without the agency of water. Iron carbide is magnetic, and significance has been attached to the general correspondence between the locations of oil in the western United States and regions of magnetic disturbance.

It seems not unlikely that some inorganic theory of this sort is necessary to explain the ultimate source of oil or of the substances which become oil, but the evidence is overwhelming that organic agencies have been mainly responsible for the principal oil pools now known.

Oil exploration. A simple geographic basis for oil exploration is the fact that the major oil fields of the world are situated between 20° and 50° north latitude, and that thus far there are no major oil areas within the tropics or within the southern hemisphere. This broad generalization may have little value when exploration is carried further. It has also been suggested that the geographic distribution of oil corresponds roughly with the average annual temperatures, or isotherms, between 40° and 70.°[25] It is thought that this present distribution of temperatures may indicate roughly the temperatures of the past when the oil was accumulated; and the inference is drawn that there was some sort of limitation of areal deposition within these temperature limits. If this be true, the only reasons why the southern hemisphere is not productive are the relatively small size of the land areas and the lack of exploration to date.

In approaching broadly the problem of oil exploration, the geologist considers in a general way the kinds and conditions of rocks which are likely to be petroliferous or non-petroliferous. Schuchert[26] summarizes these conditions for North America as follows:

1. The impossible areas for petroliferous rocks.

   (*a*) The more extensive areas of igneous rocks and especially those of the ancient shields; exception, the smaller dikes.

   (*b*) All pre-Cambrian strata.

   (*c*) All decidedly folded mountainous tracts older than the Cretaceous; exceptions, domed and block-faulted mountains.

   (*d*) All regionally metamorphosed strata.

   (*e*) Practically all continental or fresh-water deposits; relic seas, so long as they are partly salty, and saline lakes are excluded from this classification.

   (*f*) Practically all marine formations that are thick and uniform in rock character and that are devoid of interbedded dark shales, thin-bedded dark impure limestones, dark marls, or thin-bedded limy and fossiliferous sandstones.

   (*g*) Practically all oceanic abyssal deposits; these, however, are but rarely present on the continents.

2. Possible petroliferous areas.

   (*a*) Highly folded marine and brackish water strata younger than the Jurassic, but more especially those of Cenozoic time.

   (*b*) Cambrian and Ordovician unfolded strata.

   (*c*) Lake deposits formed under arid climates that cause the waters to become saline; it appears that only in salty waters (not over 4 per cent?) are the bituminous materials made and preserved in the form of kerogen, the source of petroleum; some of the Green River (Eocene) continental deposits (the oil shales of Utah and Colorado) may be of saline lakes.

3. Petroliferous areas.

   (*a*) All marine and brackish water strata younger than the Ordovician and but slightly warped, faulted, or folded; here are included also the marine and brackish deposits of relic seas like the Caspian, formed during the later Cenozoic. The more certain oil-bearing strata are the porous thin-bedded sandstones, limestones, and dolomites that are interbedded with black, brown, blue, or green shales. Coal-bearing strata of fresh-water origin are excluded. Series of strata with disconformities may also be petroliferous, because beneath former erosional surfaces the top strata have induced porosity and therefore are possible reservoir rocks.

(*b*) All marine strata that are, roughly, within 100 miles of former lands; here are more apt to occur the alternating series of thin and thick-bedded sandstones and limestones interbedded with shale zones.

The extent to which marine or brackish water conditions of sedimentation are requisite to the later formation of oil, as is suggested in the above quotation, has long been a debatable question. It may be noted that certain oil shales formed in fresh water basins contain abundant organic matter which is undoubtedly suitable for the generation of oil and gas, and that these shales on distillation yield oil essentially like that obtained from oil shales of marine origin; that certain important oil-bearing sands of the younger Appalachian formations were laid down in waters which are believed to have been only slightly saline; that natural gas is present in fresh water basins; and that it has not been demonstrated that salt in appreciable amounts is necessary for the geologic, any more than for the artificial, distillation of oil. Most of the great oil fields have been in regions of marine or other saline water deposits, but it has not been proved that this is a necessary condition. White[27] says: "At the present stage of our knowledge, fresh-water basins appearing otherwise to meet the requirements should be wildcatted without prejudice."

The principal oil-bearing horizons in any locality are comparatively few, and it is ordinarily easy to determine by stratigraphic methods the presence or absence of a favorable geologic horizon. By knowing the succession and thicknesses of the beds in a given region it is possible to infer from surface outcrops the approximate depth below the surface at which the desired horizon can be found. To do this, however, the conditions of sedimentation, the initial irregularities of the beds, the structural conditions, including unconformities, and other factors must be studied.

In exploration for oil the determination of the existence and location of the proper horizon is but an initial step. For instance, the oil of the Midcontinent field of the United States is in the beds of the Pennsylvanian, which are known to occupy an enormous area extending from Illinois and Wyoming south to the Gulf of Mexico. This information is clearly not sufficiently specific to limit the location of drill holes. Sometimes seepages of oil or showings of gas near the surface are sufficient basis for localizing the drill holes.[28] Commonly, however, it is necessary to find some structural feature in the nature of a dome or anticline which suggests proper trapping conditions for an oil pool. This is accomplished by geologic and topographic mapping of the surface. Levels and contours are run and outcrops are platted. As the outcrops are usually of different geologic horizons, it is necessary to select some one or more identifiable beds as horizon markers, and to map their elevations at different points as a means of determining the structural contours of the beds. When several key horizons are thus used, their elevations must be reduced to the elevations of one common horizon by the addition or subtraction of the intervals between them. For instance, knowing the succession, an outcrop of a certain sandstone may indicate that the marking horizon is 200 feet below, and the structural contour is then drawn accordingly. Observations of strike and dip at the surface are helpful; but where the beds are but slightly flexed, small irregularities in deposition may make strike and dip observations useless in determining major structures. It is then necessary to have recourse to the elevations of the marking horizons.

In the selection of key horizons, knowledge of the conditions of sedimentation is very important. For example, some of the oil fields occur in great delta deposits, where successive advances and retreats of the sea have resulted in the interleaving of marine and land deposits. The land-deposited sediments usually show great variations in character and thickness laterally and vertically; and a given bed is likely to thin out and disappear when traced for a short distance, rendering futile its use as a marker. The marine sediments, on the either hand, show a much greater degree of uniformity and continuity, and a bed of marine limestone may extend over a large area and be very useful as a key horizon.

Over large areas outcrops and records of previously drilled water and oil wells may not be sufficient to give an indication of structure; it then becomes necessary to secure cross sections by drilling shallow holes to some identifiable bed, and to determine the structure from these cross sections, in advance of deeper drilling through a favorable structure thus located. The coöperative effort of the Illinois State Survey and private interests, cited on page 306, is a good illustration of this procedure. This method is only in its infancy, because well-drilling has not yet exhausted the possibilities of structures located from surface outcrops.

The so-called anticlinal structures, which have been found by experience to be so favorable to the accumulation of oil, are by no means symmetrical in shape or uniform in size. They may be elongated arches with equal dip on the two sides, or one side may dip and the other be nearly flat. In a territory with a general dip in one direction, a slight change in the angle, though not in the direction of dip, sometimes called an arrested dip, may cause sufficient irregularity to produce the necessary trapping conditions. In other cases the anticline may be of nearly equidimensional dome form. The largest anticlines which have been found to act as specific reservoirs are rarely more than a few miles in extent, and in many cases only a mile or two. The "closure" of an anticline is the difference between the height of a given stratum at the highest point and at the edges of the structure. A considerable number of productive anticlines are known in which the beds dip so gently as to give a closure of 20 feet or less.

After the structural outlines of beds near the surface have been determined, all possible information should be used in projecting these structures downward to the oil-producing horizons. Where a number of wells have been previously drilled in the vicinity, examination of their records may indicate certain lateral variations in the thickness of the beds between the horizon which has been mapped and the producing horizon. The effect of such lateral variations may be either to accentuate the surface structure, or to cause it to disappear entirely and thus to indicate lack of favorable trapping conditions. The possibility of several oil-producing beds, at different depths—a not uncommon condition in many fields—should also be kept in mind.

As already indicated, anticlines are not always essential to make the necessary trapping conditions. In the Beaumont field of Texas, for instance, it has been shown that irregular primary deposition of sediments differing in porosity both vertically and horizontally allowed the oil to migrate upward irregularly along the porous beds and parts of beds, and to be trapped between the more impervious portions of the beds.

Further questions to be considered in the exploration of an area are the content of organic matter in the sediments which may have served as a source of oil, the presence of impervious caprocks or of variations in porosity sufficient to retain the oil, the thickness of sediments and the extent to which they have undergone differential stresses, the amount of erosion and the possibilities that oil, if formed, has escaped from the eroded edges of porous strata, and, where carbonaceous beds are present, their degree of carbonization, and many other similar matters.

Each field in fact has its own "habit," determined by the interaction of several geologic factors. This habit may be learned empirically. Geologists have often gone wrong in applying to a new district certain principles determined elsewhere, without sufficient consideration of the complexity and relative importance of the sundry geologic factors which in the aggregate determine the local habit of oil occurrence.

Geographically associated fields characterized by similarity of oil occurrence, age, and origin, are known as *petroliferous provinces*. The factors entering into the classification of fields are so numerous that more precise definition of a petroliferous province is hardly yet agreed upon.

The part played by the economic geologist in oil exploration and development is a large one for the obvious reasons given above. Probably no other single division of economic geology now employs so large a number of geologists. Practically no large oil company, or large piece of oil exploration and development, is now handled without geologic advice. Quoting from Arnold:[29]

It ought to be as obvious that exploration with the drill should be preceded by careful geologic studies as it is that railroad construction should be based on surveys. These studies should include such subjects as topography, stratigraphy, structure, and surface evidence of petroleum in the regions to be tested. The work divides itself into two stages—preliminary reconnaissances and detailed surveys.

The preliminary reconnaissance should consist in procuring all the available published and hearsay evidence regarding the occurrence of oil or gas seepages or hydrocarbon deposits in the region; in making preliminary geologic surveys to determine from which formations the oil is to come and the areal distribution of these formations; in determining those general regions in which the surface evidence is supposed to be most favorable for the accumulation of hydrocarbons; and in determining the best routes and methods of transportation.

The second stage includes detailed geologic surveys of those regions where the surface evidence indicates that petroleum is most likely to be found and the location of test holes at favorable points. By working out the surface distribution and structure of the formations it is usually possible to select the areas offering the best chances of success. Geology should always be the dominant factor in determining the location of test holes, although modifications to meet natural conditions must sometimes be made.

OIL SHALES

One of the sources of oil which is likely to become important in the future is oil shales,—that is, shales from which oil product can be extracted by distillation. These have already been referred to on previous pages. Such shales are now mined only in Scotland and in France to a relatively small extent, but there are immense reserves of these shales in various parts of the world which are likely to be drawn upon when commercial conditions require it. In the United States alone it is estimated that the oil shales are a potential source of oil in amounts far greater than all the natural petroleum of this hemisphere.[30] The solution of the problem of extraction of oil from shales is fairly well advanced technically, and the problem has now become principally one of cost. In order to recover any large amount of oil from this source, operations of stupendous magnitude, approximately on the scale of the coal industry, must be established. As long as there are sufficient supplies of oil concentrated by nature to be drawn upon, it is unlikely that oil shale will furnish any considerable percentage of the world's oil requirements. With the great increase in world demand for oil, however, which may very possibly outstrip the available annual supply in the future, and particularly with the increase in the United States demand relative to domestic supplies, exhaustive surveys of the situation are being made with a view to development of oil shales when warranted by market conditions.

Oil shales are sedimentary strata containing decomposed products of plants and animals. Locally they grade into cannel coal, with which they are genetically related. They may be regarded as representing the kinds of sediments from which the oil of oil pools has in the main originated.

The most extensive of the oil shales of the United States are found in the Eocene beds of northwestern Colorado, northeastern Utah, and southwestern Wyoming, and in the Miocene beds of northern Nevada. The largest known foreign deposits occur in Brazil and Russia.

NATURAL GAS

Economic Features

Natural gas is used both for lighting and for fuel purposes. In the United States it has become the basis of a great industry, the value of the product ranging above that of lead and zinc. The United States is the largest producer of natural gas. Other producers are Canada, Dutch East Indies, Mexico, Hungary, Japan, and Italy. Nearly all producing oil fields furnish also some natural gas.

In the United States nearly 40 per cent of the total production of natural gas comes from West Virginia, about 17 per cent from Pennsylvania, about 17 per cent from Oklahoma, and less than 10 per cent from each of Ohio, California, Louisiana, Kansas, Texas, and several other states.

One of the recent interesting developments in this industry is the recovery of gasoline from the natural gas. This is obtained by compression and condensation of the casing-head gas from oil wells, and also, more recently, by an absorption process which is applied not only to "wet" gas from oil wells but also to so-called "dry" gas occurring independently of oil. It is a high-grade product which in recent years has amounted to about 10 per cent of the total output of gasoline for the United States.

Geologic Features

Natural gas, like oil, originates in the distillation of organic substances in sediments, and migrates to reservoirs capped by impervious strata. It is commonly, though not always, associated with oil and coal. The geologic features of its occurrence have so much in common with oil that a description would essentially duplicate the above account of the geologic features of oil.

ASPHALT AND BITUMEN

Economic Features

Asphalt and bitumen are not used as energy resources, but they have so much in common with oil in occurrence and origin that they are included in this chapter.

Asphalt and bitumen find their main use in paving. Other important uses are in paints and varnishes, in the manufacture of prepared roofing, for various insulating purposes, and in substitutes for rubber.

Nearly the entire world's supply of natural asphalt comes from the British Island of Trinidad and from Venezuela. Both of these deposits are under United States commercial control probably affiliated with Dutch-English interests. Prior to the war about half the product went to Europe and half to the United States. Large amounts of asphaltic and bituminous rock, used mainly in paving, are normally produced in Alsace, France, and in Italy. Prior to the war both the Alsatian and Italian deposits were under German commercial control. Their output is practically all consumed in Europe.

The United States takes a large part in the world's trade in natural asphalt, by importation from Trinidad and Venezuela, and by some reëxportation chiefly to Canada and Mexico. The United States also produces some natural asphalt and bituminous rock for domestic consumption. Deposits of natural asphaltic material are widely distributed through the United States, but commercial production is limited to a few localities in Kentucky, Texas, Utah, Colorado, Oklahoma, and California.

The asphalt manufactured from petroleum constitutes a much larger tonnage than natural asphalt though it does not enter so largely into world trade. The manufactured product is largely but not exclusively in American control. Large amounts are made in this country and will no doubt be made for the next decade, from oil produced in the southwestern states and in Mexico. At the present time as much or more asphalt is made in the United States from Mexican as from domestic crude oil. The refineries are located near the Gulf coast so that exports can avoid overland shipments. The relative merits of natural asphalt and asphalt manufactured from oil may be subject to some discussion; but it is perfectly clear that the manufactured material is sufficient, both in quantity and variety, to make the United States entirely independent and have an exportable surplus.

Geologic Features

Natural asphalt and similar products are in the main merely the residuals of oil and gas distillation accumulated by nature under certain conditions already described in connection with oil (pp. 140-144). In some cases the asphaltic material is found as impregnations of sediments, and appears to have remained in place while the lighter organic materials were volatilized and migrated upward. In other cases it occurs in distinct fissure veins; the fissures and cavities apparently were once filled with liquid petroleum, which has subsequently undergone further distillation. The original liquid character of some of these bitumens is shown by occasional fragments of unworn "country rock" imbedded in the veins. The effect of surface waters, carrying oxidizing materials and sulphuric acid, is believed to have contributed to the drying out and hardening of these veins or dikes.

Asphalts and bitumens include a wide variety of hydrocarbon materials, such as gilsonite, grahamite, elaterite, ozokerite, etc., which are used for somewhat different purposes. The deposits of the United States show much variety in form, composition, age, and geologic associations. The important Kentucky deposits occur as impregnations of Carboniferous sandstones at the base of the Coal Measures of that state.

The Trinidad asphalt comes from the famous "pitch lake," which is a nearly circular deposit covering about a hundred acres 150 feet above sea level, and which is believed to fill the crater of an old mud volcano. The so-called pitch consists of a mixture of bitumen, water, mineral and vegetable matter, the whole inflated with gas, which escapes to some extent and keeps the mass in a state of constant ebullition. The surface of the lake is hard, and yet the mass as a whole is plastic and tends to refill the excavations. The lake is believed to be on the outcrop of a petroleum-bearing stratum, and the pitch to represent the unevaporated residue of millions of tons of petroleum which have exuded from the oil-sands. The pitch is refined by melting,—the heat expelling the water, the wood and other light impurities rising, and the heavy mineral matter sinking to the bottom.

The asphalt of Venezuela is similar in nature, but the pitch "lake" is here covered with vegetation and the soft pitch wells up at certain points as if from subterranean springs.

FOOTNOTES:

[17] For more detailed treatment of international coal movements before the war and of coal movements within the United States, see the U. S. Geological Survey's *World Atlas of Commercial Geology*, Pt. 1, 1921, pp. 11-16.

[18] Campbell, Marius R., The coal fields of the United States: *Prof. Paper 100-A, U. S. Geol. Survey*, 1917, pp. 5, 6, 7.

[19] Compiled from tables quoted by White, David, The petroleum resources of the world: *Annals Am. Acad. Social and Political Sci.*, vol. 89, 1920, pp. 123 and 126.

[20] White, David, *loc. cit.*, p. 113.

[21] See Arnold, Ralph, Petroleum resources of the United States: *Econ. Geol.*, vol. 10, 1915, p. 707.

[22] White, David, Late theories regarding the origin of oil: *Bull. Geol. Soc. Am.*, vol. 28, 1917, p. 732.

[23] McCoy, A. W., Notes on principles of oil accumulation: *Jour. Geol.*, vol. 27, 1919, pp. 252-262.

[24] White, David, Genetic problems affecting search for new oil regions: *Mining and Metallurgy, Am. Inst. of Min. Engrs.*, No. 158, Sec. 21, Feb., 1920.

[25] Mehl, M. G., Some factors in the geographic distribution of petroleum: *Bull. Sci. Lab., Denison Univ.*, vol. 19, 1919, pp. 55-63.

[26] Schuchert, Charles, Petroliferous provinces: *Bull. 155, Am. Inst. Mining and Metallurgical Engrs.*, 1919, pp. 3059-3060.

[27] Loc. cit., p. 20.

[28] Seepages or residual bituminous matter near the surface may be due to upward escape of oil material through joints in the rocks capping a reservoir, and productive pools may be found directly below such showings. In other regions similar surface indications may mean that the stratum in the outcrop of which they are found is oil-bearing; but accumulations of oil, if present, may be several miles down the dip, at places where the structural conditions have been favorable. In still other cases the seepage may have been in existence for such a long time as to exhaust the reservoir. It must also be remembered that gas seeps are common in sloughs and marshes where vegetation is decaying, and may be of no significance in the search for petroleum.

[29] Arnold, Ralph, Conservation of the oil and gas resources of the Americas: *Econ. Geol.*, vol. 11, 1916, pp. 321-322.

[30] Oil shales may also be made to yield large quantities of fuel and illuminating gas, and of ammonia (see pp. 101-102).

CHAPTER IX

MINERALS USED IN THE PRODUCTION OF IRON AND STEEL (THE FERRO-ALLOY GROUP)

GENERAL FEATURES

Iron and steel and their alloys are the most generally used of the metals. The raw materials necessary for their manufacture include a wide variety of minerals.

Iron is the principal element in this group; but in the manufacture of iron and steel, manganese, chromium, nickel, tungsten, molybdenum, vanadium, zirconium, titanium, aluminum, uranium, magnesium, fluorine, silicon, and other substances play important parts, either as accessories in the furnace reactions or as ingredients introduced to give certain qualities to the products.

Nearly all parts of the world are plentifully supplied with iron ores for an indefinite period in the future, but their abundant use has thus far been confined mainly to the countries bordering the North Atlantic,—the United States, Germany, and England,—which, possessing ample coal supplies, have had the initiative to develop great iron and steel industries. China has abundant coal, moderate quantities of iron ore, and a large population, but a low per capita consumption of iron and steel products. Development of its iron and steel industry is just beginning. Japan has neither coal nor iron in sufficient quantities, and hence the Japanese effort in recent years to control the mineral resources of China and other countries. As a result of the war Germany has been largely deprived of its iron ores, and France may assume somewhat the rank in iron ore production once held by Germany. Sweden and Spain have been considerable producers of iron ore, but both lack coal, with the result that their ores have been largely exported to England and Germany. With increase of per capita consumption in outlying parts of the world, iron and steel industries are beginning to develop locally on a small scale, as in India, South Africa, and Australia. Russia has had sufficient supplies of coal and iron, but the stage of industrial development in that country has not called for great expansion of its iron and steel industry.

There has been a tendency for iron and steel manufacture to become concentrated at a comparatively few places on the globe favored by the proper combinations of coal, iron, transportation, proximity to consuming populations, initiative and capacity to take advantage of a situation, and other factors. Even though on paper conditions may seem to be favorable in outlying territories for the development of additional plants, this development is often held back by competition from the established centers. On the west coast of the United States, there are raw materials for an iron and steel industry and there has been discussion for years as to the possibilities of starting a successful large scale steel industry. The consuming power of the local population for all kinds of iron and steel would seem to be great enough to warrant such action. However, the demand is for an extremely varied assortment of iron and steel products; and to start an industry, making only a few of the cruder products such as pig iron and semi-finished forms, would not meet this demand. All varieties of finishing plants and associated factories would also need to be started in order to meet the situation. This would require large capital. Furthermore the local demand for some of the accessory finished products might not warrant the establishment of the accessory plants.

Throughout the history of the iron and steel business there has been a marked tendency for the iron ore to move to regions of coal production rather than for the coal to move to the iron ore regions. The coal or energy factor seems ultimately to control. This is due in considerable part to the fact that coal furnishes the basis of a great variety of industries for which iron ore is only one of the feeders, and which are so interrelated that it is not always easy to move the iron and steel industry to a spot near the sources of iron ore where iron and steel alone could be produced.

In regard to iron ore supplies of proper grade and quantity, the United States is more nearly self-sufficing than any of its competitors. It imports minor amounts of ore from Cuba and Canada, and even from Chile and Sweden, to border points, in the main merely because these imported ores can compete on a price basis with the domestic ores. The entire exclusion of these ores, however, would make comparatively little difference in the total volume of our iron and steel industry; though it would probably make some difference in distribution, to the disadvantage of plants along the coast. There is only one kind of iron ore in which the United States has anything approaching deficiency, and that is ore extremely low in phosphorus, adapted to making the so-called low-phosphorus pig which is needed for certain special steels. Ordnance requirements during the war put a premium on these steels. While some of these extremely low-phosphorus ores are mined in the United States, additional quantities have been required from Spain and Canada and to a lesser extent from North Africa and Sweden. Also the Spanish pyrite, imported ordinarily for its sulphur content, on roasting leaves a residue of iron oxide extremely low in phosphorus which is similarly used. The elimination of pyrite imports from Spain during the war, therefore, was a considerable contributing factor to the stringency in low-phosphorus iron ores. War experience showed that the United States was dependent on foreign sources for 40 per cent or upwards of its needs in this regard. Certain developments in progress, notably the project for concentration of siliceous eastern Mesabi Range ores, make it likely that future domestic production will more nearly be able to meet the requirements.

The equivalent of 15 per cent of the iron ore mined in the United States is exported as ore to Canadian ports on the Great Lakes and in the form of crude iron and steel products to many parts of the world. England and Germany are almost the sole competitors in the export trade.

When we turn to the minerals used for making the alloys of iron and as accessories in the manufacture of iron, it appears that no one of the principal iron and steel producing countries of the world is self-supporting, but that these "sweeteners" must be drawn in from the far corners of the earth. The importance of these minor constituents is altogether out of proportion to their volume. For instance, only fourteen pounds of manganese are necessary in the making of a ton of steel, yet a ton of steel cannot be made without manganese. The increasing specialization in iron and steel products, and the rapidly widening knowledge of the qualities of the different alloys, are constantly shifting the demand from one to the other of the ferro-alloy minerals. Each one of the ferro-alloy minerals may be regarded as being in the nature of a key mineral for the iron and steel industry, and the control of deposits of these minerals is a matter of international concern. Control is not a difficult matter, in view of the fact that the principal supplies of practically every one of the alloy minerals are concentrated in comparatively few spots on the globe,—as indicated on succeeding pages.

Nature has not endowed the United States, nor in fact the North American continent, with adequate high-grade supplies of the principal ferro-alloy minerals,—with the exception of molybdenum, and with the exception of silica, magnesite, and fluorspar, which are used as accessories in the process of steel making. With plenty of iron ore and coal, and with an iron and steel capacity amounting to over 50 per cent of the world's total, the United States is very largely dependent on other countries for its supplies of the ferro-alloy minerals. The war brought this fact home. With the closing of foreign sources of supplies, it looked at one time as if our steel industry was to be very greatly hampered; and extraordinary efforts were made to keep channels of importation open until something could be done in the way of development, even at excessive cost, of domestic supplies. The result of war efforts was a very large development of domestic supplies of practically all the ferro-alloy minerals; but in no case, with the exceptions noted above, did these prove sufficient to meet the total requirements. This development was at great cost and at some sacrifice to metallurgical efficiency, due to the low and variable grades of the raw materials. With the post-war reopening of importation much of the domestic production has necessarily ceased, and large amounts of money patriotically spent in the effort to meet the domestic requirements have been lost. These circumstances have resulted in the demand in Congress from producers for direct financial relief and in demand for protective tariffs, in order to enable the new struggling industries to exist, and to permit of development of adequate home supplies. Such tariffs might be beneficial to these particular domestic industries if wisely planned; but also, in view of the limited amounts of these particular ores in this country, their general low grade, and the high cost of mining, tariffs might very probably hasten exhaustion of our limited supplies and might handicap our metallurgical industries both in efficiency and cost (see pp. 365-366, 393-394).

IRON ORES

Economic Features

Technical and commercial factors determining use of iron ore minerals. Popularly, an iron ore is an iron ore, and there is little realization of its really great complexity of composition and the difficulty of determining what is or is not a commercial ore. Percentage of iron is of course an important factor; but an ore in which the iron is in the mineral hematite is more valuable than one with an equivalent percentage of iron which is in the form of magnetite. Substances present in the ore in minor quantities, such as phosphorus, sulphur, and titanium, have a tendency to make the iron product brittle, either when it is cold or when it is being made, so that excessive amounts of these substances may disqualify an ore. Excessive quantities of silica, lime, or magnesia may make the ore undesirable. Where an acid substance, like silica, is balanced by basic constituents like lime and magnesia, considerable amounts of both may be used. Excessive moisture content may spoil an ore because of the amount of heat necessary to eliminate it in smelting.

The metallurgical processes of the iron and steel industry are essentially adapted to the principal grades of ore available. The cheapest of the steel-making processes, called the acid Bessemer process, requires a very low-phosphorus ore (usually below .050 per cent in the United States and below .030 per cent in England.) The basic open-hearth processes, making two-thirds of the steel in the United States, allow higher percentages of phosphorus, but not unlimited amounts. The basic Bessemer (Thomas) process, used for the "minette" ores of western Europe and the Swedish magnetites, may use an ore with any amount of phosphorus over 1.5 per cent. The phosphatic slag from this process is used as fertilizer. The supply of low-phosphorus Bessemer ore in the United States is at present limited as compared with that of the non-Bessemer ores, with the result that steel-plant construction for many years past has been largely open-hearth. The open-hearth process is favored also because it allows closer control of phosphorus content in the steel.

Small but increasing amounts of steel are also made in the electric furnace; for the most part, however, this process is more expensive than the others, and it is used principally for special alloy steels.

Iron ores are seldom so uniform in quality that they can be shipped without careful attention to sampling and grade. In the Lake Superior region the ores are sampled daily as mined, and the utmost care is taken to mix and load the ore in such a way that the desired grades can be obtained. Ordinarily a single deposit produces several grades of ore. When ores are put into the furnace for smelting the mixtures are selected with great care for the particular purpose for which the product is to be used. The mixture is compounded as carefully as a druggist's prescription. An ore salesman, after ascertaining the nature of the iron and steel products of a plant, has to use great skill in offering particular ores for sale which not only will meet the desired grade in regard to all elements, but also will meet competition in price. In some respects, the marketing of different grades of iron ore is as complex as the marketing of a miscellaneous stock of merchandise. With ores, as with merchandise, custom and sentiment play their part,—with the result that two ores of identical grade mineralogically and chemically may have quite a different vogue and price, simply because of the fact that furnace men are used to one and not to the other and are not willing to experiment.

The geologist is ordinarily concerned merely with finding an ore of as good a general grade as possible; but he often finds to his surprise that his efforts have been directed toward the discovery of something which, due to some minor defect in texture, in mineralogical composition, or in chemical composition, is difficult to introduce on the market. There is here a promising field, intermediate between geology (or mineralogy) and metallurgy, for the application of principles of chemistry, metallurgy, and mineralogy, which is occupied at the present time mainly by the ore salesman. Both the mineralogist and metallurgist touch the problem but they do not cover it. With increasingly precise and rapidly changing metallurgical requirements, this field calls for scientific development.

Geographic distribution of iron ore production. Iron ores are widely distributed over the world, but are produced and smelted on a large scale only in a few places where there is a fortunate conjunction of high grades, large quantity, proximity of coal, cheap transportation to markets, and manufacturing enterprise. Over 90 per cent of the iron ore production of the world is in countries bordering the North Atlantic basin. The United States produces about 40 per cent, France about 12 per cent, England about 10 per cent, Germany before the war 15 to 20 per cent, and Spain, Russia, and Sweden each about 5 per cent. Lesser producing countries are Luxemburg, Austria-Hungary, Cuba, Newfoundland, and Algeria; and insignificant amounts are produced in many other parts of the world. Of the world's iron and steel manufacturing capacity, the United States has about 53 per cent, Germany 16 per cent, England 14 per cent, France 10 per cent, the remainder of Europe (chiefly Russia, Austria-Hungary, and Belgium) 7 per cent. The absence of important iron ore production and of iron and steel manufacture either in the southern hemisphere or in any of the countries bordering the Pacific is a significant feature, when we remember what part iron plays in modern civilization. Japan, however, is beginning to develop a considerable iron and steel industry, which promises to use a large amount of ore from China, Manchuria, and Korea, and possibly to compete in American Pacific Coast markets.

In the United States about 85 per cent of the production, or one-third of the world's production, comes from the Lake Superior region, a large part of the remainder from the Birmingham district, Alabama, and smaller quantities from the Adirondacks. For the rest of the North American continent, the only largely producing deposit is that at Belle Isle, Newfoundland, which is the basis of the iron industry of eastern Canada. Cuba supplies some ore to the east coast of the United States.

In Europe there are only three large sources of high-grade iron ore which have heretofore been drawn on largely,—the magnetite deposits of northern Sweden, the hematites and siderites of the Bilbao and adjacent districts of northern Spain, and the magnetite-hematite deposits of southern Russia. The first two of these ores have been used to raise the percentage of iron in the low-grade ores which are the principal reliance of western Europe. The Swedish ores have also been necessary in order to raise the percentage of phosphorus and thus make the ores suitable for the Thomas process; on the other hand the Spanish ores and a small part of the Swedish material have been desired because of their low phosphorus content, adapted to the acid

Bessemer process and to the manufacture of low-phosphorus pig. The Russian ores have largely been smelted in that country.

The largest of the western European low-grade deposits is a geographic and geologic unit spreading over parts of Lorraine, Luxemburg, and the immediately adjacent Briey, Longwy, and Nancy districts of France. The ores of this region are called "minette" ores. This unit produces about a fourth of the world's iron ore. Low-grade deposits of a somewhat similar nature in the Cleveland, Lincolnshire, and adjacent districts of England form the main basis for the British industry. There is minor production of iron ores in other parts of France and Germany, in Austria-Hungary, and in North Africa (these last being important because of their low phosphorus content).

Comparison of figures of consumption and production of iron ores indicates that the United States, France, Russia, and Austria-Hungary are self-supporting so far as quantity of materials is concerned. Certain ores of special grades, and ores of other minerals of the ferro-alloy group required in steel making, however, must be imported from foreign sources; this matter has been discussed above. Great Britain and Germany appear to be dependent on foreign sources, even under pre-war conditions, for part of the material for their furnaces. During the war there was considerable development of the low-grade English ores, but this does not eliminate the necessity for importing high-grade ores for mixture. Belgium produces a very small percentage of her ore requirements and is practically dependent on the Lorraine-Luxemburg field.

The principal effect of the war on iron ore production was the occupation of the great French mining and smelting field by the Germans, thereby depriving the French of their largest source of iron ore. Since the war the situation has been reversed, France now possessing the Lorraine field, which formerly supplied Germany with 70 per cent of its iron ore. As the German industrial life is largely based on iron and steel manufacture, the problem of ore supplies for Germany is now a critical one. It has led to German activity in Chile and may lead to German developments in eastern Europe and western Asia, particularly in the large and favorably located reserves of southern Russia. It seems likely, however, that arrangements will also be made to continue the export of ore from the Lorraine field down the Rhine to the principal German smelting centers. France needs the German coal for coking as badly as Germany needs the French iron ore. The Rhine valley is the connecting channel for a balanced movement of commodities determined by the natural conditions. These basic conditions are likely in the long run to override political considerations.

The Lake Superior deposits, the Swedish magnetites, the Spanish hematites, and the Russian ores carry 50 to 65 per cent of metallic iron. The Birmingham deposits of southeastern United States, the main British supplies, and the main French and German supplies contain about 35 per cent or less. It is only where ores are fortunately located with reference to consuming centers that the low-grade deposits can be used. For outlying territories only the higher-grade deposits are likely to be developed, and even there many high-grade deposits are known which are not mined. The largest single group not yet drawn on is in Brazil. Others in a very early stage of development are in North Africa and Chile.

World reserves and future production of iron ore. The average rate of consumption of iron ore for the world in recent years has been about 170 million tons per year. At this rate the proved ore reserves would last about 180 years. If it be assumed that consumption in the future will increase at about the same rate as it has in the past, the total measured reserve would still last about a century. These calculations of life, however, are based only on the known reserves; and when potential reserves are included the life is greatly increased. And this is not all; for beyond the total reported reserves (both actual and potential), there are known additional large quantities of lower-grade ores, at present not commercially available, but which will be available in the future,—to say nothing of expected future discoveries of ores of all grades in unexplored territories. Both geological inference and the history of iron ore exploration seem to make such future discoveries practically certain. Iron ore constitutes about 4 per cent of the earth's shell and it shows all stages of concentration up to 70 per cent. Only those rocks are called "iron ores" which have a sufficiently high percentage of iron to be adapted to present processes for the extraction of iron. When economic conditions demand it, it may be assumed that iron-bearing rocks not now ordinarily regarded as ores may be used to commercial advantage, and therefore will become ores.

Not only is an indefinitely long life assured for iron ore reserves as a whole, but the same is true of many of the principal groups of deposits.

The question of practical concern to us, therefore, is not one of total iron ore reserves, but one of degrees of ***availability*** of different ores to the markets which focus our requirements for iron.

The annual production of ore from a given district is roughly a measure of that ore's ability to meet the competitive market, and therefore, of its actual immediate or past availability. Annual production is the net result of the interaction of all of the factors bearing on availability. It may be argued that there are ores known and not yet mined which are also immediately available. On the whole, they seem to be less available than ores actually being produced; otherwise general economic pressure would require their use and actual production.

In considering the future availability of iron ores, it is obvious that tables of past production afford only a partial basis for prediction. Presumably districts which have produced largely in the past may be expected to continue as important factors. In these cases production has demonstrated availability. Continued heavy production may thus be expected from the ores of the Lake Superior region, from the Clinton hematites of Alabama, from the ores of the Lorraine-Luxemburg-Briey district, from the Cleveland ores of England, from the Bilbao ores of Spain, from the high-grade magnetites of northern Sweden, and (assuming political stability) from the ores of southern Russia.

Similarly, also, recent increases in production from certain districts are probably significant of increased use of such ores in the future. Among these developments are the increasing production of Swedish ores and their importation into England and Germany, and the increasing use of Clinton hematites and Adirondack magnetites in the United States. Low-grade ores from the great reserves of Cuba are being mined and brought to the east coast of the United States in increasing amounts, and it is highly probable that they will take a larger share of the market. A similar project in Chile, which lay dormant during the war because of restricted shipping facilities, is expected in the near future to yield important shipments to the United States. In none of these cases will production be limited in the near future by ore reserves. Increased production and use of iron ores are also to be looked for in Newfoundland, North Africa, China, India, Australia, and South Africa.

On the commercial horizon are ores of still newer districts, the availability of which may not be read from tables of production. Their availability must be determined by analysis and measurement of the factors entering into availability. Availability of iron ore is determined by percentage of iron, percentages of impurities, percentages of advantageous or deleterious minor constituents, physical texture, conditions for profitable mining, adaptability to present furnace practice, distance from consuming centers, conditions and costs of transportation, geographical and transportational relation to the coal and fluxes necessary for smelting, trade relations, tariffs and taxes, inertia of invested capital, and other considerations. All of these factors are variable. A comparison of ores on the basis of any one of these factors or of any two or three of them is likely to be misleading. A comparison based on the quantitative consideration of all of the several factors seems to be made practically impossible by the difficulty of ascertaining accurately the quantitative range and importance of each factor, and by the difficulty of integrating all of the factors even if they should be determined. However, their combined effect is expressed in the cost of bringing the product to market; and comparison of costs furnishes a means of comparing availability of ores. A high-grade ore, cheaply mined and favorably located with reference to the points of demand, will command a relatively high price at the point of production. The same ore so located that its transportation costs are higher will command a lower price; or it may be so located that the costs of mining and bringing it to places where it can be used are so high that there is no profit in the operation. There are known high-grade iron ores which, because of cost, are not available under present conditions.

The availability of an ore, then, depends on its relation to a market,—whether, after meeting the cost of transportation, it can be sold at prevailing market prices at the consuming centers, and can still leave a fair margin of profit for the mining operation. The price equilibrium between consuming centers affords a reasonably uniform basis against which to measure availability of ores.

Figures of cost are obtainable as a basis for comparison of availability of iron ores of certain of the districts, but not enough are at hand for comparison of the ores of all districts. Careful study of costs has demonstrated the availability in the near future of the Brazilian high-grade Bessemer hematites; and projects which are now under way for exportation to England and the United States will doubtless make this enormous reserve play an important part in the iron industry. Iron ore is known but not yet mined in many parts of the western United States and western Canada. With the increasing population along the west coast of North America, projects for smelting the ore there are becoming more definite. Establishment of smelters on the west coast would make available a large reserve of ore (see also, however, p. 155).

The list of changes now under way or highly probable for the future might be largely extended. The use of iron and steel is rapidly spreading through populous parts of the world which have heretofore demanded little of these products. This increased use is favoring the development of local centers of smelting, which will make available other large reserves of iron ore. The growth of smelting in India, China and Australia illustrates this tendency.

Iron ore reserves are so large, so varied, and so widely distributed over the globe, that they will supply demands upon them to the remote future. Reserves become available and valuable only by the expenditure of effort and money. Ores are the multiplicand and man the multiplier in the product which represents value or availability. Iron ore can be made available, when needed, almost to any extent, but at highly varying cost and degree of effort. The highest grade ores, requiring minimum expenditure to make them available, are distinctly limited as compared to total reserves. Any waste in their utilization will lead more quickly to the use of less available ores at higher cost. One of the significant consequences of the exhaustion of the highest grade reserves will be an increased draft upon fuel resources for the smelting of the lower grade ores. Availability of iron ores is limited, not by total reserves, but by economic conditions.

Geologic Features

Iron rarely exists in nature as a separate element. It occurs mainly in minerals which represent combinations of iron, oxygen, and water, the substances which make up iron rust. Very broadly, most of the iron ores might be crudely classified as iron rust. In detail this group is represented by several mineral varieties, principal among which are hematite (Fe_2O_3), magnetite (Fe_3O_4), and limonite (hydrated ferric oxide). Iron likewise combines with a considerable variety of substances other than oxygen; and some of these compounds, as for instance iron carbonate (siderite), iron silicate (chamosite, glauconite, etc.), and iron sulphide (pyrite), are locally mined as iron ores. While an ore of iron may consist dominantly of some one of the iron minerals, in few cases does it consist exclusively of one mineral. Most ores are mixtures of iron minerals.

Fully nine-tenths of the iron production of the world comes from the so-called hematite ores, meaning ores in which hematite is the dominant mineral, though most of them contain other iron minerals in smaller quantities. About 5 per cent of the world's iron ores are magnetites, and the remainder are limonites and iron carbonates.

Iron ores are represented in nearly all phases of the metamorphic cycle, but the principal commercial values have been produced by processes of weathering and sedimentation at and near the surface.

Sedimentary iron ores. Over 90 per cent of the world's production of iron ore is from sedimentary rocks. The deposits consist in the main either of beds of iron ore which were originally deposited as such and have undergone little subsequent alteration, or of those altered portions of lean ferruginous beds which since their deposition have been enriched or concentrated sufficiently to form ores. A minor class of iron ores in sediments consists of deposits formed by secondary replacement of limestones by surface waters carrying iron in solution.

1. Deposits of the first class,—originally laid down in much their present form,—are usually either oölitic, *i. e.*, containing great numbers of flat rounded grains of iron minerals like flaxseeds, or consist in large part of fossil fragments of sea shells, replaced by iron minerals. The Clinton ores of the Birmingham district, the Wabana ores of Newfoundland, the minette ores of the Lorraine district in central Europe, and the oölitic ores of northern England are all of these types. Their principal iron mineral is hematite, although the English ores also contain

considerable iron carbonate or siderite. The cementing or gangue materials are chiefly calcite and quartz, in variable proportions.

The large reserves of high-grade hematite in the Minas Geraes district of Brazil are also original sediments, but lack the oölitic texture.

An insignificant proportion of the world's iron is obtained from "bog ores," which are sedimentary deposits of hydrated iron oxide in swamps and lakes. These ores have been used only on a small scale and chiefly in relatively undeveloped countries. They are of particular interest from a genetic standpoint in that they show the nature of some of the processes of iron ore deposition as it is actually going on today.

None of the ores of this class, with the exception of the iron carbonates, have undergone any considerable surface enrichment since their primary deposition. Neither, with the exception of the Brazilian ores, have they undergone any deep-seated metamorphism. The shapes, sizes, and distribution of the deposits may be traced back to the conditions of original deposition. In England and western Europe the principal deposits have been only slightly tilted by folding. In the United States the Clinton ores have partaken in the Appalachian folding. In Brazil, the ores have undergone close folding and anamorphism.

2. Deposits of the second class, which owe much of their value to further enrichment since deposition, are represented by the hematite ores of the Lake Superior district. These may be thought of as the locally rusted and leached portions of extensive "iron formations," in which oxidation of the iron, and the leaching of silica and other substances by circulating waters, have left the less soluble iron minerals concentrated as ores. The Lake Superior iron formations now consist near the surface mainly of interbanded quartz (or chert) and hematite, called *jasper* or *ferruginous chert* or *taconite*. These are similar in composition to the leaner iron ores of Brazil, called *itabirite*, but differ in that the silica is in the form of chemically deposited chert, rather than fragmental quartz grains.

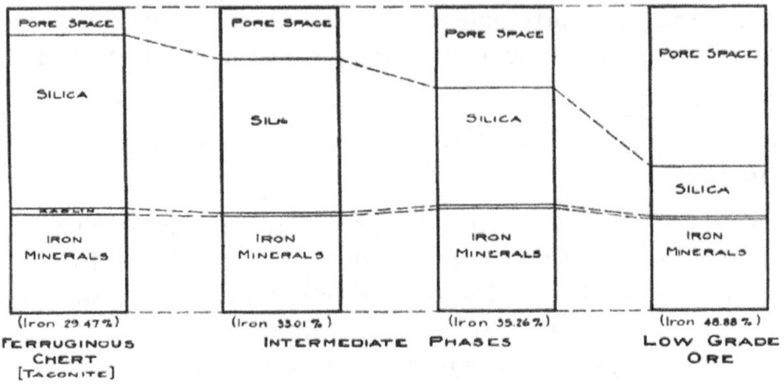

FIG. 11. Alteration of Lake Superior iron formation to iron ore by the leaching of silica.ToList

When originally deposited the iron was partly hematite (perhaps some magnetite) and largely in the form of iron carbonate (siderite) and iron silicate (greenalite), interbanded with chert. The original condition is indicated by the facts that deep below the surface, in zones protected from weathering solutions, siderite and greenalite are abundant, and that they show complete gradation to hematite in approaching the surface. The ore has been concentrated in the iron formation almost solely by the process of leaching of silica by surface or meteoric waters, leaving the hematite in a porous mass. Figure 11 illustrates this change as calculated from analyses and measurements of pore space. During this process a very minor amount of iron has been transported and redeposited. In short, the Lake Superior iron ores are residual deposits formed by exactly the same weathering processes as cause the accumulation of clays, bauxites, and the oxide zones of sulphide deposits. The development of an iron ore rather than of other materials as an end-product is due merely to the peculiar composition of the parent rock. The solution of silica on such an immense scale as is indicated by these deposits has sometimes been questioned on the general ground that silica minerals are insoluble. However, there is plenty of evidence that such minerals *are* soluble in nature; and the assumption of insolubility, so often made in geologic discussions, is based on the fact that most other minerals are *more* soluble than silica minerals, and that in the end-products of weathering silica minerals therefore usually remain as important constituents. Iron oxide, on the other hand, is *less* soluble even than silica,—with the result that when the two occur together, the evidence of leaching of silica from the mixture becomes conspicuous.

The fact that these deposits are almost exclusively residual deposits formed by the leaching of silica has an important bearing on exploration. If they have been formed by the transportation and deposition of iron from the surrounding rocks, there is no reason why they should not occasionally be found in veins and dikes outside of the iron formation. As a matter of fact they do not transgress a foot beyond the limits of the iron formation. Failure to recognize the true nature of the concentration of these ores has sometimes led to their erroneous classification as ores derived from the leaching and redeposition of iron from the surrounding rocks.

The distribution and shapes of ore deposits of this class are far more irregular and capricious than those of the primary sediments, as would be expected from the fact that their concentration has taken place through the agency of percolating waters from the surface, which worked along devious channels determined by a vast variety of structural and lithological conditions. The working out of the structural conditions for the different mines and districts constitutes one of the principal geologic problems in exploration. These conditions have been fully discussed in the United States Geological Survey reports, and are so various that no attempt will be made to summarize them here.

One of the interesting features of the concentration of Lake Superior iron ores is the fact that it took place long ago in the Keweenawan period, preceding the deposition of the flat-lying Cambrian formations, at a time when the topography was mountainous and the climate was arid or semi-arid. These conditions made it possible for the oxidizing and leaching solutions to penetrate very deeply, how deeply is not yet known, but certainly to a depth below the present surface of 2,500 feet. At present the water level is ordinarily within 100 feet of the surface, and oxidizing solutions are not going much below this depth. This region, therefore, furnishes a good illustration of the intermittent and cyclic character of ore concentration which is now coming to be recognized in many ore deposits.

Subsequent changes far beneath the surface have folded, faulted, and metamorphosed some of the Lake Superior iron ores but have not enriched them. The same processes have recrystallized and locked together the minerals of some of the lean iron formations, making them hard and resistant, so that subsequent exposure and weathering have had little effect in enriching them to form commercial ores.

The weathering of limestones containing minor percentages of iron minerals originally deposited with the limestones may result in the residual concentration of bodies of limonite or "brown ores" associated with clays near the surface. This process is similar in all essential respects to the concentration of the Lake Superior ores. Such limonitic ores are found rather widely distributed through the Appalachian region and in many other parts of the world. Because of the ease with which they can be mined and smelted on a small scale they have been used since early times, but have furnished only a very small fraction of the world's iron.

3. In a third class of sedimentary ores, the iron minerals are supposed to have been introduced as replacements of limestones subsequent to sedimentation. Such ores are not always easy to discriminate from ores resulting primarily from sedimentation. This class is represented by the high-grade deposits of Bilbao, Spain, Austrian deposits, and by smaller deposits in other countries. The Bilbao ores consist mainly of siderite, which near the surface has altered to large bodies of oxide minerals. They occur in limestones and shales and are not associated with igneous rocks. The deposits are believed to have been formed by ordinary surface waters carrying iron in solution, and depositing it in the form of iron carbonate as replacements of the limestones. The original source of the iron is believed to have been small quantities of iron minerals disseminated through the ordinary country rocks of the district. The action of surface waters, in thus concentrating the iron in certain localities which are favorable for precipitation, is similar to the formation of the lead and zinc ores of the Mississippi valley, referred to in the next chapter. Deposits formed in this manner may be roughly tabular and resemble bedded deposits, or they may be of very irregular shapes.

The sedimentary iron ores in general evidently represent an advanced stage of katamorphism, and illustrate the tendency of this phase of the metamorphic cycle toward simplification and segregation of certain materials. The exact conditions of original sedimentation present one of the great unsolved problems of geology, referred to in Chapter III.

Iron ores associated with igneous rocks. About five per cent of the world's production of iron ore is from bodies of magnetite formed in association with igneous rocks. These are dense, highly crystalline ores, in which the iron minerals are tightly locked up with silicates, quartz, and other minerals, suggestive of high temperature origin. The largest of these deposits is at Kiruna in northern Sweden; in fact this is the largest single deposit of high-grade ore of any kind yet known in the world. Here the magnetite forms a great tabular vertical body lying between porphyry and syenite. In the Adirondack Mountains of New York and in the highlands of New Jersey, magnetites are interbedded and infolded with gneisses, granites, and metamorphic limestones. In the western United States there are many magnetite deposits, not yet mined, at contacts between igneous intrusives and sedimentary rocks, particularly limestones (so-called "contact-metamorphic" deposits). The ores of the Cornwall district of Pennsylvania and some of the Chilean, Chinese, and Japanese ores are of the same type.

Magnetites containing titanium, which prevents their use at the present time, are known in many parts of the world as segregations in basic igneous rocks. They are actually parts of the igneous rock itself (p. 34). Among the large deposits of this nature are certain titaniferous ores of the Adirondacks, of Wyoming, and of the Scandinavian peninsula.

In all of these cases, it is clear that the origin of the ores is in some way related to igneous processes, and presumably most of the ores are deposited from the primary hot solutions accompanying and following the intrusion of the igneous rocks; but thus far it has been difficult to find definite and positive evidence as to the precise processes involved. None of these deposits have undergone any important secondary enrichment at the surface. Their sizes, shapes, and distribution are governed by conditions of igneous intrusion, more or less modified, as in the Adirondacks, by later deformation.

Iron ores due to weathering of igneous rocks. A small part of the world's iron ores, less than 1 per cent of the total production, are the result of surface alteration of serpentine rocks. These ores are mined principally in Cuba (Fig. 12). Here they have been developed on a plateau-like area on which erosion is sluggish. The process of formation has been one of oxidation of the iron minerals and leaching of most of the other constituents, leaving the iron concentrated near the surface in blanket-like deposits. The minerals of the original rock contained alumina, which, like the iron, is insoluble under weathering conditions, and hence the Cuban iron ores are high in alumina. They also contain small quantities of nickel and chromium which have been concentrated with the iron. A large part of the iron minerals, especially where close to the surface, have been gathered into small shot-like nodules called *pisolites*. It is thought that the solution and redeposition of the iron by organic acids from plant roots may be at least a contributing cause in the formation of this pisolitic texture.

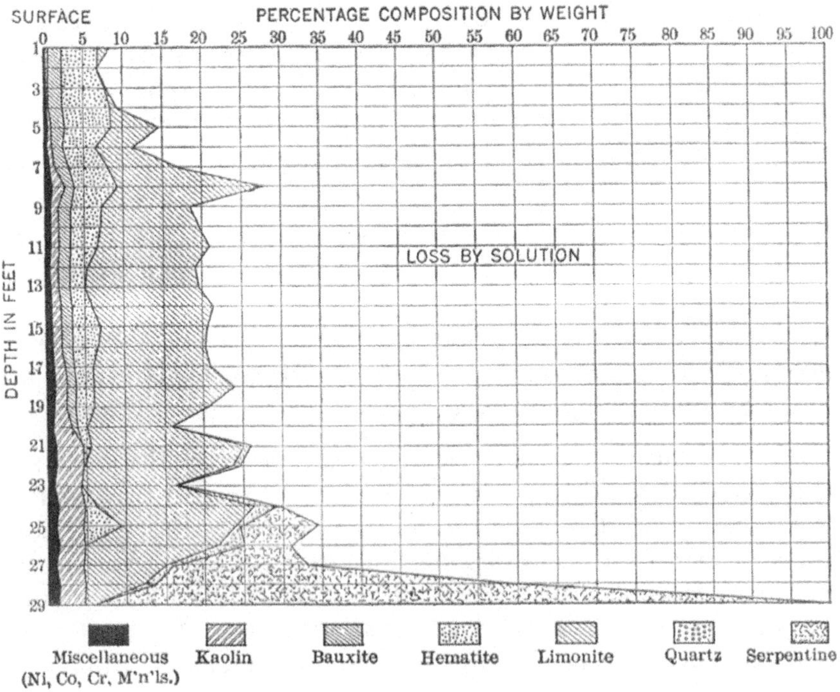

FIG. 12. Representing in terms of weight the mineralogical changes in the katamorphism of serpentine rock to iron ore, on the assumption that alumina has remained constant, eastern Cuba.ToList

The Cuban iron ores are similar in their origin to *laterites*, which are surface accumulations of clay, bauxite, and iron oxide minerals, resulting from the weathering of iron-bearing, commonly igneous, rocks. The typical laterites carry more clay and bauxite than the Cuban iron ores, but this is due merely to the fact that the original rocks commonly carry more materials which weather to clay. In fact the Cuban iron ores are themselves, broadly speaking, laterites.

Iron ores due to weathering of sulphide ores. A relatively minute portion of the world's iron ore comes from the "gossans" or "iron caps" over deposits of iron sulphides. The gossans are formed by oxidation and leaching of other minerals from the deposits, leaving limonite or hematite in concentrated masses (see pp. 46-47).

MANGANESE ORES

ECONOMIC FEATURES

Manganese ores are used mainly in the manufacture of steel, the alloys spiegeleisen and ferromanganese being added to the molten steel after treatment in the Bessemer converter and open-hearth furnace in order to recarburize and purify the metal. The alloy ferromanganese is also used in the production of special manganese steels. Manganese ore is used in relatively small amounts in dry batteries, in the manufacture of manganese chemicals, in glass making, and in pigments. Steel uses 95 per cent of the total manganese consumed, batteries and chemicals 5 per cent. On an average each ton of steel in the United States requires 14 pounds of metallic manganese, equivalent to 40 pounds of manganese ore.

With manganese ores, as with iron ores, the percentage of minor constituents,—phosphorus, silica, sulphur, etc.,—determines to a large extent the manner of use. Low-grade manganese ores, ranging from 10 to 35 per cent in manganese, 20 to 35 per cent in iron, and containing less than 20 per cent of silica, are used mainly in the production of the low-grade iron-manganese alloy called *spiegeleisen* or *spiegel* (16 to 32 per cent manganese). The higher-grade ores, ranging from 35 to 55 per cent in manganese, are used mainly in the production of the high-grade alloy called *ferromanganese* or *ferro*, in which the manganese constitutes 65 to 80 per cent of the total. To a very limited extent manganese is smelted directly with iron ores, thus lessening the amount to be introduced in the form of alloys; this, however, is regarded as wasteful use of manganese, since its effectiveness as so used is not very great. Steel makers usually prefer to introduce manganese in the form of ferromanganese rather than as spiegel. On the other hand, the ores of the United States as a whole are better adapted to the manufacture of spiegel. With the shutting off of foreign high-grade supplies during the war, resulting in the increased use of local ores, it became necessary to use larger amounts of the spiegel which could be made from these ores. Metallurgists stated that it was theoretically possible to substitute spiegel for the higher grade alloy up to 70 per cent of the total manganese requirement, but in actual practice this substitution did not get much beyond 18 per cent.

The principal manganese ore-producing countries in normal times are Russia, India, and Brazil. Relatively little ore is used in these countries, most of it being sent to the consuming countries of Europe and to the United States. The Indian ore has been used largely by British steel plants, but much of it also has gone to the United States, Belgium, France, and Germany. The Russian ore has been used by all five of these countries, Germany having a considerable degree of commercial control and receiving the largest part; a small quantity is also used in Russia. Brazilian ore has gone mainly to the United States, and in part to France, Germany, and England.

Smaller amounts of manganese ore have been produced in Germany, Austria-Hungary, Spain, and Japan. This production has had little effect on the world situation. That produced in Austria-Hungary and Germany is used in the domestic industry. That from Spain and Japan is in large part exported.

The highest grade of manganese ore comes from the Russian mines, especially those in the Caucasus region. Most of the ore used for the manufacture of dry batteries and in the chemical industry, where high-grade ores are required, has come from Russia. By far the larger part of the Russian production, however, has gone into steel manufacture. Indian and Brazilian ores have likewise been used mainly in the steel industry. Some Japanese ore also is of high grade and is used for chemical and battery purposes.

Nature has not endowed the United States very abundantly with manganese ores, and such as are known are widely scattered, of relatively small tonnage, and of a wide range of grade. The principal producing districts are the Philipsburg district of Montana and the Cuyuna Range of Minnesota; there are also scattering supplies in Virginia, Arizona, California, and many other states. The use of domestic ores has sometimes been unsatisfactory, because of frequent failure of domestic producers to deliver amounts and grades contracted for. It has been, on the whole, cheaper, easier, and more satisfactory for the large consumers to purchase the imported ore, which is delivered in any desired amount and in uniform grades, rather than to try to assemble usable mixtures from various parts of the country.

Before the European War, the United States produced only 1 to 2 per cent of its needed supply of manganese, the rest being imported mainly from India, Russia, and Brazil, in the form of ore, and from England in the form of ferromanganese (about half of the total requirement). The partial closing of the first two and the fourth of these sources of supply under war conditions made it necessary to turn for ore to Brazil and also to Cuba, where American interests developed a considerable industry in medium-grade ores. At the same time steps were taken to develop domestic resources; and with the high prices imposed by war conditions, the domestic production, both of high- and low-grade ore, was increased largely, but still was able to supply only 35 per cent of the total requirements of manganese.

At the close of the war sufficient progress had been made—in the discovery of many new deposits in the United States, in the use of low-grade domestic ores, which before had not been able to compete with imported ores, and in the increased use of spiegel, allowing wider use of low-grade ores,—to demonstrate that, if absolutely necessary, and at high cost, the United States in another year or two could have been nearly self-sufficing in regard to its manganese requirements. The release of shipping from war demands resulted immediately in larger offerings of foreign manganese ore and of ferromanganese from England, at prices which would not allow of competition from much of the domestic or Cuban ore production or from the domestic manufacture of alloys. The result was a rather dramatic closing down of the manganese industry, with much financial loss, the passage of a bill for reimbursement of producers, and a demand on the part of the producers, though not of consumers, for a protective tariff. In the questions thus raised it is desirable that geologists and engineers professionally connected with the industry thoroughly understand the basic facts; for they are liable to be called upon for advice, not only on questions relating to domestic supplies affected by possible future foreign policies, but on the formulation of the policies themselves. Conservation, cheaper steel, and future trade relations of the United States all require consideration, before action is taken to protect this one of several similarly situated mineral industries, in the effort to make the country self-supporting. These questions are further dealt with in Chapters XVII and XVIII.

Manganese production was also developed during the war in the Gold Coast of West Africa, in Costa Rica, in Panama, in Java, and elsewhere; but with the possible exception of Java and Chile, none of these sources are likely to be factors in the world situation. The war-developed manganese production of Italy, France, Sweden, and United Kingdom is also unlikely to continue on any important scale.

GEOLOGIC FEATURES

Like iron ores, manganese ores consist principally of the oxides of manganese (pyrolusite, psilomelane, manganite, wad, and others), and rarely the carbonate of manganese (rhodochrosite). They are similar in their geologic occurrence to many of the iron ores and are often mixed with iron ores as manganiferous iron ores and ferruginous manganese ores.

The higher grade manganese ores are of two general types. Those of the Caucasus district in Russia are sedimentary beds, oölitic in texture, which were originally deposited as rather pure manganese oxides, and which have undergone little secondary concentration. They are mined in many places in much the same manner as coal. Those of India and Brazil are chiefly surface concentrations of the manganese oxides, formed by the weathering of underlying rocks which contain manganese carbonates and silicates. The origin of the primary manganese minerals in the Indian and in some of the Brazilian deposits is obscure. In others of the Brazilian ores, the manganese was deposited in sedimentary layers interbedded with siliceous "iron formations," and the whole series has subsequently been altered and recrystallized.

The manganese ores of Philipsburg, Montana, the principal large high-grade deposits mined in the United States, were derived by surface weathering from manganese carbonates which form replacements in limestone near the contact with a great batholith of granodiorite. The primary manganese minerals probably owe their origin to hot magmatic solutions, as suggested by the close association of the ores with the igneous rock, the presence of minerals containing chlorine, fluorine, and boron, and the development in the limestone of dense silicates and mineral associations characteristic of hot-water alteration. The manganese ores are mined principally in the oxidized zone. Rich silver ores are found below the water table, but mainly in veins independent of the manganese deposits.

At Butte, Montana, a little high-grade manganese material has been obtained from the unoxidized pink manganese carbonate, which is a common mineral in some of the veins. It is associated with quartz and metallic sulphides and is similar in origin to the copper ores of the same district (pp. 201-202).

The lower-grade and the more ferruginous manganese ores are of a somewhat similar origin to the principal high-grade ores, in that they represent surface concentrations of the oxides from smaller percentages of the carbonates and silicates in the rocks below. Deposits of this nature have been derived from a wide variety of parent rocks—from contact zones around igneous intrusions, from fissure veins of various origins, from calcareous and clayey sediments, and from slates and schists. The manganese and manganiferous iron ores of the Cuyuna district of Minnesota, the largest source of low-grade ores in this country, were formed by the action of weathering processes on sedimentary beds of manganese and iron carbonates constituting "iron formations." The process is the same as the concentration of Lake Superior iron ores described elsewhere.

Manganese, like iron, is less soluble than most of the rock constituents, and tends to remain in the outcrop under weathering conditions. To some extent also it is dissolved and reprecipitated, and is thus gathered into concretions and irregular nodular deposits in the residual clays. In some cases it is closely associated with iron minerals; in others, due to its slightly greater solubility, it has been separated from the iron and segregated into relatively pure masses. With manganese, as with iron, katamorphic processes are responsible for the concentration of most of the ores. The ores are in general surface products, and rarely extend to depths of over a hundred feet.

CHROME (OR CHROMITE) ORES

ECONOMIC FEATURES

The principal use of chrome ores is in the making of the alloy ferrochrome (60 to 70 per cent chromium), used for the manufacture of chrome, chrome-nickel, and other steels. These steels have great toughness and hardness, and are used for armor-plate, projectiles, high-speed cutting tools, automobile frames, safe-deposit vaults, and other purposes. Chrome ore is used also both in the crude form and in the form of bricks for refractory linings in furnaces, chiefly open-hearth steel furnaces; and as the raw material for bichromates and other chemicals, which are used in paints and in tanning of leather. In the United States in normal times about 35 per cent of the total chromite consumed is used in the manufacture of ferrochrome, and about 35 per cent for bichromate manufacture, leaving 30 per cent for refractory and other purposes.

In the higher commercial grades of chrome ore the percentage of chromic oxide is 45 to 55 per cent, but under war conditions ore as low as 30 per cent in Cr_2O_3 was mined. Recovery of chrome from slags resulting from the smelting of chromiferous iron ores was one of the war-time developments.

The principal chromite-producing countries in normal times are New Caledonia, and Rhodesia (controlled by French and British interests), and to a somewhat lesser extent Russia and Turkey (Asia Minor). Small amounts of chromite are mined in Greece, India, Japan, and other countries. The Indian deposits in particular are large and high-grade but have been handicapped by inadequate transportation. The production of chrome ore in New Caledonia, Rhodesia, Russia, and Turkey has usually amounted to more than 90 per cent of the total world's production. The ore from New Caledonia has been used by France, Germany, England, and to some extent by the United States. Rhodesian ore has been used by the United States and the principal European consumers. Latterly more Rhodesian ore has gone to Europe and more Caledonian ore to the United States. The Russian ore has been in part used in Russia and in part exported, probably going mainly to France and Germany. The Turkish ore has been exported to the United States, England, and Germany; it probably supplied most of Germany's chromite requirements during the war.

During the war the United States was temporarily an important producer, as were also Canada, Brazil, Cuba, and to a minor degree Guatemala.

The richest chrome ore mined at present comes from Guatemala, but the mines are relatively inaccessible. The New Caledonian, Rhodesian, Russian, Turkish, and Indian ores are also of high grade. The ores mined in the United States, Canada, Brazil, Cuba, Greece, and Japan are of lower grade.

The use of domestic chromite supplies in the United States presents much the same problem as does manganese. The ore bodies are small, scattered, and of a generally law grade. War-time experience showed that they could be made to meet a large part of the United States requirements, but at high cost and at the risk of early exhaustion of reserves. California and Oregon are the principal sources, and incidental amounts have been produced in Washington, Wyoming, and some of the Atlantic states. With the resumption of competition from foreign high-grade ores at the close of the war, the domestic mining industry was practically wiped out; the consequences being financial distress, partial direct relief from Congress, and consideration of the possibilities of a protective tariff,—which in this case would have to be a large one to accomplish the desired results (see Chapters XVII and XVIII).

Geologic Features

The principal chrome mineral is chromite, an oxide of chromium and iron. Chromite is a common minor constituent of basic igneous rocks of the peridotite and pyroxenite type. In these rocks it occurs both as disseminated grains, and as stringers, and large irregular masses which probably represent magmatic segregations. Alteration, and weathering of the parent rock, forming first serpentine and then residual clays, make the chromite bodies progressively richer and more available, by leaching out the soluble constituents of the rock leaving the chromite as residual concentrates. All the important chromite deposits of the world are associated in somewhat this manner with serpentine or related rocks. They are formed in the same way as the lateritic iron ores of Cuba, and from the same sort of rocks (pp. 171-173). Chromite is very insoluble, and the mechanical breaking down of deposits and transportation by streams frequently forms placers of chrome sands and gravels. Such placers have not been worked to any extent.

Katamorphic processes give the important values to chromite deposits.

NICKEL ORES

Economic Features

The principal use of nickel is in the manufacture of nickel steel, the most important of all alloy steels. Ordinary nickel steels carry about 3-½ per cent nickel. Nickel is used in all gun and armor-plate steels, and in practically all other good steels except tool steels. It is also extensively alloyed with other metals, particularly with copper to form the strong non-corrosive metal (monel metal) used for ship propellers and like purposes. Nickel is also used for electroplating, for nickel coins, for chemicals, etc. Of the total production about 60 per cent is used in steels, 20 per cent in non-ferrous alloys and 20 per cent in miscellaneous uses. The ores mined range from 2 to 6 per cent in metallic nickel.

Canada (Sudbury, Ontario) produces over three-fourths of the world's nickel and is likely to have an even greater share of the future production. The French supply from New Caledonia is second in importance, and minor amounts are produced in Norway and in several other countries. The control and movement of the Canadian and New Caledonian supplies are the salient features of the world nickel situation. Nickel leaves the producing countries mostly as matte. Canadian matte has been refined mainly in the United States, but the tendency is toward refining a larger proportion in Canada. In Europe there are refineries in France, England, Belgium, Germany, and Norway, which normally treat the bulk of the New Caledonian and some of the Canadian production. Small quantities of New Caledonian matte or ore are also refined in Japan, and during the war considerable amounts came to the United States.

The United States now produces perhaps 10 per cent of its normal requirements of nickel from domestic sources, principally as a by-product of copper refining. However, the United States has a large financial interest in the Canadian deposits, and refines most of the matte produced from Sudbury ores in a New Jersey refinery. Shipments to Europe of Canadian nickel refined in the United States have been a feature of the world's trade in the past.

The nickel-bearing iron ores of Cuba, consumed in the United States, constitute a potential nickel supply of some importance, if processes of preparation become commercially perfected.

Known supplies of nickel in Canada and New Caledonia are ample for a considerable future, and geologic conditions promise additional discoveries at least in the former field. The probable reserves of the Sudbury district have been estimated to be fully 100,000,000 tons, which would supply the world's normal pre-war requirements for about a hundred years.

In recent years the British and Canadian governments have taken an active interest in the nickel industry. They organized a joint commission for its investigation, the report[31] of which furnishes the most comprehensive view of the world nickel situation yet available. The British government has directly invested in shares of the British-American Nickel Company, and has negotiated European contracts for sale of nickel for this company. The Canadian government has exerted some pressure toward larger refining of nickel matte in Canada.

Geologic Features

The principal ore minerals are the nickel sulphides and arsenides (particularly pentlandite, but also millerite, niccolite, and others), which are found at Sudbury intergrown with the iron and copper sulphides, pyrrhotite and chalcopyrite; and the hydrated nickel-magnesium silicates (garnierite and genthite), which are products of weathering. The richer ores of Canada contain about 5 or 6 per cent of nickel, the New Caledonian ores less than 2 per cent. The Sudbury ores carry also an average of about 1.5 per cent of copper.

Nickel, while present in the average igneous rock in greater amounts than copper, lead, or zinc, is apparently not so readily concentrated in nature as the other metals and is rarely found in workable deposits. The few ore bodies known have been formed as the result of unusual segregation of the nickel in highly magnesian igneous rock of the norite or gabbro type, at the time of its solidification or soon after; and in some cases, in order to produce the nickel ore, still further concentration by the agency of weathering has been necessary. Thus there are two main types of deposits.

The first, the sulphide type, is represented by the great ore bodies of the Sudbury district. These are situated in the basal portions of a great norite intrusive, and are ascribed to segregation of the sulphides as the rock solidified. To some extent the segregation was aided by mineralizing solutions following the crystallization of the magma, but in general there is little evidence that the ores were deposited from vagrant solutions of this kind (see pp. 34-35). These ores owe their value to primary concentration; secondary transportation and reprecipitation by surface waters has not been important. A small amount of the green arsenate, annabergite or "nickel bloom," has been developed by oxidation at the surface.

The second, the garnierite or "lateritic" type of nickel ores, is somewhat more common and is represented by the deposits of New Caledonia. In this locality the original rock is a peridotite, relatively low in nickel, which has been altered to serpentine. Weathering has concentrated the more resistant nickel at the expense of the more soluble minerals, and has produced extensive blanket deposits of clay, which in their lower portions contain nickel in profitable amounts. Similar processes, working on material of a somewhat different original composition, have produced the nickel-bearing and chrome-bearing iron ores of Cuba (pp. 171-173).

TUNGSTEN (WOLFRAM) ORES

Economic Features

The principal use of tungsten is in the making of high speed tool steels. It is added either as the powdered metal or in the form of ferrotungsten, an alloy containing 70 to 90 per cent of tungsten. Tungsten is also used for filaments in incandescent lamps, and in contacts for internal combustion engines, being a substitute for platinum in the latter use. Of late years tungsten alloys have also been used in valves of airplane and automobile engines.

The average grade of tungsten ores mined in the United States is less than 3 per cent of the metal; before smelting they are concentrated to an average grade of 60 per cent tungsten oxide.

Germany through its smelting interests controlled the foreign tungsten situation prior to the war; two-thirds of its excess output of ferrotungsten was consumed by England and the balance principally by the United States and France. Other consumers in the main satisfied their requirements by imports of tool steel from these four countries.

The bulk of the tungsten ore consumed in Europe prior to 1914 came from British possessions; these were principally the Federated Malay States, Burma, Australia, and New Zealand. The United States, Portugal, Bolivia, Japan, Siam, Argentina, and Peru were also producers. The great demand for tungsten created by the war added China to the list of important producers and greatly increased the production from Burma and Bolivia. Smelting works were established in England and those of the United States and France were greatly enlarged. England is at present in a position to dominate the world tungsten situation. The question of control of the ores obtainable in China, Korea, Siam, Portugal, and western South America is likely to be an important one for the future.

Of the annual pre-war world production, the United States used about one-fifth. Three-fourths of this requirement was met by domestic production. The balance was obtained by importation, chiefly from Germany, from Portugal and Spain, and from England, both of concentrates and of ferrotungsten.

To the considerable demand for high speed tool steels occasioned by munitions manufacture, production in the United States responded quickly. Supplies of tungsten came chiefly from California, Colorado, Arizona, Nevada, and South Dakota. At the same time importation largely increased, chiefly from the west coast of South America and the Orient. Consumption reached a half of the world's total. Considerable amounts of ferrotungsten were exported to the Allies.

The end of the war created a possible tungsten shortage in this country into a tungsten surplus. In so far as actual domestic consumption is concerned there has been a return to something like pre-war conditions, as the only known new use to which tungsten may be put—the manufacture of die steel—does not involve the use of any large amount of ferrotungsten. The richer mines of the two chief tungsten-producing districts in the United States have shown impoverishment and at present no important new deposits are known. The grade of the producing deposits is on an average low. The domestic production of tungsten ore will doubtless decrease, owing to the importation of cheaper foreign ores, unless a high tariff wall is erected. Importation from the Orient and the west coast of South America should continue in reduced amounts, depending upon the ability of domestic manufacturers to obtain and hold foreign markets for ferrotungsten and high speed tool steel. In the commercial control of tungsten ores the United States has at present a strong position, second only to that of England.

Geologic Features

Tungsten ores contain tungsten principally in the form of the minerals scheelite (calcium tungstate), ferberite (iron tungstate), hübnerite (manganese tungstate), and wolframite (iron-manganese tungstate). All these minerals are relatively insoluble and have high specific gravity, and as a consequence they are frequently accumulated in placers, along with cassiterite and other stable, heavy minerals. A large part of the world's tungsten production in the past has been won from such deposits. Placers are still important producers in China, Siam, and Bolivia, although in these countries vein deposits are also worked.

With the exhaustion of the more easily worked placer deposits, increasing amounts of tungsten are being obtained from the primary or fixed deposits. These are found almost exclusively in association with granitic rocks, and have a variety of forms. The most productive deposits are in the form of veins, cutting the granites and the surrounding rocks into which the granites were intruded, and containing quartz, metallic sulphides, and in some cases minerals of tin, gold, and silver. The deposits of the two most important districts in the United States, in Boulder County, Colorado, and at Atolia, California, are of this general nature. The close association of such deposits with plutonic igneous rocks, and the characteristic mineral associations (see pp. 37-41) suggest strongly that the deposits were formed by hot solutions deriving their material from a magmatic source.

Other tungsten deposits, which only recently became of importance, are of the contact-metamorphic type—in limestones which have been invaded by hot aqueous and gaseous solutions near the borders of granitic intrusions. In these occurrences the tungsten mineral is almost invariably scheelite, and is associated with calcite, garnet, pyroxene, and other silicates. A magmatic origin of the tungsten is probable. Some of the deposits of the Great Basin area and of Japan are of this nature, and it is believed that important deposits of this type may be discovered in many other countries.

Tungsten is likewise found in original segregations in igneous rocks and in pegmatite dikes, but these deposits are of comparatively small commercial importance.

In some tungsten deposits a hydrated oxide called tungstite has been formed as a canary-yellow coating at the surface. On the whole, however, tungsten minerals are very resistant to weathering, and in all their deposits secondary concentration by chemical action at the surface has not played any appreciable part. The disappearance of tungsten minerals from alluvial materials which are undergoing laterization, which has been described in Burma,[32] seems to indicate that the tungsten is dissolved in surface waters to some extent; but in the main it is probably carried completely out of the vicinity and not reprecipitated below.

MOLYBDENUM ORES

Economic Features

The main use of molybdenum is in the manufacture of high-speed tool steels, in which it has been used as a partial or complete substitute for tungsten. Its steel-hardening qualities are more effective than those of tungsten, but it is more difficult to control metallurgically. It has been used in piston rods and crank shafts for American airplanes. Its use in tool steel is mainly confined to Europe, where its metallurgical application is in a more advanced stage than in the United States. Molybdenum is added to steel either as powdered molybdenum or in the form of ferromolybdenum, an alloy containing 60 to 70 per cent of the metal. Molybdenum chemicals are essential reagents in iron and steel analysis and other analytical work; they are also used as pigments. Molybdenum metal has been used to a small extent in incandescent lamps and as a substitute for platinum in electric contacts and resistances.

Molybdenum ores range from considerably less than 1 per cent to about 5 per cent in molybdenum.

The world's principal sources of molybdenum ores in approximate order of importance are the United States, Canada, Norway, Australia, Korea, Austria, Peru, and Mexico.

About half of the world's supply is produced in the United States. Production of molybdenum in this country practically began in 1914. Most of the production has come from Colorado and Arizona. It is believed that the United States contains reserves more than sufficient to meet any possible future demand. Thus far the demand has not kept up with capacity for production. The principal consuming countries are England, France, and Germany.

Geologic Features

The chief ore minerals are molybdenite (molybdenum sulphide) and wulfenite (lead molybdate). The larger part of the world's production is from the molybdenite ores. Molybdenite occurs principally in association with granitic rocks,—in pegmatite dikes, in veins, and in contact-metamorphic deposits,—in all of which associations its origin is traced to hot solutions from the magma. It is frequently present as an accessory mineral in sulphide deposits containing ores of gold, copper, silver, lead, and zinc. At Empire, Colorado, one of the principal producing localities, it is found in veins, associated with pyrite, and filling the interstices between brecciated fragments of a wall rock composed of alaskite (an acid igneous rock). In molybdenite deposits secondary concentration has not been important.

Wulfenite is rather common in the upper oxidized zone of deposits which contain lead minerals and molybdenite. It is probably always secondary. Deposits of wulfenite have been worked on a small scale in Arizona.

VANADIUM ORES

Economic Features

Vanadium is used mainly in steel, to which it gives great toughness and torsional strength. Vanadium steels are used in locomotive tires, frames, and springs, in those parts of automobiles that must withstand special bending strains, in transmission shafts, and in general in forgings which must stand heavy wear and tear. Vanadium is also used in high-speed tool steels, its use materially reducing the amount of tungsten necessary. It is added in the form of ferrovanadium, carrying 35 to 40 per cent of vanadium. Another use of vanadium is in chrome-vanadium steels for armor-plate and automobiles. Minor amounts are used in making bronzes, in medicine, and in dyeing.

The low-grade ores of the United States range from 1 to 8 per cent of vanadium oxide, the general mean being nearer the lower figure. The high-grade ores of Peru contain from about 10 to as high as 50 per cent of the oxide; the roasted ore as shipped averages about 35 to 40 per cent.

Two-thirds of the world's supply of vanadium comes from Peru, where the mines are under American control. The concentrates are all shipped to the United States and some of the ferrovanadium is exported from this country to Europe. The Germans during the war supplied their needs for vanadium from the minette iron ores in the Briey district in France, and presumably the French will in the future utilize this source. An unrecorded but small quantity is obtained by the English from lead-vanadate mines in South Africa. There are some fairly large deposits of vanadium minerals in Asiatic Russia, which may ultimately become an important source.

The United States supplies less than one-half of its normal needs of vanadium, from southwestern Colorado and southeastern Utah. The grade of these deposits is low and the quantity in sight does not seem to promise a long future. Through its commercial control of the Peruvian deposits, the United States dominates the world's vanadium situation.

GEOLOGIC FEATURES

The Minasragra vanadium deposit of Peru contains patronite (vanadium sulphide) associated with a peculiar nickel-bearing sulphide and a black carbonaceous mineral called "quisqueite," in a lens-shaped body of unknown depth, enclosed by red shales and porphyry dikes. The origin is unknown. The patronite has altered at the surface to red and brown hydrated vanadium oxides.

The deposits of Colorado and Utah are large lens-shaped bodies containing roscoelite (a vanadium-bearing mica) in fissures and brecciated zones and replacing the cementing materials of flat-lying sandstones. Locally the sandstones contain as much as 20 per cent of the roscoelite. The deposits contain small amounts of fossil wood which may have been an agent in the precipitation of the vanadium. There is considerable doubt as to their origin, but it is generally supposed that they represent concentrations by surface waters of minute quantities of material originally scattered through the surrounding sediments; it has also been suggested that certain igneous dikes in this region may have had some connection with the mineralization. Deposits of carnotite, a potassium-uranium vanadate, which have been worked for their content of uranium and radium and from which vanadium has been obtained as a by-product, are found as impregnations of the sandstone in these same localities (p. 265).

There are other deposits containing small amounts of vanadium which are not at present available as ores. Vanadinite, a lead-vanadate, and descloizite, a vanadate of copper or lead, are found in the oxide zones of a number of lead and copper deposits in the southwestern United States and Mexico. Titaniferous iron ores, extensive deposits of which are known in many places, usually contain a small percentage of vanadium.

Outside of the Peruvian deposit, the affiliations of which are doubtful, the vanadium deposits of economic importance owe their positions and values mainly to the action of surface processes, rather than to igneous activity.

ZIRCONIUM ORES

Economic Features

The oxides of zirconium have high refractory properties which make them useful for refractory bricks and shapes for furnace linings, for chemical ware, and for other heat, acid, and alkali resisting articles. For these purposes they find a limited market. Experimental work seems to show possibilities of a very considerable use of zirconium as a steel alloy; indeed, results are so suggestive that during the war the government conducted an active campaign of investigation with a view to using it in ordnance and armor steel. For such purposes the alloy ferrozirconium is used, which carries 25 to 35 per cent zirconium metal.

The principal known deposits of zirconium ores, in order of commercial importance, are in Brazil, in India, and in the United States (Pablo Beach, Florida). The Brazilian and Indian deposits are also the principal sources of monazite (pp. 288-289). The United States controls one of the important Brazilian deposits. Germany before the war controlled the Indian deposits, and is reported to have taken much interest in the development of zirconium steels. During the war German influence in India was effectively broken up. The use of zirconium has been in an experimental state, and known sources of supply have been ample for all requirements.

Geologic Features

The zirconium silicate, zircon, is a fairly common accessory constituent of granitic rocks and pegmatite veins. From these rocks it is separated by weathering, disintegration, and stream transportation, and, having a high specific gravity, it becomes concentrated in placers. The deposits of southern India, of the coast of Brazil, and of Pablo Beach, Florida, all contain zircon along with ilmenite, garnet, rutile, monazite, and other insoluble, heavy minerals, in the sands of the ocean beaches. Smaller deposits of zircon-bearing sands exist in rivers and beaches in other parts of the United States and in other countries, but none of these deposits has thus far proved to be of commercial importance.

The largest and most important zirconium deposits are on a mountainous plateau in eastern Brazil and are of a unique type, entirely different from those just described. They contain the natural zirconium oxide, baddeleyite or brazilite, mixed with the silicate, the ore as produced carrying about 80 per cent zirconia (ZrO_2). The ores consist both of alluvial pebbles and of extensive deposits in place. The latter are associated with phonolite (igneous) rocks, and seem to owe their origin to the agency of hot mineralizing solutions from the igneous rocks.

TITANIUM ORES

ECONOMIC FEATURES

Titanium is sometimes used in steel manufacture to take out occluded gases and thus to increase the strength and wearing qualities. Its effect is to cure certain evils in the hardening of the molten steel, and it is not ordinarily added in amounts sufficient to form a definite steel alloy. Aluminum is frequently used in place of titanium. Titanium is added in the form of ferrotitanium, containing either about 15 per cent titanium and 6 to 8 per cent carbon, or about 25 per cent titanium and no carbon. Titanium compounds are also used in pigments, as electrodes for arc-lights, and by the army and navy for making smoke-clouds.

The United States has domestic supplies of titanium sufficient for all requirements. Production has come chiefly from Virginia. Additional quantities have been imported from Canada and Norway. The recently developed deposits of Pablo Beach, Florida, may produce important amounts of titanium minerals along with the output of zircon and monazite.

Geologic Features

The principal titanium minerals are rutile (titanium oxide) and ilmenite (iron titanate). These minerals are formed mainly under high temperatures, either during the original solidification of igneous rocks, or as constituents of the pegmatites which follow the crystallization of the main igneous masses. The Virginia production comes from pegmatite dikes cutting through gabbros, syenites, and gneisses. The deposits contain rutile in amounts as high as 30 per cent of the mass, but averaging 4 or 5 per cent, in addition to varying amounts of ilmenite. Titaniferous magnetites, formed in many basic igneous rocks by the segregation of certain iron-bearing materials into irregular masses, contain large quantities of ilmenite which are not commercially available under present metallurgical processes.

Rutile and ilmenite both have high specific gravity and are little affected by weathering. Consequently they are not decomposed at the surface, but when carried away and subjected to the sorting action of streams and waves, they form placer deposits. Both of these minerals are recovered from the sands at Pablo Beach, Florida.

MAGNESITE

Economic Features

The most important use of magnesite is as a refractory material for lining furnaces and converters. It is also used in the manufacture of Sorel cement for stucco and flooring, in making paper, in fire-resisting paint, in heat insulation, and as a source for carbon dioxide. Small amounts are used in Epsom salts and other chemicals.

As taken from the ground the ore consists principally of the mineral magnesite or magnesium carbonate, with minor impurities (1 to 12 per cent) of lime, iron, silica, and alumina. In making magnesite bricks, it is calcined or "dead-burned" to drive out the carbon dioxide.

Austria-Hungary and Greece are the large European producers of magnesite and Scotland supplies a little. Most of the European production is consumed in England and the Central European countries, but part has been sent to America. Outside the United States there are American supplies in Canada, and recent developments in Venezuela and Mexico (Lower California).

Magnesite is produced in considerable quantities in the United States, in California and Washington. Some material is imported from Canada, and a small amount comes from Scotland as return cargo for ballast purposes.

Before the war only about 5 per cent of the United States requirements of magnesite were met by domestic production. The country was practically dependent on imports from various European countries; chiefly from Austria-Hungary and Greece The Austrian magnesite (controlled in large part by American capital) was considered especially desirable for lining open-hearth steel furnaces, because of the presence of a small percentage of iron which made the material slightly more fusible than the pure mineral. When the shipments from this source were discontinued during the war and prices rose to a high figure, experiments were made with American magnesite, and the deposits on the Pacific Coast were developed on a large scale. A process of treatment was perfected by which the Washington magnesite was made as desirable for lining furnaces as the Austrian material. At the same time large amounts were imported from Canada and Venezuela and lesser amounts from Lower California.

Under the high prices which prevailed during the war, dolomite was to some extent substituted for magnesite. Dolomite, which may be thought of as a magnesite rock high in lime, occurs in large quantities close to many points of consumption. It is cheaper but less satisfactory than magnesite, and is not likely to be used on any large scale.

While the United States has undoubtedly sufficient reserves of magnesite to supply the domestic demands for many years, the mines are far from the centers of consumption and it is expensive to transport the material. Since the war, magnesite shipped from Canada and overseas has again replaced the American product in the eastern market to some extent. The Canadian magnesite is of lower grade than the domestic and European magnesite and is consequently less desirable. Deposits in Venezuela are also expected to furnish some material for the eastern furnaces, in competition with those of Austria and Greece. Austrian magnesite, however, will be likely to dominate the market in the future if delivered at anything like pre-war prices. This situation has led to agitation for a protective tariff on magnesite.

Geologic Features

Magnesite, as noted above, is the name of a mineral, the composition of which is magnesium carbonate. The principal magnesite deposits are of two types, of different modes of origin and of somewhat different physical characteristics.

The large magnesite deposits of Austria and of Washington, as well as those of Quebec, occur as lenses in beds of dolomite (calcium-magnesium carbonate). They are in fairly close proximity to igneous rocks, and magnesia-bearing solutions issuing from these rocks are believed to have dissolved out the calcium carbonate of the dolomite and replaced it with magnesium carbonate. In these deposits the material is coarsely crystalline and forms fairly large, continuous bodies, which are worked by quarrying. The Washington deposits closely resemble marble, and had sometimes been mistaken for that rock until war-time needs resulted in their more thorough investigation.

The commoner type of magnesite deposits is represented by those of Greece, California, Venezuela, and many other countries. These consist of veins and replacements in serpentine. The original rock was a highly magnesian igneous rock of the peridotite type, which is very unstable under weathering conditions, and rapidly alters to serpentine. Magnesite is formed both by this process and by the further breaking down of the serpentine itself. The processes are those of katamorphism. Under these circumstances the magnesite is characteristically fine-grained or massive, and occurs in veins, lenses, and irregular bodies in cavities and fractured zones. It is usually worked by open cuts.

Magnesite is also reported to occur in sedimentary beds in which it was primarily deposited in its present form and has not undergone later alteration. Such deposits are not important commercially.

FLUORSPAR

ECONOMIC FEATURES

The chief use of fluorspar is as a flux in the manufacture of open-hearth steel. Minor uses are in chemical and enameling industries, in the smelting of copper, lead, and iron, and in the manufacture of the ferro-alloys in the electric furnace.

In order to be used in steel-making, the fluorspar after being concentrated should contain at least 85 per cent calcium fluoride and less than 4 per cent silica. Chemical and enameling industries require material with 95 to 98 per cent calcium fluoride and less than 1 per cent silica.

The chief foreign producer of fluorspar is Great Britain, and much of this product comes to the United States. Canada produces a small amount, some of which also comes to the United States. Several thousand tons are produced yearly in Germany and France, and are largely consumed there.

The production of fluorspar in the United States is several times that of any other country. The ore mined comes principally from the southern Illinois and western Kentucky field, and is used largely for fluxing purposes in open-hearth steel furnaces. Minor amounts are produced in Colorado, New Mexico, and other states.

The United States has sufficient supplies of fluorspar to meet all its own demands for this material. Small amounts, however, are imported for use in eastern furnaces because the material can be brought over from England very cheaply. The domestic fluorspar is suitable for practically all purposes for which fluorspar is used except for lenses in optical instruments. For this use very small quantities of material imported from Japan have been used, but recently fluorspar of a grade suitable for optical purposes has been found in Illinois, Kentucky, New Hampshire, and other states. For fluxing purposes domestic fluorspar is superior to the foreign product.

GEOLOGIC FEATURES

Fluorspar is the trade name for the mineral fluorite, which is composed of calcium fluoride. This is a common mineral in veins and replacements which carry ores of zinc, lead, silver, gold, copper, and tin. It is formed under a variety of conditions, but is always ascribed to solutions coming from nearby igneous rocks.

The large fluorspar deposits of Illinois and Kentucky contain fluorite with calcite, barite, and metallic sulphides, in wide veins filling fissures in limestones and sandstones and replacing the fissure walls. Into these sediments there are intruded certain peridotite dikes. The fluorite and associated minerals were probably deposited by hot solutions bringing the material from some large underlying igneous mass of which the dikes are offshoots.

In the western United States many metalliferous deposits carry large amounts of fluorite, which is treated as a gangue or waste mineral, but which could be profitably extracted if there were local markets. In England, fluorite is obtained in this manner as a by-product from lead and zinc mines.

SILICA

Economic Features

Silicon and its oxide, silica, find important applications in the manufacture of iron and steel. Silicon, like manganese, is an important constituent of many steels, the alloy ferrosilicon being added to deoxidize and purify the metal and thus to increase its tensile strength. Like titanium, it is added chiefly for its curative effect rather than as a useful ingredient. On an average 4 pounds of 50 to 55 per cent ferrosilicon are used in the United States for each ton of steel produced. A higher grade of ferrosilicon (80 to 85 per cent) is used for certain special steels, and during the war considerable quantities were used in making hydrogen gas for balloons. Lower grades (10 to 15 per cent silicon) are practically a high silicon pig iron.

Silica has an important use in the form of silica brick or "ganister" for lining furnaces and converters in which acid slags are formed. For this purpose siliceous rocks, chiefly quartzites and sandstones, are ground up, mixed with lime as a binder, and fused and pressed into bricks and shapes. For the most satisfactory results the rock should contain 96 per cent or more of silica, and very little of the alkali materials, which increase the fusibility.

In addition to its applications to the iron and steel industry, silica finds an almost universal use in a wide variety of structural and manufacturing operations. The extensive use of sand and gravel—composed chiefly of silica—for road materials and railway ballast is well known. In construction work silica is used in the form of stone, sand-lime brick, cement, mortar, concrete, etc. Large quantities of sand, or silica, are used for molds in foundries, for abrasives, for the manufacture of glass, for filters, and for a great variety of other purposes which readily suggest themselves (see pp. 84, 267).

For most uses of silica there are local supplies available. For certain purposes requiring material of a particular chemical composition or texture, however, satisfactory deposits are known in only a few places. For example, the material for silica refractories is obtained in the United States chiefly from certain regions in Pennsylvania, Missouri, and Wisconsin. The United States has ample domestic supplies of silica for practically all requirements.

Ferrosilicon of the higher grades is manufactured principally in electric furnaces at Niagara Falls. The capacity is ample to meet all demands, but cheap ferrosilicon from Canada also enters United States markets.

GEOLOGIC FEATURES

Silicon and oxygen, making up the compound silica, are the two most abundant elements in the earth's crust, and quartz (SiO_2) is a very abundant mineral. The processes of weathering and transportation everywhere operative on the surface of the earth tend to separate quartz from other materials, and to concentrate it into deposits of sand. Katamorphism is primarily responsible for most of the deposits of silica which are commercially used. Anamorphism—cementing and hardening the sands into sandstones and quartzites—has created additional value for certain uses, as in refractories, building stones, and abrasives (see pp. 84, 267).

FOOTNOTES:

[31] Report of the Royal Ontario Nickel Commission. Printed by order of the Legislative Assembly of Ontario, Toronto, 1917.

[32] Campbell, J. Morrow, Tungsten deposits of Burma and their origin, *Econ. Geol.*, vol. 15, 1920, p. 511.

CHAPTER X

COPPER, LEAD, AND ZINC MINERALS

COPPER ORES

ECONOMIC FEATURES

The electrical industry is the largest consumer of copper. The manufacture of brass, bronze, and other copper alloys constitutes another chief use for the metal. Considerable quantities of copper sheets, tubes, and other wares are used outside of the electrical industry, as for instance in roofing, plumbing, and ship bottoms. Copper is also used in coinage, particularly in China, where it is the money standard of the working population.

The average grade of all copper ores mined in the United States in recent years has been about 1.7 per cent metallic copper. Ores containing as low as 0.6 per cent have been mined in the Lake Superior country, and bonanza deposits containing 20 to 60 per cent have been found and worked in some places, notably in Alaska and Wyoming. The lower-grade ores, carrying 1 to 3 per cent copper, are usually concentrated before smelting, while the richer ores, carrying 3 to 5 per cent or more, are generally smelted direct. Many of the ores contain values in gold and silver, and also in lead and zinc. An average of about 40c. worth of gold and silver per ton is obtained from all the copper ores of the United States.

In other countries the average grade of copper ores mined is somewhat higher than in the United States,—where large scale operations, particularly the use of steam-shovel methods on extensive bodies of disseminated or "porphyry" copper ores, as well as improvements in concentrating and metallurgical processes, have made possible the use of low-grade ore.

The principal sources of copper are the North American continent, Chile and Peru, Japan, south and central Africa, Australia, and Spain and Portugal. Smaller quantities are produced in Russia, Germany, Norway, Cuba, Serbia, and a number of other countries.

The United States normally produces nearly two-thirds of the world's copper and consumes only about one-third. In addition the great bulk of the South American, Mexican, and Canadian crude copper comes to the United States for refining. Through financial interests abroad and by means of refining facilities, the United States controls a quantity of foreign production which, together with the domestic production, gives it control of about 70 per cent of the world's copper. No other country produces one-sixth as much copper as the United States.

England, because of production in the British Empire (mainly Africa and Australia) and British financial control of production in various foreign countries, is not dependent upon the United States for supplies of raw copper. Japan, Spain, Portugal, and Norway are able to produce from local mines enough copper for their own needs and for export. But France, Italy, Russia, Germany, and the rest of Europe normally are dependent upon foreign sources, chiefly the United States. South America, Mexico, Canada, Africa, and Australia are exporters of copper. The control of these countries over their production in each case is political and not financial, except in the case of Canada, where about half the financial control is also Canadian. It is in these countries and in Spain that the United States and England have financial control of a large copper supply.

Before the war German interests had a considerable control over the American copper industry through close working arrangements with electrolytic refineries. Germany was the largest foreign consumer of copper, and German companies bought large quantities of the raw copper in the United States, Canada, Mexico, and South America, had it refined, and sold the finished material in both the American and foreign markets. During the war this control was broken up.

In view of the importance of copper metal as a raw material, particularly in the electrical industry, the strength of the United States in copper as a key resource ranks even above its control of petroleum.

In the United States in recent years about 40 per cent of the annual production of copper has come from Arizona, chiefly from the Bisbee, Globe, Ray-Miami, Jerome, and Morenci-Metcalf districts; about 18 per cent has come from the Butte district of Montana; about 12 to 15 per cent from Keweenaw Point,

Michigan; and about 12 per cent from Bingham, Utah. From 3 to 5 per cent of the country's output comes from each of the states of New Mexico, Nevada, Alaska, and California. All other states together produce only a little over 2 per cent of the total.

The so-called "porphyry" coppers in Utah, Arizona, Nevada, and New Mexico, described below, are the source of about 35 per cent of the present production of the United States. The deep mines of Butte and Michigan are responsible for about 30 per cent of the production, and the ore bodies of Arizona (other than porphyry) and of Alaska produce about 25 per cent.

Reserves of copper ore are such as to give no immediate concern about shortage, nor to indicate any large shift in the distribution of production in the near future. Development is on the whole considerably in advance of present demands. The principal measured reserves are in the so-called porphyry coppers of the United States and Chile. In the United States the life of these reserves now estimated is approximately 25 years. The reserves of the Chile Copper Company are the largest of any known copper deposit in the world, and the Braden copper reserve (also in Chile) is among the largest. For the deep mines of the United States, the developed reserves have a life of perhaps only five years, but for most of these mines the life will be greatly extended by further and deeper development. The porphyry coppers, because of their occurrence near the surface and the ease with which they may be explored by drilling, disclose their reserves far in advance. The deep mines are ordinarily developed for only a few years in advance of production.

Geologic Features

The principal copper minerals may be classified into the sulphide group, the oxide group, and native copper. Native copper, mined in the Lake Superior region, is the source of 8 to 10 per cent of the world's copper supply. The oxide group of minerals—including the copper carbonates, azurite and malachite; the silicate, chrysocolla; the oxide, cuprite; the sulphates, chalcanthite and brochantite; and some native copper associated with these minerals—probably supplies another 5 per cent. The remaining 85 per cent is derived from the sulphide group. Of the sulphide group by far the most important mineral is chalcocite (cuprous sulphide), which supplies the bulk of the values in the majority of the mining camps of the western hemisphere. Locally, as at Butte, enargite (copper-arsenic sulphide) is of great value. Other minerals of considerable importance in some districts are chalcopyrite and bornite (copper-iron sulphides), tetrahedrite (copper-antimony sulphide), and covellite (cupric sulphide). Very commonly the copper sulphides are associated with large quantities of the iron sulphide, pyrite, as well as with varying amounts of lead and zinc sulphides and gold and silver minerals.

The principal copper ores originate in the earlier stages of the metamorphic cycle, in close association with igneous activity. Katamorphism or weathering, in place, has played an important part in enriching them. The processes of transportation and sedimentary deposition, which have done so much toward making valuable iron ore deposits, have contributed little to the formation of copper ores.

Copper deposits associated with igneous flows. The copper ores of the Lake Superior district, and of a few small deposits in the eastern United States, contain small percentages of native copper in pre-Cambrian volcanic flows or in sediments between the flows. The ore bodies have the form of long sheets parallel to the bedding, the copper and associated minerals filling amygdaloidal openings and small fissures in the flows, and replacing conglomeratic sediments which lie between the flows. The copper was probably deposited by hot solutions related to the igneous rocks, either issuing from the magmas or deriving heat and dissolved material from them. Secondary concentration has not been important. There is practically none of it near the present erosion surface; but it appears in one part of the district near an older erosion surface covered by Cambrian sediments, suggesting a different climatic condition at that time.

The Kennecott copper deposits of Alaska have a number of resemblances to the Lake Superior copper deposits, suggesting similarity in origin. The Kennecott deposits occur exclusively in limestone, which rests conformably on a tilted surface of igneous flows ("greenstones") not unlike those of Lake Superior. The flows carry native copper and copper sulphides in minutely disseminated form and in amygdules, but apparently not in quantities sufficiently concentrated to mine. The flows are supposed to be the original source of the copper now in the limestone. The primary copper mineral in the limestone is chalcocite, in exceptionally rich and solid masses, showing no evidence of having replaced earlier sulphides. It is regarded as a product of primary deposition, under the influence of hot solutions related in some way to the igneous flows; but whether the solutions were magmatic, originating in the lavas or below, or whether they were meteoric waters rendered hot by contact with the extrusives, and thereby made effective in leaching copper from them, is not clear. The oxidation of the Kennecott copper ores is not extensive. It presents an interesting feature, in that since glacial time the ground has been frozen and the moisture is now present in the form of ice. The oxidation clearly took place before glacial time. Abundant fragments of both the oxide and the sulphide ores are mined from the lateral moraine of a nearby glacier. This is a good illustration of the cyclic nature of secondary concentration which is coming to be recognized in so many camps.

The Boleo copper deposits of Lower California occur in volcanic tuffs and associated conglomerates of Tertiary age. They have certain peculiar mineralogic associations—the ores containing large quantities of all the common copper oxide minerals, and a number of rare oxide minerals of copper, lead, silver, and cobalt, together with gypsum, sulphur, and much iron and manganese oxide. The copper oxides and carbonates are in places gathered into rounded concretions called "boleos" (balls). Sulphides are present in the lowest beds and may represent the form in which the copper was originally deposited. The copper-bearing beds have been much silicified, and it has been suggested that mineralization was accomplished by hot-spring waters, probably of igneous origin. These deposits have a few marked similarities to the Lake Superior copper ores.

Copper veins in igneous rocks. A second group of copper ores in igneous rocks is made up of deposits in distinct fissure veins and as replacements along such veins. The chief deposits of this type are at Butte, Montana—which is, from the standpoint of both past and present production, the greatest single copper district in the world. Here a large batholith of Tertiary granite was intruded by porphyry dikes; and faulting, accompanying and following the intrusions of the dikes, developed numerous fissures. The fissures were mineralized with copper sulphides and arsenides, iron sulphides, and locally with zinc sulphide and manganese carbonate,—all in a matrix of quartz. At the same time the wall rocks were extensively mineralized and altered; the fissure veins grade off into the wall rock, and in fact the larger part of the ore is simply altered granite with disseminated sulphides. The solutions which deposited the ores are inferred to have been hot from the nature of the wall-rock alterations, from the presence of hot-water minerals like fluorite, cassiterite, and others, and from the general association of the ores in time and place with the porphyry intrusions. The solutions are believed to have originated from the porphyry and possibly from other intrusives.

In the Butte district, and in the great majority of copper sulphide vein ores throughout the world, secondary concentration by surface waters has played a considerable part in developing ores of commercial value. Near the surface the copper is leached out and carried down by waters containing various solvents, particularly sulphuric acid from the oxidation of pyrite. A leached zone is formed containing the ordinary products of rock weathering,—rusty quartz and clay, sometimes black with manganese oxides. A small part of the copper remains in this zone as oxides, carbonates, and silicates. Below the oxidized and leached zone there is evidence of deposition of a large amount of secondary copper sulphide in the form of chalcocite. This is supposed to have been formed by the leaching of copper from above as soluble copper sulphate, and its precipitation below by iron and other sulphide minerals which the solutions meet on their downward course—a reaction which has been demonstrated experimentally. It was formerly supposed that most of the chalcocite was of this origin; but as chalcocite is found in important amounts with enargite and chalcopyrite to great depths (now 3,500 feet), where the veins are still rich and strong, it begins to appear that much of the chalcocite is of primary origin.

The fissures along which the Butte ores occur are in three main sets, which in order of age strike roughly east-west, northwest-southeast, and northeast-southwest. Two-thirds of the ore is in the first set, about 30 per cent in the second, and the remainder in the third. The mineralization of the several vein systems cannot be discriminated, and it is thought that it was accomplished as a more or less continuous and progressive process. There is some evidence, also, that the fracturing in the several fracture systems was likewise a nearly continuous progressive process, contemporaneous with the ore deposition, and perhaps developing under a single great shear which caused more or less simultaneous and overlapping systems of fractures in the various directions.

"**Porphyry coppers.**" Another type of copper deposits in igneous rocks is the disseminated or "porphyry" deposits. The term "porphyry" as commonly used includes true porphyries, monzonites, granites, and other igneous rocks. Ores of this type are represented by the great deposits of Bingham, Utah; Ray, Miami, and the New Cornelia mine of Arizona; Ely, Nevada; Santa Rita, New Mexico; Cananea, Sonora, Mexico; northern Chile; and many other districts of importance. They form the greatest known reserves of copper ore. These deposits contain copper minerals, usually in the marginal portions of acid porphyries, in many irregular, closely spaced veins, and in minute seams and spots disseminated through the mass of the rock. In the Ray and Miami and other districts the mineralization has spread largely through adjacent schists, but these deposits are included with the porphyry copper deposits in commercial parlance. The porphyry deposits are of an undulating blanket form of considerable areal extent and shallow depth. At the surface is a leached and weathered zone, often containing more or less of the oxides, carbonates, and silicates of copper, ranging in thickness up to 1,000 feet, but averaging 200 feet or less. Below this is a zone carrying copper in the form of chalcopyrite, enriched by chalcocite deposition from above, ranging in thickness up to 400 feet. The ore in this zone varies from one-half of 1 per cent to 6 per cent of copper and ordinarily averages between 1 and 2 per cent. The use of ore of this grade is made possible by the large quantities and by the cheap and efficient mining and metallurgical practices. The ore body grades below into a zone characterized by lean chalcopyrite, which is supposed to represent original or primary deposition from hot waters associated with the porphyry intrusion. This primary ore, or protore, was clearly formed after the solidification of the igneous rocks, though soon after, by solutions from igneous sources which followed fractured and shattered zones.

Copper in limestone near igneous contacts. Another great group of copper deposits occurs as replacements of limestone adjacent to porphyry or granitic intrusives. This type is illustrated by some of the deposits at Bingham, Utah, and at Bisbee, Arizona.

The primary deposition was of chalcopyrite and other copper sulphides, together with garnet, diopside, and other minerals known to have required high temperature in their formation. The ore fills fissures and replaces extensive masses of the limestone. It is likely to show a fairly sharp contact on the side toward the intrusive, and to grade off into the country rock on the other side with numerous embayments and irregularities. These deposits have been enriched by weathering in the same manner as indicated above for the porphyry coppers, but to highly varying degrees. In the Bisbee deposits large values were found in the weathered zone, and secondary sulphide enrichment below this zone is also important. In the Bingham camp, on the other hand, the weathered zone is insignificant and most of the ore beneath is primary. The weathering of the silicated limestone gangue results in great masses of clay which are characteristic features of the oxide zones of these deposits.

Copper deposits in schists. Other copper deposits, as at Jerome, Arizona, in the Foothill and Shasta County districts of California, at Ducktown, Tennessee, etc., are irregular lenticular bodies in schists and other rocks, but all show relationship to igneous rocks. The Rio Tinto ores of Spain and Portugal, which belong in this group, have been referred to on page 108.

In the Jerome or Verde district of central Arizona, folded pre-Cambrian greenstones and sediments were invaded by masses of quartz-porphyry, and after further deformation, rendering many of the rocks schistose, were intruded by an augite-diorite. Contact metamorphism along both the quartz-porphyry and the diorite contacts was practically lacking. The ore bodies were formed as irregular pipe-like replacements of the schists, being localized in one case by a steeply pitching inverted trough of impervious diorite, and in other cases by shear zones which favored vigorous circulation. A later series of small diorite or andesite dikes cut the ore bodies. The primary ores consist of pyrite, chalcopyrite, and other sulphides, with large amounts of jaspery quartz and some calcite and dolomite. They were clearly formed by replacement of the schists particle by particle, as shown by the frequent preservation of the schist structure in a banding of the sulphide minerals, the residual shreds of unreplaced schist material in the ores, and the usual gradual transition from unreplaced schists to those completely replaced by massive sulphides. The localization of the most important mineralization in an inverted trough is good evidence that the solutions came from below, and the nature of the mineral associations suggests an origin through the work of hot waters associated with igneous intrusives. The diorite, being most closely related in time and space with the ore bodies, seems the most logical source of the ore materials.

Secondary concentration of the Jerome ores has proceeded along the general lines previously outlined (pp. 46-50, 202). Here again the evidence is clear that the ores were concentrated in an earlier period, in this case in pre-Cambrian times, probably during the long interval required for the base-leveling of the pre-Cambrian mountains. Since Cambrian times the deposits have been for the most part buried by later sediments. Some of the deposits are still protected by this overlying blanket and mining has not yet reached the zone of altogether primary sulphides. Others have been faulted up and again exposed by erosion; but since being uncovered, steep slopes and rapid erosion have apparently favored the scattering of the copper rather than its concentration and enrichment. In the United Verde Mine, oxidizing conditions at present prevail to the bottom of the chalcocite zone.

The very large reserves of the Katanga copper belt of the Belgian Congo are in the form of tabular masses in schistose and highly metamorphosed Paleozoic sediments. The ore bodies are roughly parallel to the bedding, but in instances follow the schistosity which cuts across the bedding. They consist dominantly of the oxide minerals, though in several ore bodies sulphides have been shown by diamond-drilling. The ores have a high content of cobalt and also carry precious metals. The origin of the deposits is not known, but has been ascribed to granitic masses intrusive into the schists.

Sedimentary copper deposits. In the later phases of the metamorphic cycle, the agencies of transportation (in solution) and sedimentary deposition have resulted in some low-grade deposits of copper sulphides in sedimentary rocks. Deposits of this type are found in the Rocky Mountain region, where they are referred to as the "Red Beds" coppers, but are of no commercial importance. Similar deposits in Germany, the Mansfield copper-bearing shales, have been worked for some time, and during the war were Germany's main source of copper. On Keweenaw Point, Michigan, deposits of native copper formed in this manner in the

"Nonesuch" beds have been worked on a commercial scale. Other copper ores on Keweenaw Point are replacements of conglomerate beds between igneous flows, and are of a different origin already described (p. 200).

While much of the copper of sedimentary beds gives evidence that it was deposited from solution in cracks and as replacements of the wall rocks, often through the agency of abundant organic material in the beds, and while also comparatively little of this copper can be identified as having been deposited in detrital flakes or fragments along with the other mineral fragments, there is, nevertheless, considerable evidence that some of these deposits were formed essentially during the sedimentation of the enclosing beds and as incidents to this process. Such evidence consists of a close limitation of the copper to certain beds, its wide and uniform distribution within these beds, its absence in similar beds near at hand, the absence of evidence of feeding and escape channels of the kind which would be necessary in case the solutions were introduced long afterward, and often a minute participation of the copper minerals in the minor structures of bedding, false-bedding, and ripple-marks, which would be difficult to explain as due to secondary concentration.

The Corocoro copper deposits of Bolivia occur in beds of sandstone with no igneous rocks in the vicinity. However, they are all closely associated with a fault plane, igneous rocks occur at distances of a few miles, and the general mineralization is coextensive with the belt of igneous rocks; the deposits are therefore ascribed to a magmatic source rather than to sedimentary processes. Toward the surface the copper is in part in the form of sulphides, somewhat altered to oxide minerals, and farther down it is entirely native copper, associated with gypsum. This is the only district outside of Lake Superior where native copper has been mined on an important scale.

General comments. In general, the commercially prominent copper deposits show a close relationship to igneous rocks in place, time, and origin. Seldom do the ores extend more than 1,000 feet away from the igneous rock.

The common downward order in sulphide deposits is: first, a weathered zone, originally formed mainly above the water table, consisting above of a leached portion and below of oxides and carbonates of copper in a gangue of quartz or clay; second, a zone of secondary sulphide enrichment, characterized by chalcocite coatings, chalcopyrite, and pyrite, with a gangue of quartz and igneous rock or limestone; and third, a zone of primary deposition with similar gangue, characterized by chalcopyrite, and at Butte by enargite and chalcocite. The oxide zone as a whole may be rich or lean in values, depending on the nature of the associated gangue material and country rock. When these are more soluble than the copper—as is commonly the case in limestone—the copper may be residually concentrated, notwithstanding the fact that much copper originally present has been carried off in solution. When the associated gangue and country rock are less soluble than the copper—as is common with quartz and igneous rocks—the oxide zone is likely to be depleted of values.

The zones formed by weathering and secondary enrichment are extremely irregular, both in distribution and depth, in any one deposit, and they overlap and grade into one another in a very complex fashion. In many places the primary zone is too lean to be mined to commercial advantage; but in other places, as at Butte, and in the limestone deposits of Bingham, the primary ores are of considerable importance.

When evidence of secondary sulphide enrichment was first recognized there was a tendency to magnify its effectiveness, and to assume that in most cases the values were due to this process; that the primary zones would be found to be valueless. In recent years the emphasis is being somewhat changed because of the recognizing in many camps of rich primary zones. While some chalcocite is clearly the result of secondary enrichment from above, other chalcocite seems to have been related closely to the primary deposition. The quantitative discrimination of the two is a matter of great difficulty.

It has come to be recognized that the zonal arrangement caused by enrichment from the surface has been imposed usually on a zonal arrangement caused by the primary hot solutions and not related to the surface but to the source of the solutions. In some districts, as illustrated by Butte and Bingham, the copper-bearing minerals seem to have been deposited nearest the igneous source, while the lead, zinc, gold, and silver minerals have been deposited farther away,—suggesting the cooling of the solutions with increasing distance from the igneous source. The further investigation of this primary zonal arrangement promises interesting results with a practical bearing on exploration and development.

One of the newer features of the investigation of copper deposits has been the recognition of the cyclic nature of the secondary concentration. This process has been related not only to the present erosion surface, but to older surfaces now partly buried under later rocks. Ransome's[33] summary of conditions at the Ray-Miami camp has a somewhat general application.

Supergene enrichment has generally been treated as a continuously progressive process. There is considerable probability, however, that it is essentially cyclic, although the cyclic character may not be patent in all deposits. A full development of the cycle can take place only under a certain equilibrium of a number of factors, including climate, erosion, topography, and character of rock. The essential fact appears to be that as enrichment progresses and chalcocite increases the process of enrichment becomes slower in action, and erosion may, in some circumstances, overtake it. With the removal of some of the protecting zone of chalcocite the protore is again exposed to oxidation and a second cycle of enrichment begins.

Although much of the enriched ore is now below ground-water level, it probably was once above that level, and enrichment is believed to have taken place mainly in the zone of rock above any general water table.

Where the old erosion surface roughly coincides with the present erosion surface, the deposits follow more or less the topography. Where the old erosion surface pitches below later sediments, the ores pitch with it, and therefore do not follow the present topography. The recognition of the cyclic nature of secondary concentration is obviously of great significance in exploration and development.

Although a vast amount of study has been devoted to the origin and enrichment of copper deposits, and although the general conditions and processes are pretty well understood, the results thus far have been largely qualitative rather than quantitative.

LEAD ORES

ECONOMIC FEATURES

The most prominent uses of lead are in the manufacture of alloys, such as type-metal, bearing metal, shot, solder, and casting metal; as the oxide, red lead, and the basic carbonate, white lead, in paints; for lead pipe, cable coverings, and containers of acid active material; and in lead compounds for various chemical and medical uses. Of the lead consumed in the United States before the war about 38 per cent was utilized in pigments, 30 per cent in alloys other than shot, 15 per cent in pipe, 10 per cent in shot, and 7 per cent in all other uses. During the war much larger quantities were used in munitions, such as shot and shrapnel.

The lead content of commercial ores varies widely. It ranges from as low as .25 per cent in the Joplin district of Missouri, to about 15 per cent in the Broken Hill deposits of Australia, and over 20 per cent in the Bawdwin mines of Burma. In the Cœur d'Alene district of Idaho and the southeastern district of Missouri, the two greatest lead producers in the United States, the average grades are about 10 per cent and about 3-½ per cent respectively. The grade of ore which may be profitably worked depends not only upon the economic factors,—such as nearness to consuming centers, and the price of lead,—but also upon the amenability of the ore to concentration, the content of other valuable metals, and the fact that lead is very useful in smelting as a collector of gold and silver.

Most lead ores contain more or less zinc, and lead is obtained as a by-product of most zinc ores. Argentiferous lead ores form one of the principal sources of silver, and also yield some gold. Lead and copper are produced together from certain ores. Thus the separation of many ores into hard and fast classes, as lead, or zinc, or copper, or silver, or gold ores, cannot be made; in some of the mineral resource reports of the United States Geological Survey the statistics of these five metals are published together.

The main sources of lead ore, named in order of their importance, are the United States, Australia, Spain, Germany, and Mexico, which account for over 80 per cent of the world's production. Most of the countries of Europe outside of Spain and Germany produce small amounts of lead, but are largely dependent on imports. Spain exports argentiferous lead and pig lead mainly to England and France, with minor quantities to other countries of Europe and to Argentina. Before the war Germany, which was the largest European consumer, utilized all its own production of lead ores and imported an additional 10 per cent of the world's ores for smelting, as well as considerable amounts of pig lead. Its principal deposits were those of Silesia; under the Peace Treaty they may possibly be lost to Poland, leaving German smelters largely dependent on imports. Australia before the war normally shipped lead concentrates and pig lead to England and also to Belgium, Germany, and Japan. England, the second largest European consumer, before the war had insufficient smelting capacity within the British Empire and was partly dependent on foreign-smelted lead. During the war, however, England contracted for the entire Australian output, and enlarged its smelting capacity accordingly. This may mean permanent loss to Belgium, which had depended mainly on the Australian ores for its smelting industry before the war.

In North Africa there is a small but steady production of lead, most of which goes to France. Recent developments in Burma have shown large reserves of high-grade lead-zinc-silver-copper ores, and this region may be expected to become an important producer. There are also large reserves of lead in the Altai Mountains of southwestern Siberia and in the Andes Mountains of South America.

England, through control of Australian and Burman lead mines and smelters, domestic smelting facilities, and some financial control in Spain, Mexico, and elsewhere, and France, through financial control of Spanish and North African mines and Spanish, Belgian, and domestic smelters, have adequate supplies of lead.

The United States produces about a third of the world's lead and twice as much as any other country. Normally the domestic production is almost entirely consumed in this country. Mexico sends large quantities of bullion and ore to the United States to be smelted and refined in bond. Mexican lead refined and exported by the United States equals in amount one-sixth of the domestic production. Small quantities of ore or bullion from Canada, Africa, and South America are also brought into the United States for treatment.

Through domestic production, smelting facilities for Mexican ore, and commercial ownership in Mexico and elsewhere, the United States controls over 45 per cent of the world's lead production. Before the war Germany, through the "Lead Convention" or International Sales Association, and through smelting and selling contracts with large producing mines, practically controlled the European lead market as well as exports from Mexico and the United States and from Australia. During the war German foreign influence was practically destroyed.

In the United States about a third of the production of lead comes from southeastern Missouri and about a fourth from the Cœur d'Alene district of Idaho. The five states, Missouri, Idaho, Utah, Colorado and Oklahoma, produce about nine-tenths of the country's total output. Reserves of lead ore are not large in proportion to demand, contrasting in this regard with zinc ore.

Geologic Features

The principal lead mineral is the sulphide, galena, from which the great bulk of the world's lead is derived. Cerussite (lead carbonate) and anglesite (lead sulphate) are mined in some places in the upper part of sulphide deposits, and supply a small fraction of the world's output.

The ores of lead are of two general classes:

The first class, the so-called "soft" lead ores, nearly free from copper and precious metals, and commonly associated with zinc ores, are found in sedimentary beds independent of igneous intrusion. They are of world-wide distribution, were the first to be extensively exploited, were at one time the dominant factor in world production of lead, and at present produce about 30 per cent of the world's total. They are represented by the deposits of the Mississippi Valley, of Silesia, and some of the Spanish deposits. The general description of the origin of the zinc ores of the Mississippi Valley on pp. 216-218 applies to this class of lead ores. It should be noted, however, that in the principal United States lead-producing district, that of southeastern Missouri, the lead ores occur almost to the exclusion of the zinc ores, and are more disseminated through the limestone than is characteristic of the zinc ores. Ores of this type have been found extending only to shallow depths (not over a few hundred feet), and because of the absence of precious metals their treatment is comparatively simple.

The second class consists of ores more complex in nature, which are found in association with igneous rocks, and which usually contain some or all of the metals, zinc, silver, gold, copper, iron, manganese, antimony, bismuth, and rare metals, with various gangue minerals among which quartz, siderite, and silicates are important. Today these ores are the source of about 70 per cent of the world's lead. They are represented by the lead deposits of the Rocky Mountain region (Cœur d'Alene, Idaho; Leadville, Colorado; Bingham, Utah; etc.); of Broken Hill, New South Wales; of Burma; and of many other places. They are all related to the earlier stages of the metamorphic cycle and occur in close genetic association with igneous activity. They include deposits in the body of igneous rocks,—in the form of well-defined veins, replacements along zones of fissuring and shearing, and disseminated masses,—as well as veins and replacements in the rocks, particularly in limestones, adjoining igneous intrusions. The deposits present a wide variety of shapes depending on the courses of the solutions by which they were formed. The materials of the ore minerals are believed to have been derived from the igneous rocks and to have been deposited by hot solutions. The source of the solutions—whether magmatic or meteoric—presents the same problems which have been discussed elsewhere (pp. 41-42). The ores are frequently mined to great depths. Because of their complexity they require involved processes of treatment to separate out the values.

Ores of this nature have already been referred to in the discussion of the copper ores of Bingham and Butte, and will be referred to in connection with the zinc-lead-silver ores of Leadville, Colorado. Special reference may be made here to the Cœur d'Alene district of Idaho, which is the second largest producer of lead in the United States.

The Cœur d'Alene deposits are almost unique in that they contain galena as vein-fillings and replacements in quartzite, with a gangue of siderite (iron carbonate). Quartzite (instead of limestone) is an unusual locus of replacement ores, and siderite is an unusual gangue. These ores are believed to owe their origin to acid igneous intrusives, because of the close association of the ores with some of these intrusives, and because of the content of high-temperature minerals. Some of the ore bodies are found far from intrusives, but it is supposed that in such cases further underground development may disclose the intrusives below the surface. Secondary concentration has been insignificant.

In general, weathering of lead ores at the surface and secondary sulphide enrichment below are not so extensive as in the case of copper and zinc. Galena is fairly stable in the oxide zone, and even in moist climates it is found in the outcrop of many veins. Weathering removes the more soluble materials and concentrates the lead sulphide with the residual clay and other gangue. In some districts cerussite and a little anglesite are also found in the oxide zone. The carrying down of lead in solution and its deposition below the water table as a secondary sulphide is not proved on any extensive scale. In this respect it contrasts with zinc; and when the two minerals occur together, lead is likely to be more abundant in the oxide zone, and zinc in the sulphide zone below. Such a change in composition with depth is also found in some cases as the result of primary vertical variations in the mineralization.

ZINC ORES

Economic Features

Zinc metal has commonly gone under the name of "spelter." Brass and galvanized iron contain zinc as an essential ingredient. Of the total United States zinc consumption in normal times, about 60 per cent is used in galvanizing iron and steel objects to protect them from rust, 20 per cent is used in the manufacture of brass and other alloys, 11 per cent goes into the form of rolled sheets for roofing, plumbing, etc., 1 per cent is employed in desilverizing lead bullion, and the remaining 8 per cent is used for pigments, electrodes, and other miscellaneous purposes. During the war the use in brass-making was greatly increased.

The zinc content of the ores mined today ranges from a little over 1-½ per cent in the Joplin district of Missouri, to 25 per cent and higher in some of the deposits of the Cœur d'Alene and other western camps, and over 40 per cent in certain bonanzas in British Columbia and Russia. The ores usually contain both zinc and lead in varying proportions, and sometimes gold, silver, and copper are present. Of the zinc produced in the United States, about 73 per cent is obtained from ores containing zinc as the principal element of value, about 25 per cent from zinc-lead ores, and 2 per cent from copper-zinc and other ores. The average grade of the straight zinc ores is about 2-½ per cent.

Of the world's zinc ore, the United States produces in normal times about one-third, Germany about one-fifth, Australia about 15 per cent, Italy, North Africa and Spain each about 5 per cent. The remaining 15 to 20 per cent comes from a large number of scattered sources, including Japan, East Asia, Norway and Sweden, Canada, Mexico, Austria, France, Greece, Siberia, and Russia. In the near future the Bawdwin mines of Burma will probably be increasingly important producers. Large reserves of zinc also exist in the Altai Mountains of southwestern Siberia, and in the Cordilleran region of South America. In short, zinc is one of the most widely distributed of metallic resources; there is consequently less necessity for great international movements than in the case of many other commodities.

The smelting of zinc concentrates is in general carried on close to the points of consumption and where skilled labor is available, rather than at the mines,—although smelters to handle part of the output have recently been built in Australia and in Burma. In Europe the great smelting countries have been Germany and Belgium, and to a lesser extent England and France. Before the war these four countries with the United States produced over nine-tenths of the world's spelter. Belgium did principally a custom business, and a large part of its exports went to England. Australian and Tasmanian zinc ores were the basis of the Belgian and English smelting industries, and also supplied about one-third of the German requirements. Since the war England has contracted to take practically the entire Australian output. This fact, in connection with war-time destruction of Belgian smelters, leaves the future of the Belgian zinc industry in some doubt. Germany may possibly lose to Poland its richest zinc mines, those of Silesia. German activity in the rich deposits of Mexico is to be expected. France controls the deposits of North Africa and satisfies a considerable part of its requirements from that source. Smaller movements of zinc include exports from Italy to England, and a complex interchange among the lesser producers of Europe. English and French zinc-smelting capacity was expanded during the war, and the industry in these countries is in a strong position. Japan also developed a considerable smelting industry during the war, importing ores from eastern Asia and Australia.

The United States normally smelts and consumes all its large production of zinc ores and does not enter foreign markets to any extent. Small amounts of zinc concentrates are brought in from Mexico and Canada to be smelted in bond. During the war,—when the Allies were cut off by enemy operations from the customary Belgian and German supplies of spelter, and by shortage of ships from Australian zinc ores,—Australian, Spanish, Italian, and other ores were imported into the United States, and large quantities of spelter were exported from this country to Europe. Mine and smelter capacities were greatly increased, over-production ensued, and with the cessation of hostilities many plants were obliged to curtail or cease operations. The United States has now about 40 per cent of the zinc-smelting capacity of the world. For the present at least the capacity is far in excess of the domestic requirements.

Before the war German control of the international zinc market was even stronger than in the case of lead. The German Zinc Syndicate, through its affiliations, joint share-holdings, ownership of mines and smelters, and especially through smelting and selling contracts, controlled directly one-half of the world's output of zinc and three-fourths of the European production. It regulated the Australian exports by means of long-term contracts, and had considerable influence in the United States. To some extent it was able to so manipulate the market that zinc outside the syndicate was also indirectly controlled. During the war political jurisdiction was used by the Allied countries to destroy this German influence.

In the United States the principal zinc-producing regions are the Joplin and adjacent districts of Missouri, Oklahoma, Kansas, and Arkansas, furnishing about one-third of the country's output; the Franklin Furnace district of New Jersey, and the Butte district of Montana, each yielding about one-sixth of the total supply; and the Upper Mississippi Valley district of Wisconsin, Iowa, and Illinois, the Leadville district of Colorado, and the Cœur d'Alene district of Idaho, each producing between one-tenth and one-twentieth of the total. Smaller quantities are produced in Tennessee, New Mexico, Nevada, and several other states.

Reserves of zinc are ample for the future. They are now developed considerably in advance of probable requirements, a fact which causes keen competition for markets and renders zinc-mining more or less sensitive to market changes.

GEOLOGIC FEATURES

The most important mineral of zinc is the sulphide, sphalerite or "zinc blende." The minerals of the oxide zone are smithsonite (zinc carbonate) and calamine (hydrous zinc silicate), which yield minor amounts of zinc and are especially productive at Leadville, Colorado. Zincite (zinc oxide) and willemite (zinc silicate) are the important minerals in the deposits of Franklin Furnace, New Jersey. The association of most deposits of zinc with more or less lead has been noted.

The ores of zinc are of two general classes, corresponding to the two classes of lead ores (pp. 211-212). Zinc ores of the first type are in veins and replacements in sedimentary rocks at shallow depths, independent of igneous association, and are supposed to have been formed by cold solutions. They are found in the Mississippi Valley, in Silesia, and in many of the smaller European deposits. They were formerly the leading zinc-producers, and now produce about 45 per cent of the world's total. Zinc ores of the second type consist of veins and replacements related to intrusive rocks, sometimes extending to considerable depths, and of more complex composition. They include most of the deposits of the American Cordilleran region (Butte, Cœur d'Alene, Leadville, etc.), of Franklin Furnace, of Australia, of Burma, and of many other places.

The zinc-lead ores of the type found in the Mississippi Valley are of special interest, in that they are sulphide ores of an origin apparently independent of igneous agencies. These ores occur as fissure-fillings and replacements, mainly in nearly flat-lying Paleozoic limestones and dolomites—the Bonne Terre dolomitic limestone of southeastern Missouri, the Boone formation of southwestern Missouri and Oklahoma, the Galena dolomite of Wisconsin and Illinois. They are variously associated with a gangue of dolomite, calcite, quartz, iron pyrite, barite, and chert. Not infrequently they are spread out both in sheets and in disseminated form along carbonaceous layers within or at the base of the limestone.

The source of the primary sulphides has been a subject of much discussion. All are agreed that they were first deposited with the sediments in minutely dispersed form, through the agency of the organic contents of the sediments, and that such deposition was somewhat generally localized by estuarine conditions which favored the accumulation of organic remains. Many years ago, before the evidence of estuarine deposition was recognized, Chamberlin suggested an ingenious hypothesis for the northern Mississippi Valley,—that the organic material had been localized by ocean's currents forming something in the nature of a Sargasso sea. Differences of opinion become acute, however, when the attempt is made to name the precise sedimentary horizon, out of several available horizons, in which for the most part this primary concentration occurred. Judging from the organic contents of the several beds, the primary source may have been below, within, or above the present ore-bearing horizons. If the ore came from the lower horizons, it was introduced into its present situation by an artesian circulation, for which the structural conditions are favorable. If the ore was derived from overlying horizons, downward moving solutions accompanying erosion did the work. If the primary source was within the horizon of present occurrence of the ores, both upward and downward moving waters may have modified and transported them locally. For each of these hypotheses a plausible case can be made; but much of the evidence can be used interchangeably for any one of them. In spite of the wealth of data available, it is astonishingly difficult to arrive at a conclusion which is exclusive of other possibilities. Without attempting to argue the matter in detail the writer merely records his view, based on some familiarity with these districts, that, on the whole, the evidence favors the accumulation of these deposits by downward moving meteoric solutions during the weathering of overlying strata; but that it is by no means certain that a part of the ores has not been derived from lower horizons. The great area of the producing districts in comparison with their depth, the uniform association of the ore-bearing zone with the surface regardless of geologic horizon uncovered by erosion, the failure of the ores to extend in quantity under cappings of later formations, and the known efficacy of oxidizing waters in local downward transfers of zinc and lead, seem to suggest concentrating agencies which are clearly related to surface conditions.

It is of interest to note that in many places in the limestones of Missouri and Virginia, and elsewhere in the Paleozoic rocks, there are sinks of limonite and clay near the surface, which are likewise believed to have originated through downward movement of waters deriving their mineral contents from the erosion and stripping of overlying sediments. Still further, the primary deposition of Clinton iron ores in many parts of the Mississippi Valley and eastward to the Appalachians took place in stratigraphic horizons not far removed from the horizons of lead and zinc deposition. When the peculiar conditions controlling the deposition of the Clinton ores are understood (see pp. 52-53) it is entirely possible that they may throw some light on the genesis of the lead and zinc ores.

Since the ores were introduced into essentially their present locations, secondary concentration has produced an oxide zone of clay, chert, and iron oxide, with varying amounts of zinc carbonate, zinc silicate, lead sulphide, and rarely lead carbonate. This zone is obviously developed above water level, and is seldom as much as 100 feet thick. Zinc, and to a less extent lead, have been taken into solution as sulphates, with the aid of sulphuric acid resulting from the oxidation of the associated pyrite. Zinc has been carried away from the weathered zone in solution faster than lead, leaving the lead more or less concentrated near the surface. Some of the zinc carried down has been redeposited secondarily as zinc sulphide. Evidences of this secondary sulphide enrichment can be seen in many places; yet certain broad quantitative considerations raise a doubt as to whether this process has been responsible for the main portion of the values of the sulphide zone. If downward secondary enrichment had been a dominant process, it might be expected that the ores would be richer in places where erosion had cut away more than half the limestone formation carrying the ore, than in places where it had barely cut into the formation. This is not the fact,—which suggests that erosion in its downward progress has carried a large part of the zinc completely out of the vicinity.

Zinc ores of this same general character are also found in Paleozoic rocks (Knox dolomite) in Virginia and Tennessee. Their manner of occurrence suggests the same problem of origin as in Missouri and Wisconsin, but no decisive evidence of their source has been discovered.

Of the zinc ores associated with igneous intrusions, those of the Butte and Cœur d'Alene districts are described in connection with copper and lead ores on pp. 201-203, 208, and 212-213.

Zinc constitutes about 75 per cent by weight of the recoverable metals of the Leadville district of Colorado. About two-thirds of the zinc occurs as the sulphide and about one-third as the carbonate resulting from weathering of the sulphide. The zinc sulphide is associated with lead, iron, and copper sulphides and gold and silver minerals. In the oxide zone the zinc carbonate is associated with oxides and carbonates of various metals, including those of lead, copper, iron, and manganese. The iron and manganese oxides are mined in considerable tonnage as a flux. It is an interesting fact that, although mining has been carried on in this district for upwards of forty years, only within the last decade has the existence of zinc ores in the oxide zone been recognized. This has been due largely to the fact that the iron and manganese oxides effectively stain and mask the zinc carbonate.

The Leadville ores occur as replacements and vein-fillings in a gently faulted and folded Carboniferous limestone, in deposits of a general tabular shape, parallel to the bedding but with very irregular lower surfaces. The limestone is intruded by numerous sheets of porphyry, mainly parallel to the bedding but sometimes cutting across it, against the under sides of which most of the ore occurs. The primary sulphides are believed to be genetically related in some fashion to these porphyries. The older view was that the agents of deposition were aqueous solutions from the surface above, which derived their mineral content chiefly from the porphyries. Later views favor solutions coming directly from the porphyries or deeper igneous sources. While in form and association these ores are characteristic igneous contact deposits, they lack the high-temperature silicates which are so distinctive of many ores of this type.

The zinc ores of Franklin Furnace, New Jersey, belong in the group associated with igneous agencies, but have certain unique features. They consist of willemite and zincite, together with large amounts of franklinite (an iron-manganese oxide) and silicates, in a pre-Cambrian white crystalline limestone near its contact with a coarse-grained granite-gneiss. The origin of the ores is obscured by later shearing and metamorphism, but it seems best explained by replacement of the limestone by heated solutions coming from the granitic mass. The view has also been advanced that the ores originated in the limestone before the advent of the igneous rocks. Secondary concentration is not apparent.

FOOTNOTES:

[33] Ransome, Frederick Leslie, The copper deposits of Ray and Miami, Arizona: *Prof. Paper 115, U.S. Geol. Survey*, 1919, pp. 12-13.

CHAPTER XI

GOLD, SILVER, AND PLATINUM MINERALS

GOLD ORES

ECONOMIC FEATURES

The principal and most essential use of gold is as a standard of value and a medium of exchange. Gold has been prized since the earliest times because of its luster, color, malleability, and indestructibility, and has long been used as a trading medium. At present little of the metal is actually circulated from hand to hand. Stocks of gold, however, accumulated by governments and banking interests, form the essential foundation of paper currency and of the vast modern system of credit relations. In the settlement of international trade balances considerable quantities of gold frequently move from debtor to creditor nations. Although the amounts thus shipped are frequently great in value, they are very small in volume. It is interesting to note that the entire accumulated gold stocks of the world's governments—about nine billion dollars—cast in a solid block, with the horizontal dimensions of the Washington monument, would be only about 12 feet high.

Other uses of gold are in dentistry, and in the arts for jewelry, gilding, and other forms of ornamentation. Consumption for these purposes has been increasing of late years and now takes a third or more of the world's annual production. In the United States, before war-time restrictions were adopted, the consumption for jewelry and similar uses exceeded the consumption in coinage. Since the war it has exceeded the total domestic production of gold. An interesting problem for the future is how an adequate supply of gold is to be distributed between monetary uses and the arts. The curve of increase in the requirements of the arts indicates that, unless there is greatly increased production, all the world's gold will be necessary for the arts in a comparatively few years. To retain it for monetary purposes would require government restrictions.

Of all the mineral commodities, gold has played perhaps the most important and certainly the most romantic part in the world's history. The "lure of gold" has taken men to the remotest corners of the globe. It has been the moving force in the settlement and colonization of new countries, in numerous wars, and in many other strenuous activities of the human race.

About two-thirds of the annual gold production of the world comes from the British Empire—from South and West Africa, Australasia, Canada, and India. A single colony, the Transvaal, produces about 40 per cent of the world's total. British capital, which seems to have a particular affinity for investments in gold mines, controls not only the larger part of the output from the colonies, but also important mines in Siberia, Mexico, South America, and the United States.

Russia, Mexico, and Japan have small gold production. The chief deposits of Russia are those of Siberia, which have had an important output and have apparent great possibilities of increase. Other foreign districts are numerous and widely scattered, but, with the exception of Colombia and Korea, no one of them yields 1 per cent of the world's gold.

French interests control about a tenth of the production of the Transvaal, and minor supplies in Mexico and South America—in all about 6 per cent of the world's production. Germany and Austria control less than 1 per cent of the total gold production. German interests formerly had extensive holdings in South Africa and Australia, but during the war this control was eliminated.

The United States, the second largest gold-producing country, supplies about 20 per cent of the world's total. Commercially it controls production of another 5 per cent in foreign countries, chiefly in Canada, Mexico, South America, and Korea. About one-fourth of the United States production comes from California. Other producing states in order of importance are Colorado, Alaska, South Dakota, Nevada, Arizona, Montana, and Utah. These eight states supply 95 per cent of the country's output, and most of the remainder is obtained from other western states.

International movements of gold depend chiefly upon its use in the settlement of trade balances, and are not governed by the considerations which control ordinary mineral commodities. Imports and exports vary with changing foreign trade balances.

Large amounts of gold normally go to London, because Great Britain requires all gold produced in the colonies to be sent to England; but since England ordinarily has an unfavorable balance of trade, much of this gold is reëxported. The United States up to a few years ago was also a debtor nation, and more gold was exported than was imported. During the war, however, this country became the greatest of the creditor nations and imports of gold, chiefly from Europe, were several times the exports.

The total world's gold production up to 1920 has been upwards of 19 billions of dollars, of which about 10 billions have gone into the arts or been hidden and lost, leaving 9 billions in monetary reserve.

At the present writing the United States government holds an unusually large fraction of the world's gold reserve, about 28 per cent or 2 billion dollars,—an amount equal to two-thirds of the aggregate production of the United States to date. Other large stocks of gold are held, in order, by Great Britain, France, and Russia, these three with the United States holding over a half of the world's total gold reserve. Germany has about 1-½ per cent of the total reserve, and, with its tremendous debt and no sources of new production, is of course in a particularly unfavorable position.

The total amount of gold now (1920) accounted for by governments as money is not more than 10 per cent of the value of the notes and currency issued against this gold. Before the war it was 60 per cent. In the United States the pre-war percentage was 99-½ per cent. Since the war it has been 45 per cent. The ratio of gold to currency is now so small that the gold standard is hardly a physical fact, but is to be regarded rather more as a profession of faith. Notwithstanding the recent falling off in gold production, an increment of approximately 350 million dollars is potentially available each year to be added to the gold reserves. Whether this increment, or a larger increment which may come from new discoveries, is sufficient to maintain a reasonable proportion between gold stocks and the necessary normal increase in paper currency, has been, and doubtless will continue to be, a subject of vigorous discussion and speculation.

During and immediately following the war, the gold production of the world showed rather an alarming progressive decrease. About 1915 the group of three greatest producers—South Africa,

United States, and Australia,—reached the acme of its production, and output then fell off. Simultaneously there was a marked decrease of production in many of the less important districts. This general decline was due in considerable part to the fact that during the war the price of gold was fixed and its use restricted to monetary purposes. The price of gold, which is itself the standard of value, could not rise to offset growing mining costs and to maintain profits, as was the case with iron, copper, and the other metals,—with the result that the margin of profit in gold mining became so small as materially to affect exploration and production. Another important cause of decreased production was the actual exhaustion of certain mines, and the lowering of the grades of ore available in many others. New discoveries did not supply these deficiencies. In the United States, for instance, physical conditions of one kind or another were responsible for lessening of production from Alaska, Cripple Creek, and California. Minor causes included conflicts in California between agricultural and mining interests over water rights, and a succession of dry seasons which did not afford enough water for the working of placers; and in Alaska difficulties due to litigation over the oil-flotation process of recovering gold from its ores. As a result of all these conditions, many of the smaller mines were closed down, others continued operations only by curtailing exploration and by mining solely the richest and most accessible ore bodies, and there was a general discouragement and lack of inducement to engage in gold mining.

The gold situation has become a matter of great concern to the various governments, since national financial stability and the confidence of the public in the national credit are based largely upon the acquirement of an adequate gold reserve. Both in England and in the United States, committees of experts have been appointed to make exhaustive investigations and present recommendations for measures to stimulate production. The report of the joint committee from the United States Bureau of Mines and Geological Survey gives a comprehensive review of the conditions in the gold-mining industry.[34]

During the war there was vigorous demand by gold miners both in the United States and South Africa for a bonus on gold. These demands received serious consideration on the part of the governments, but were denied on the general ground of the doubtful adequacy of such a measure to meet the situation, and the danger of upsetting the gold standard of value. In the United States, for instance, a bonus of $10 per ounce was asked for. It did not appear likely that this could increase the annual production from the United States by more than 10 per cent, in face of the physical conditions being met in gold mining. The bonus would have had to be paid on all the gold mined, which would make the increment of production very expensive; to secure an added production of ten million dollars would have cost in the neighborhood of forty millions. Ten millions is only one-third of 1 per cent of the gold reserve already held by this country, and it would obviously have taken a long time for this small increase in annual production to make itself felt in the size of the gold reserves.

Since the war gold has gone to a considerable premium in England, due to the action of the British government in establishing a "free" market,—that is, abandoning the restriction that gold marketed in London should be offered to the government or the Bank of England at the fixed statutory price for monetary purposes. With the pound sterling at a considerable discount outside of England, other countries could afford to bid, in terms of British currency, far above the British mint price. The result is that the South African miner of gold receives a premium due to depreciation of sterling exchange, while the American miner still receives the regular mint price. The agitation for a bonus therefore continues in the United States. However, with the removal of war-time restrictions gold has been allowed to go to the arts, the demand from which is already equal to one-third of the world's gold production, is rapidly increasing, and is temporarily acute due to the accumulation of requirements resulting from war restrictions. This situation has a general tendency to improve the position of the gold miner, though the outlook is still far from bright.

It is an interesting fact that India is absorbing a good half of the free gold. India, in regard to its demand for precious metals and stones, has been described as "an abyss from which there is no return." This is an important contributing cause of the shortage of gold in the rest of the world.

Looking forward to the future, it seems that increased exploration, which is resulting from the present premium on gold, is likely to bring in new reserves to increase production.

Because of lack of important discoveries in recent years, there is pessimism in some quarters as to the possibilities of large increase of production; but, considering the history of gold discoveries, and the amount of ground still to be explored both areally and vertically, this pessimism does not seem to be wholly justified from the geologic standpoint. Curves representing the world's gold production in past years show periods of increasing annual production as new fields are discovered, followed by periods of decreasing production when no new ore bodies are coming in to replace dwindling reserves. It is entirely possible that in recent years the gold-mining industry has been merely in one of these temporary stagnant periods. There are many regions, both in the vicinity of worked-out lodes and in unsettled and poorly explored countries, where gold may still be discovered; there may be far greater resources of this metal still covered up than all those which man has thus far uncovered. A single new deposit or district may make a great difference in the world's production, as suggested by the experience of the past. Regions which are especially attractive for exploration and the discovery of new deposits are in Siberia and South America, which in the opinion of many engineers may eventually rival South Africa. Mexico, with the establishment of a stable government, should also have a greatly increased production.

Geologic Features

The principal gold mineral is native or metallic gold. This occurs in nature in small scales, crystals, and irregular masses, and also in microscopic particles mechanically mixed with pyrite and other sulphides. Chemically, gold is very inactive and combines with but few other elements. A small part of the world's supply is obtained from the gold-silver tellurides—calaverite, sylvanite, krennerite, and petzite.

Gold deposits are of two general classes—placers, and veins or lodes.

Placers, which are in general the more easily discovered and more easily worked deposits, have in the past been the chief source of the world's gold supply. It is estimated that in the first twenty-seven years of the modern era of gold-mining, beginning with the discovery of gold in California in 1848, 87 per cent of the world's production was obtained from placers. At present the placers of recent geologic age supply a tenth to a fifth of the gold, and ancient or fossil placers in the Transvaal supply another two-fifths. In the United States about a fourth of the gold production comes from placers, mainly from California and Alaska.

Placers are detrital or fragmental sediments containing the ore in mechanical fragments, which are derived from the erosion and transportation of solid-rock veins or lodes, sometimes called the "mother lode." During the process of transportation and deposition there is more or less sorting, because of differing density of the mineral fragments, resulting in the segregation or concentration of the ore minerals in certain layers or channels. Gold, because of its weight, tends to work down toward bedrock, or into scoured or excavated portions of stream channels. In a few cases it is carried in some quantity to the sea and concentrated in beach sands. The processes are not unlike the mechanical concentration of ores by crushing and water sorting. Seldom, however, do the processes go far enough in nature to produce an ore which can be used directly without some further mechanical sorting. Ore minerals concentrated in placers are those which resist abrasion and chemical solution during the processes of weathering and transportation, and which have a density sufficiently high so that they are partially sorted out and concentrated from the accompanying quartz and other minerals. To warrant their recovery they must also be of such high intrinsic value that it pays to mine small quantities. The most important of such minerals are gold, tin, platinum, and the precious stones. Iron, copper, lead, and zinc minerals are often somewhat concentrated as placers, but their intrinsic value is not high enough to warrant the attempt to recover them in the large amounts necessary to make them commercially available.

Placers are forming now and have formed at all stages of the earth's history. Early placers may be reworked and further concentrated by renewal of the proper erosional and transportational conditions. Old placers may be buried beneath younger rocks, cemented, and more or less recrystallized. "Fossil" placers of this kind are best represented by deposits in the Black Hills of South Dakota and probably by the South African gold deposits.

In the Witwatersrand deposits of South Africa, the gold is concentrated in the lower parts of large conglomerate and quartz sand layers of great areal extent. Pebbles of the conglomerate are mainly quartz and quartzite. The gold, in particles hardly visible to the eye, is in a sandy matrix and is associated with chloritoid, sericite, calcite, graphite, and other minerals. The origin of the gold deposits of this district is not entirely agreed on, but the evidence seems on the whole to favor their placer origin. Some investigators of these ores believe them to have been introduced into the conglomerate and sand by later solutions, possibly by hot solutions related to certain diabase intrusions that cut the beds.

In the vein or lode or hard-rock deposits, the gold is mainly metallic gold, and to a minor extent is in the form of gold tellurides. It is usually closely associated with iron pyrite in a matrix or gangue of quartz. Seldom is a gold deposit free from important values in other minerals. About 84 per cent of the gold mined in vein or lode deposits of the United States is associated with silver minerals, the combined value averaging about $6 per ton; about 13 per cent comes from copper ores which have an average yield of gold and silver of 50c. per ton; and 3 per cent comes from zinc and lead ores, with an average gold and silver yield ranging from $1 to $6 per ton. The geologic occurrence of gold in the copper, lead, and zinc ores has already been referred to in the discussions of these ores.

Reference will be made here only to the vein deposits in which gold, with silver, constitutes the principal values. Because of their common gangue of quartz these are often called "dry" or "siliceous" ores. Their principal occurrence is in distinct fissure veins in igneous rocks, with more or less replacement of the wall rock. The igneous rocks are commonly acid intrusives of a granite or porphyry type, less commonly intrusives of gabbro and diabase and surface lavas of rhyolite and basalt. In a few cases the ores are contact-metamorphic deposits of the type described under copper ores. In still rarer cases they are in pegmatites. Gold is commonly associated with minerals and wall-rock alterations indicating deposition by hot solutions, which are inferred to have come from the igneous rocks.

Because of the resistant nature both of ore minerals and gangue, weathering and secondary concentration have had little effect in enriching gold deposits. So far as there has been any noticeable effect on the gold content of the ores, it has been due to the leaching out of other constituents, principally pyrite and other sulphides, leaving the gold present in slightly larger proportions. Locally there is evidence of solution of gold in weathered zones and its deposition in the sulphide zones below. Solution is believed to be accomplished by chloride solutions, and is favored by the presence of manganese which delays precipitation. The precipitating agent below may be ferrous sulphate, various sulphides, native metals, or organic matter.

Of the vein or lode gold ores in the United States some of the most productive and best known have the following geologic features:

The California gold belt extends north and south along the west slope of the Sierra Nevada Mountains. The ore is in a series of parallel and overlapping veins striking with the trend of the range, associated with granodiorite intrusives in schist and slate. There is no pronounced secondary concentration. These deposits are the source of most of the great placer deposits of California, hence the name "Mother Lode" applied to a part of them. The principal ore deposits are somewhat removed from the main mass of intrusive which forms the crest of the Sierra Nevada range, and are more closely related to the smaller similar intrusive masses farther down the slope. The gangue is mainly quartz.

At Juneau, Alaska, great dikes of albite-diorite intrude greenstones and schists, and low-grade gold ores occur in shattered portions of the diorite. These ores were mined on a great scale at the Treadwell Mine.

Another famous low-grade deposit is the Homestake Mine in the Black Hills of South Dakota, where pre-Cambrian slates and schists of sedimentary origin are impregnated with gold, associated with quartz, dolomite, calcite, pyrite, and other minerals. The origin is supposed to have some connection with intrusives into the schists; but the relations of the ores to intrusives, both in age and in place, present many puzzling questions which make conclusion as to origin very difficult.

In the Cripple Creek district of Colorado, a volcanic neck two or three miles in diameter breaks through pre-Cambrian granites, gneisses, and schists. The volcanic rocks consist mainly of tuffs and breccias cut by basic dikes. The ore bodies are in fissures and sheeted zones, principally in the granitic rocks, but associated with these dikes. The ore is mainly gold telluride, in a gangue of quartz together with pyrite and a variety of minerals characteristic of hot-water solutions. Also the wall rocks have the characteristic hot-water alterations. There is slight enrichment near the surface.

At Goldfield, Nevada, native gold is found in surface igneous flows of a dacite type, which have undergone extensive hydrothermal alterations characterized by the development of alunite (a potassium-aluminum sulphate), quartz, and pyrite. The ore fills fissures to some extent, but is mainly a replacement of the wall rock. Association with typical hot-water minerals and hydrothermal alterations of the wall rock are again believed to indicate the origin of the ores through ascending hot solutions from a deep source.

One of the interesting features of this occurrence is the abundance of alunite. Sulphate minerals are commonly formed by oxidizing solutions. The abundant presence, therefore, of a sulphate mineral with minerals of a primary deep-seated source has led to much discussion of origin. The hypothesis was developed that these minerals result from the interaction of deep-seated sulphide-bearing solutions with surface oxidizing solutions.[35] It may be noted that in recent years other sulphate minerals have been occasionally regarded as primary, including gypsum, anhydrite, barite, and others. It has been suggested that if igneous emanations contain free oxygen and sulphur, or sulphur dioxide, it would be expected that as they become cool sulphur trioxide would be formed which would result in the sulphate at suitable temperature.[36]

Other deposits containing gold are discussed in connection with silver on following pages.

SILVER ORES

ECONOMIC FEATURES

Silver has two important uses—in money and in the arts. As money, it is used in the United States and Europe for subsidiary coinage,—silver coins normally circulating at more than their intrinsic value,—but its greatest monetary use is in India and China, where it has been the basis for the settlement of foreign exchange balances. In China also it is the money standard of the country. In the arts, silver is employed chiefly in the making of articles of luxury, such as jewelry and tableware. In the Orient this use is closely related to its use as money, since the natives invest their savings both in silver jewelry and silver coins. There is some consumption of silver by certain chemical industries, and quantities of increasing importance are used in the form of silver salts by the photographic and moving picture industries. It has been estimated that before 1914 about two-thirds of the new silver produced went into the arts and one-third into money. During the war, however, increasing amounts were used in coinage, and less than one-fifth of the output was used in the arts. Demands for silver for monetary purposes will probably continue to take the larger part of the world's production for some time. In this connection it may be noted that India has adopted a gold standard, but that the conservative habits of the population will doubtless continue to call for large amounts of silver.

About half of the silver production of the world comes from the dry or siliceous silver ores, which are mined solely for that metal and the associated gold; and about half of the production is obtained as a by-product in the mining of other metals, principally copper and lead. The average grades of these ores, in combined values of gold and silver, were referred to on p. 228. While the aggregate amount of silver obtained as a by-product of other ores is large, the percentage of silver in the copper or lead in any mine is ordinarily very small. Consequently the world output of silver depends to a considerable extent upon conditions in the copper- and lead-mining industries.

Of the total world output of silver, normally about 75 per cent comes from North America. Of this the United States and Mexico each produce about two-fifths and Canada one-fifth, and minor amounts are produced in Central America. In late years, political disturbances in Mexico reduced that country's production to less than half the normal figure, and the United States took the place which Mexico had held for many years as the leading silver producer. The United States and Mexican supply is obtained from the Rocky Mountain belt, and the Canadian production comes chiefly from the Cobalt, Ontario, district. Outside of North America the principal producing areas are Australia, South America (Peru and to a less extent Bolivia and Chile), Europe (chiefly from Spain, Germany, and Austria-Hungary, but with smaller amounts from all the other countries), and Japan. Thus, while there are sources of silver in many places, the great bulk of the world's output comes from North America. In the financial ownership of mines, including ownership in other countries, the United States controls over half the world's silver, Great Britain about a third, and Germany about a tenth (principally in Mexico).

All the silver mined in the United States is smelted and refined by domestic plants; and in addition much of the Canadian, Mexican, and South and Central American silver is exported to the United States as ore and base bullion, to be treated in this country. The United States is therefore the great silver-selling country of the world.

The great silver-consuming countries are India and China, and normally about a half of the world's output goes to these two countries. This major movement of silver, from America to the Far East, takes place through the London market, since England has been the chief nation trading in the Orient. The balance of the world's silver consumption is widely distributed among the countries of Europe and South America and the United States (which consumes about one-tenth of the total). For the European trade most of the silver also goes through London, which is the great clearing-house and the market where prices are fixed.

In the later years of the war and immediately after, the demands for silver were probably twice the world's output. The resulting rise in price was unprecedented. Silver actually became worth more as bullion than as currency, and in Europe much trouble was experienced because of its withdrawal from currency to be melted up. This condition was later followed by an equally striking drop in price as supply caught up with demand.

In the United States, as in many other countries, it was desired during the war to accumulate large stocks of gold as a basis of credit for the flotation of government loans, and the export of gold was prohibited. Consequently in the settlement of foreign trade balances, particularly with the nations of the Orient, very large amounts of silver bullion had to be used. Current production proved inadequate, and it was necessary to utilize the stocks of silver dollars in the United States Treasury. To this end the Pittman Silver Act, passed in April, 1918, authorized the melting down and conversion into bullion of 350,000,000 dollars out of the Treasury stock, and the retirement of a corresponding number of silver certificates and the issue of Federal Reserve bank notes. In this manner old stocks of silver, Manila dollars, etc., were called into service—though the stage was not reached, as it was in Germany, where it became necessary to melt down silver plate and ornaments. The silver used for exchange and export was to be replaced by the purchase of bullion from American producers at $1 per ounce, and its coining into new dollars. A minimum price of $1 per ounce was thus established for silver bullion.

The immediate result was to increase the price of silver at the mine; but with the continued rise in demands for silver, the price in the open market went far above this figure, the maximum being reached in 1920 when the price of silver went to $1.39 per ounce. Naturally, but little silver was then offered to the government at the fixed price of $1 under the Pittman Act. With the more recent slump in the general market for silver to a price below $1, offers to the government under the Pittman Act have been renewed.

That part of the silver production which is a by-product of copper production has been low since the war, because of the stagnation in the copper industry. The production from lead ores, on the other hand, was not handicapped by lack of demand for lead. With the restoration of order in Mexico, a presumption of large silver production in that country may be expected. Increases may probably be expected also from new mines in Burma and from Bolivia. On the whole, no large increase in world production can be assumed from present known resources. New discoveries will be necessary to make any considerable change.

Of the mine production of silver in the United States, about two-thirds of the total comes from the states of Montana, Utah, Idaho, and Nevada. Other considerable producers are Colorado, Arizona, California, Alaska, and New Mexico. All the other states together produce less than 5 per cent of the total. The most important single districts are the Butte district of Montana, the Cœur d'Alene district of Idaho, and the Tonopah district of Nevada, supplying respectively about one-fifth, one-eighth, and one-tenth of the country's total silver output.

Geologic Features

The most important mineral of silver is the sulphide, argentite or "silver glance." Other minerals which yield a minor percentage of the total silver produced are the silver-antimony sulphides, pyrargyrite or "ruby silver," stephanite or "black silver," and polybasite; the silver-arsenic sulphides, proustite or "light ruby silver" and pearcite; and the silver antimonide, dyscrasite. In the oxide zone the most abundant minerals are cerargyrite (silver chloride) and native or "horn" silver. In addition to these definite mineral forms, silver is present in many ores in an undetermined form in other sulphides, notably in galena, sphalerite, and pyrite. Silver differs from gold in that it is chemically active and forms many stable compounds, of which only the more important have been mentioned.

The fact that half the world's silver is obtained as a by-product in the mining of other metals has been referred to. In the United States about a third of the production comes from dry or siliceous ores, over a third from lead and zinc ores, and a fourth to a third from copper ores. A fraction of 1 per cent of the total is obtained as a by-product of gold placers, and all the remainder is won from lode or hard-rock deposits.

The general geologic features of the silver-bearing copper and lead ores, and of the dry or siliceous gold and silver ores, have been described on previous pages. The Philipsburg district has been referred to in connection with manganese ores, and the Bolivian tin-silver ores will be described in connection with tin. We shall consider here only a few of the more prominent districts which have been primarily silver producers.

The Cobalt district of northern Ontario is the most productive silver district in North America. The ores are found in numerous short, narrow veins, principally in pre-Cambrian sediments near a thick quartz-diabase sill. Locally they penetrate the sill. Native silver and various silver sulphides, arsenides, and antimonides are associated with minerals of cobalt, nickel, bismuth, lead, and zinc, in a gangue of calcite and some quartz. The ore is of very high grade. The ore minerals are believed to have been deposited by hot solutions emanating from deep magmatic sources after the intrusion of the diabase. The present oxidized zone is very shallow, but may have been deeper before being stripped off by glaciation; it is characterized by native silver and arsenates of nickel and cobalt in the form of the green "nickel bloom" and the pink "cobalt bloom." The silver minerals are distinctly later in origin than the cobalt and nickel in the unoxidized zone, as evidenced by the relations of the mineral individuals when seen under the microscope. This fact, together with the abundance of native silver in the oxide zone, has suggested downward concentration of the silver by surface waters; but recent studies have indicated the probability that some of the silver at least was deposited by the later ascending solutions of magmatic origin.

In the Tintic district of central Utah, Paleozoic limestones have been intruded by monzonite (an acid granitic or porphyritic igneous rock), and covered by surface flows, the flows for the most part having been removed by subsequent erosion. The sediments have been much folded and faulted, and the ore bodies occur as fissure veins which locally widen into chimneys or pipes in fracture zones, accompanied by much replacement of limestone. There is a rough zonal arrangement of the ore minerals around the intrusive, gold and copper minerals (chiefly enargite and chalcopyrite) being more prominent near the intrusive, and argentiferous galena and zinc blende richer at greater distances. Silver constitutes the principal value. The gangue is mainly fine-grained quartz or jasperoid, and barite. The water table is at unusually great depths (2,400 feet) and there is a correspondingly deep oxidized zone, which is characterized by lead and zinc oxide minerals much as at Leadville (p. 219).

The Comstock Lode at Virginia City, Nevada, on the east slope of the Sierra Nevadas, was one of the most famous bonanza deposits of gold and silver in the world. While the richer ore has all been extracted, lower-grade material is still being mined and the fissure is still being followed, in the hope of some day striking another fabulously rich ore body. The lode occupies a fault fissure parallel to the trend of the range and dipping about 40 degrees to the east, which can be traced about two and a half miles along the strike, with igneous rocks forming both hanging and foot walls. There are no sedimentary rocks in the district. The high-grade part of the vein is several hundred feet in thickness, with many irregular branches; the great thickness has been thought to be at least in part due to the tremendous pressure exerted by growing quartz crystals. The wall rocks have undergone a "propylitic" alteration, with development of chlorite, epidote, and probably sericite, much as at Butte. The ore contains rich silver sulphide minerals and native gold, in a gangue composed almost entirely of quartz. The ore was doubtless formed by hot solutions, but the exact nature of these solutions, whether magmatic or meteoric, has not been proven. The hypothesis was early developed that the ores were deposited by surface waters,—which are supposed to have fallen on the summits of the Sierra Nevadas, to have sunk to great depths where they were heated, enabling them to pick up metallic constituents from the diabase forming one wall of the ore body, and to have risen under artesian pressure along the fault plane, where loss of heat and pressure resulted in deposition. Later studies have emphasized the similarity of the ore-depositing conditions with those in other districts where the ores are believed to have come directly from magmatic sources, and this origin is now generally favored for the Comstock Lode. However, the earlier theory has not been disproved.

The Tonopah, Nevada, district is very similar to the Goldfield district (p. 230). Silver and gold are found in veins and replacements in a series of Tertiary volcanic flows and tuffs, all of which have been complexly faulted. Silver is the dominant constituent of value. The formation of fissures and faults accompanying and caused by the intrusion and cooling of lavas was first clearly shown in this district. Evidences of origin through the work of hot solutions, probably magmatic, are the close association of the ores in place and in time with the igneous rocks—ore deposition in most of the flows having taken place before the next overlying flows were put down,—the presence of fluorine, the nature of the wall-rock alterations, the fact that both hot and cold springs are found close together underground (indicating unusual sources for the hot springs), the contrast in composition between the ores and the country rock, and the general relation of these ores to a large number of similar occurrences in Tertiary lavas in the same general area.

Under weathering conditions, the silver sulphide minerals in general are oxidized to form native silver and cerargyrite, which are relatively insoluble and remain for the most part in the oxide zone. Silver is less soluble than copper and zinc, but more soluble than gold; and to some extent it is removed in solution, particularly where the oxidation of pyrite forms ferric sulphate. Farther down it may be reprecipitated as native silver, argentite, and the sulpho-salts, by organic matter or by various sulphides. The secondarily enriched ores are in a few districts, as at Philipsburg, Montana, the most valuable portions of the deposits. In other cases, sulphide enrichment does not appear to have contributed greatly to the values. The zones of oxide ores, secondary sulphide ores, and primary or protores are in most silver deposits much less regular and much less definitely marked than in the case of copper ores.

PLATINUM ORES

Economic Features

The principal uses of platinum are: as a catalytic agent in the contact process for the manufacture of sulphuric acid, and in the making of nitric acid from ammonia; for chemical laboratory utensils that must be resistant to heat and acids; for electrical contacts for certain telephone, telegraph, and electrical control instruments, and for internal combustion engines; in dental work; and for jewelry. In normal times before the war, it is estimated that in the United States the jewelry and dental industries used 75 per cent of the platinum metals consumed, the electrical industry 20 per cent, and the chemical industry 5 per cent. During the war, with the extraordinary expansion of sulphuric and nitric acid plants, these proportions were reversed and the chemical and electrical industries consumed about two-thirds of the platinum. Substitutes have been developed, particularly for the electrical uses, and the demand from this quarter may be expected to decrease.

About 90 per cent of the world's crude platinum produced annually comes from the Ural Mountains in Russia. The deposits next in importance are those of Colombia. Small amounts are produced in New South Wales, Tasmania, New Zealand, Borneo, British Columbia, United States, India, and Spain; and as a by-product in the electrolytic refining of the Sudbury, Canada, nickel ores. The extension of this method of refining to all of the Sudbury ores would create an important supply of platinum. The Colombian output has been increasing rapidly since 1911. Meanwhile the Russian production has declined; and from the best information available, it is not likely that Russia will be able to maintain production for many more years. Estimates of the life of the Russian fields are from 12 to 20 years at the pre-war rate of production.

The platinum situation is commercially controlled by buying and mine-operating agencies,—the French having, before the war, practically dominated the Russian industry, while American interests controlled in Colombia. The situation is further influenced by four large refineries, in England, Germany, United States, and France.

Before the war the United States produced less than 1 per cent of the new platinum it consumed annually. Production comes principally from California, with smaller amounts from Oregon, Alaska, and Nevada. The many efforts which have been made to develop an adequate domestic supply of this metal do not indicate that the United States can ever hope to become independent of foreign sources for its future supplies of platinum.

There is little reason to doubt that the Colombia field, commercially dominated by the United States, holds great promise for the future. The output has come largely from native hand labor, and with the installation of dredges can probably be greatly increased.

During the war, the need for platinum for war manufactures was so urgent and the production so reduced, that restrictions against its use in jewelry were put into force in all the allied countries. The United States government secured quantities of platinum which would have been sufficient for several years' use if war had continued. With the cessation of hostilities restrictions on the use of platinum were removed, and the accumulated metal was released by the government from time to time in small quantities; but the demands for platinum in the arts were so great that prices for a time tended to even higher levels than during the war. More recently supply is again approaching demand.

Geologic Features

Platinum, like gold, occurs chiefly as the native metal. This is usually found alloyed with iron and with other metals of the platinum group, especially iridium, rhodium, and palladium. Most of the platinum as used in jewelry and for electrical purposes contains iridium, which serves to harden it. Paladium-gold alloys are a substitute for platinum, chiefly in dental uses.

The original home of platinum is in basic igneous rocks, such as peridotites, pyroxenites, and dunites, where it has been found in small, scattered crystals intergrown with olivine, pyroxene, and chromite. Platinum is very dense and highly resistant to oxidation and solution. In the breaking up and washing away of the rocks, therefore, it is concentrated in small grains and scales in stream and beach placers. Of the world production of platinum over 99 per cent has been derived from placers.

The Ural Mountain deposits of Russia are gold- and platinum-bearing placers, in streams which drain areas of dunite rock containing minute quantities of native platinum. The deposits of Colombia and Australasia are placers of a similar character. In the United States small quantities of platinum are recovered from the gold-bearing gravels of California and Oregon, where the streams have come from areas of serpentine and peridotite.

A platinum arsenide, called sperrylite, is sometimes found associated with sulphide minerals in basic igneous rocks. At Sudbury, Ontario, this mineral, together with palladium arsenide, is found in the nickel ores, especially in the weathered zone where it is concentrated by removal of more soluble materials. It has also been found in the copper mines of Rambler, Wyoming. In the Yellow Pine district of southern Nevada, metallic gold-platinum-palladium ore shoots are found in association with copper and lead ores, in a fine-grained quartz mass which replaces beds of limestone near a granitic dike. No basic intrusives are known in the district. The deposit is unusual in that it has a comparatively high content of platinum (nearly an ounce to the ton), and is probably genetically related to acid intrusives. From all these deposits, only small quantities of platinum are mined.

FOOTNOTES:

[34] Report of a joint committee appointed from the Bureau of Mines and the United States Geological Survey by the Secretary of the Interior to study the gold situation: *Bull. 144, U. S. Bureau of Mines*, 1919. See also Report of Special Gold Committee to Secretary of the Treasury, February 11, 1919.

[35] Ransome, F. L., The geology and ore deposits of Goldfield, Nevada: *Prof. Paper 66, U.S. Geol. Survey*, 1909, p. 193.

[36] Butler, B. S., Loughlin, G. F., Heikes, V. C., and others, The ore deposits of Utah: *Prof. Paper 111, U.S. Geol. Survey*, 1920, p. 195.

CHAPTER XII

MISCELLANEOUS METALLIC MINERALS

ALUMINUM ORES

ECONOMIC FEATURES

Bauxite (hydrated aluminum oxide) is the principal ore of aluminum. Over three-fourths of the world's bauxite production and 65 per cent of the United States production is used for the manufacture of aluminum. On an average six tons of bauxite are required to make one ton of metallic aluminum. Other important uses of bauxite are in the manufacture of artificial abrasives in the electric furnace, and in the preparation of alum, aluminum sulphate, and other chemicals which are used for water-purification, tanning, and dyeing. Relatively small but increasingly important quantities are used in making bauxite brick or high alumina refractories for furnace-linings.

Aluminum is used principally in castings and drawn and pressed ware, for purposes in which lightness, malleability, and unalterability under ordinary chemical reagents are desired. Thus it is used in parts of airplane and automobile engines, in household utensils, and recently in the framework of airplanes. Aluminum wire has been used as a substitute for copper wire as an electrical conductor. Aluminum is used in metallurgy to remove oxygen from iron and steel, and also in the manufacture of alloys. Powdered aluminum is used for the production of high temperatures in the Thermite process, and is a constituent of the explosive, ammonal, and of aluminum paints.

Deposits of bauxite usually contain as impurities silica (in the form of kaolin or hydrous aluminum silicate), iron oxide, and titanium minerals, in varying proportions. Bauxites to be of commercial grade should carry at least 50 per cent alumina, and for the making of aluminum should be low in silica though the content of iron may be fairly high. For aluminum chemicals materials low in iron and titanium are preferred; and for refractories which must withstand high temperatures, low iron content seems to be necessary. The abrasive trade in general uses low-silica high-iron bauxites.

The only large producers of bauxite are the United States and France, which supplied in normal times before the war over 95 per cent of the world's total. Small amounts are produced in Ireland, Italy, India, and British Guiana. During the war a great deal of low-grade bauxite was mined in Austria-Hungary and possibly in Germany; but on account of the large reserves of high-grade material in other parts of the world, it is doubtful whether these deposits will be utilized in the future. Bauxites of good grade have been reported from Africa, Australia, and many localities in India. From geologic considerations it is practically certain that there are very large quantities available for the future in some of these regions.

The international movements and the consumption of bauxite are largely determined by the manufacture of aluminum, and to a lesser extent by the manufacture of abrasives and chemicals. The principal foreign producers of aluminum are France, Switzerland (works partly German-owned), Norway (works controlled by English and French capital), England, Canada, Italy, Germany, and Austria. French bauxite has normally supplied the entire European demands,—with the exceptions that Italy procures part of her requirements at home, and that the Irish deposits furnish a small fraction of the English demand.

The deposits of southern France, controlled largely by French but in part by British capital, have large reserves and will probably continue to meet the bulk of European requirements. France also has important reserves of bauxite in French Guiana.

The United States produces about half of the aluminum of the world, and is the largest manufacturer of artificial abrasives and probably of aluminum chemicals. Most of these are made from domestic bauxite. Prior to the war, the United States imported about 10 per cent of the bauxite consumed, but these imports were mainly high-grade French bauxite which certain makers of chemicals preferred to the domestic material. The small production of Guiana is also imported into the United States. Bauxite is exported to Canadian makers of aluminum and abrasives. During the war period domestic deposits were entirely capable of supplying all the domestic as well as Canadian demands for bauxite, although these demands increased to two and one-half times their previous figure. At the same time considerable amounts of manufactured aluminum products were exported to Europe, whereas aluminum had previously been imported from several European countries.

The United States production of bauxite comes mainly from Arkansas, with smaller amounts from Tennessee, Alabama, and Georgia. The reserves are large but are not inexhaustible. Most of the important deposits are controlled by the large consumers of bauxite, principally the Aluminum Company of America and its subsidiaries, though certain chemical and abrasive companies own some deposits. The Aluminum Company of America also controls immense deposits of high-grade bauxite in Dutch and British Guiana, and further exploration by American interests is under way.

With the return to normal conditions since the war, some of the domestic bauxite deposits probably can not be worked at a profit, a situation which is likely to require the development of the tropical American deposits.

GEOLOGIC FEATURES

Aluminum is the third most abundant element in the common rocks and is an important constituent of most rock minerals; but in its usual occurrence it is so closely locked up in chemical combinations that the metal cannot be extracted on a commercial scale. In the crystalline form aluminum oxide constitutes some of the most valuable gem stones. Many ordinary clays and shales contain 25 to 35 per cent alumina (Al_2O_3), and the perfection of a process for their utilization would make available almost unlimited aluminum supplies. The principal minerals from which aluminum is recovered today are hydrous aluminum oxides, the most prominent of which are bauxite, gibbsite, and diaspore—the aggregate of all these minerals going commercially under the name of bauxite.

Prior to the discovery of bauxite ores, cryolite, a sodium-aluminum fluoride obtained from pegmatites in Greenland, was the chief source of aluminum. It is only within about the last thirty-five years that bauxite has been used and that aluminum has become an important material of modern industry. Cryolite is used today to form a molten bath in which the bauxite is electrolytically reduced to aluminum.

Bauxite deposits in general are formed by the ordinary katamorphic processes of surface weathering, when acting on the right kind of rocks and carried to an extreme. In the weathering of ordinary rocks the bases are leached out and carried away, leaving a porous mass of clay (hydrous aluminum silicates), quartz, and iron oxide. In the weathering of rocks high in alumina, and low in iron minerals and quartz, deposits of residual clay or kaolin nearly free from iron oxide and quartz are formed. Under ordinary weathering conditions the kaolin is stable; but under favorable conditions, such as obtain in the weathered zones of tropical climates, it is broken up, the silica is taken into solution and carried away, and hydrous aluminum oxides remain as bauxite ores. This extreme type of weathering is sometimes called lateritic alteration (see pp. 172-173). Impurities of the bauxite ores are the small quantities of iron and titanium present in the original rocks, together with the kaolin which has not been broken up. The deposits usually form shallow blankets over considerable areas, with irregular lower surfaces determined by the action of surface waters—which work most effectively where joints or other conditions favor the maximum circulation and alteration. A certain degree of porosity in the original rock is also known to favor the alteration. A complete gradation from the unaltered rock through clay to the high-grade bauxite, with progressive decrease in bases and silica, concentration of alumina and iron oxide, and increase of moisture and pore space, is frequently evident (see Fig. 13). The bauxite is earthy, and usually shows a concretionary or pisolitic structure similar to that observed in residual iron ores (p. 172). Near the surface there may be an increase in silica,—probably due to a reversal of the usual conditions by a slight leaching of alumina, thus concentrating the denser masses of kaolin which have not been decomposed.

The Arkansas bauxite deposits, the most important in the United States, are surface deposits overlying nepheline-syenite, an igneous rock with a high ratio of alumina to iron content. The most valuable deposits are residual, and some parts have preserved the texture of the original rock, though with great increase in pore space; most of the ore, however, has the typical pisolitic structure. Near the surface the pisolites are sometimes loosened by weathering, yielding a gravel ore, and some of the material has been transported a short distance to form detrital ores interstratified with sands and gravels. The complete gradation from syenite to bauxite has been shown.

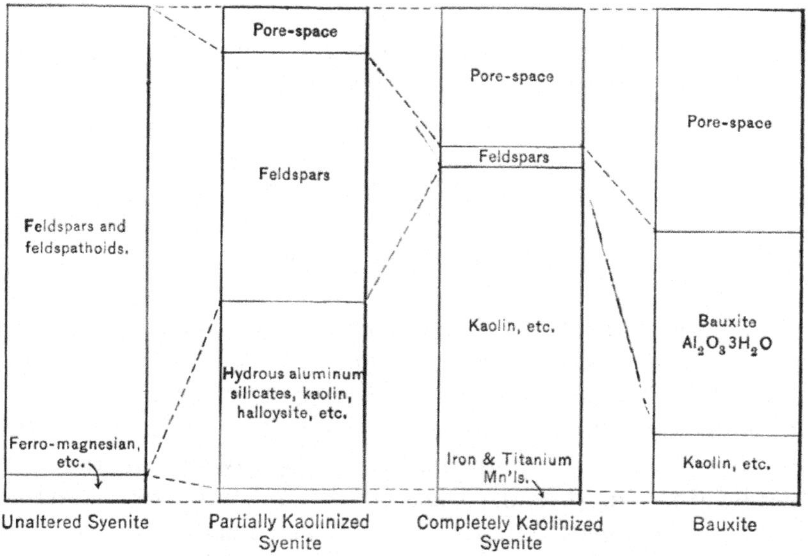

FIG. 13. Diagram showing gradation from syenite to bauxite in terms of volume. The columns represent a series of samples from a single locality in Arkansas. After Mead.ToList

In the Appalachian region of Tennessee, Alabama, and Georgia, bauxite occurs as pockets in residual clays above sedimentary rocks, chiefly above shales and dolomites. Its origin has probably been similar to that described.

The bauxite deposits of southern France occur in folded limestones, and have been ascribed by French writers to the work of ascending hot waters carrying aluminum sulphate. They present some unusual features, and evidence as to their origin is not conclusive.

At the present time bauxite is doubtless forming in tropical climates, where conditions are favorable for deep and extreme weathering of the lateritic type. The breaking up of kaolin accompanied by the removal of silica is not characteristic of temperate climates, though many clays in these climates show some bauxite. It is possible that, at the time when the bauxite deposits of Arkansas and other temperate regions were formed, the climate of these places was warmer than it is today.

In studying the origin of bauxites, it should not be overlooked that they have much in common with clays, certain iron ores, and many other deposits formed by weathering.

ANTIMONY ORES

Economic Features

Antimony is used mainly for alloying with other metals. Over one-third of the antimony consumed in the United States is alloyed with tin and copper in the manufacture of babbitt or bearing-metal. Other important alloys include type-metal (lead, antimony, and tin), which has the property of expanding on solidification; "hard lead," a lead-antimony alloy used in making acid-resisting valves; Britannia or white metal (antimony, tin, copper, zinc), utilized for cheap domestic tableware; and some brasses and bronzes, solders, aluminum alloys, pattern metals, and materials for battery plates and cable coverings. Antimony finds a very large use in war times in the making of shrapnel bullets from antimonial lead. Antimony oxides are used in white enameling of metal surfaces, as coloring agents in the manufacture of glass, and as paint pigments; the red sulphides are used in vulcanizing and coloring rubber, as paint pigments, in percussion caps, and in safety matches; and other salts find a wide variety of minor uses in chemical industries and in medicine.

Antimony ores vary greatly in grade, the Chinese ores carrying from 20 to 64 per cent of the metal. The presence of arsenic and copper in the ores is undesirable. Several of the more important antimony districts owe their economical production of that metal to the presence of recoverable values in gold. Some lead-silver ores contain small quantities of antimony, and "antimonial lead," containing 12 to 18 per cent antimony, is recovered in their smelting.

China is by far the most important antimony-producing country in the world, and normally supplies over half the world's total. Chinese antimony is exported in part as antimony crude (lumps of needle-like antimony sulphide), and in part as antimony regulus, which is about 99 per cent pure metal. France was the only other important source of antimony before the war (25 to 30 per cent of the world production), and Mexico and Hungary produced small amounts. The large demand for antimony occasioned by the war, besides stimulating production in these countries, brought forth important amounts of antimony ore from Algeria (French control) and from Bolivia and Australia (British control), as well as smaller quantities from several other countries. Of the war-developed sources, only Algeria and perhaps Australia are expected to continue production under normal conditions.

Before the war, antimony was smelted chiefly in China, England, and France, and to a lesser extent in Germany. British and French commercial and smelting interests dominated to a considerable extent the world situation, and London was the principal antimony market of the world.

During the war Chinese antimony interests were greatly strengthened, and facilities for treating the ore in that country were increased. Japan also became important as a smelter and marketer of Chinese ore, and increasing quantities of antimony were exported from China and Japan directly to the United States. English exports ceased entirely and were replaced in this country by Chinese and Japanese brands.

The United States normally consumes about one-third of the world's antimony. Before the war the entire amount was secured by importation, two-thirds from Great Britain and the rest from the Orient, France, and other European countries. Domestic production of ore and smelting of foreign ores were negligible. (These statements refer only to the purer forms of antimony; the United States normally produces considerable amounts of antimonial lead, equivalent to somewhat less than 5 per cent of the country's total lead production, but this material cannot be substituted for antimony regulus in most of its uses.)

During the war, under the stimulus of rising prices, mining of antimony was undertaken in the United States and several thousand tons of metal were produced—principally from Nevada, with smaller amounts from Alaska, California, and other western states. The great demands for antimony, however, were met chiefly by increased importation. Imports were mainly of regulus from Chinese and Japanese smelters of Chinese antimony; but about a third was contained in ores, including most of the production of Mexico which had formerly gone to England, and about 15 per cent of the Bolivian output. Antimony smelters were developed in the United States to handle these ores.

At the close of hostilities there had accumulated in the United States large surplus stocks of antimony and antimonial materials. With a very dull market and low prices, domestic mines and smelters were obliged to close down. The dependence of the United States on foreign sources of antimony and the importance of the metal for war purposes led to some agitation for a protective tariff—in addition to the present import duty of 10 per cent on antimony metal—in order to encourage home production (see pp. 365-366, 393-394).

In summary, the United States is almost entirely dependent upon outside sources for its antimony, although there are inadequately known reserves in this country which might be exploited if prices were maintained at a high level. The future of United States smelters is problematical. China, the world's chief source of antimony, at present dominates the market in this country, largely due to the low cost of production and favorable Japanese freight rates.

Geologic Features

The antimony sulphide, stibnite, is the source of most of the world's production of this metal. Antimony oxides, including senarmontite, cervantite, and others, are formed near the surface, and in some of the deposits of Mexico and Algeria they supply a large part of the values recovered. Jamesonite, bournonite, and tetrahedrite (sulphantimonides of lead and copper), when found in lead-silver deposits, are to some extent a source of antimony in the form of antimonial lead.

Stibnite is found in a variety of associations and is present in small quantities in many types of deposits. In the commercial antimony deposits, it is in most cases accompanied by minor quantities of other metallic sulphides—pyrite, cinnabar, sphalerite, galena, arsenopyrite, etc.—in a gangue of quartz and sometimes calcite. Many of the deposits contain recoverable amounts of gold and silver.

The deposits of the Hunan Province of southern China occur as seams, pockets, and bunches of stibnite ore in gently undulating beds of faulted and fissured dolomitic limestone. In the vicinity of the most important mines no igneous rocks have been observed, and the origin of the ores has not been worked out.

In the Central Plateau of France the numerous antimony deposits are stibnite veins cutting granites and the surrounding schists and sediments. An origin related in some way to hot ascending solutions seems probable.

The deposits of the National district of western Nevada, the most important war-developed antimony deposits of the United States, consist of stibnite veins with a gangue of fine-grained drusy quartz, cutting through flows of rhyolite and basalt. They are intimately related to certain gold- and silver-bearing veins, and all are closely associated with dikes of rhyolite, which were the feeders to the latest extrusion in the district. The wall rocks have undergone alteration of the propylitic type. These relations, and the presence of the mercury sulphide, cinnabar, in some of the ores (see pp. 258-259), suggest an origin through the work of ascending hot waters or hot springs. These waters probably derived their dissolved matter from a magmatic source, and worked up along vents near the rhyolite dikes soon after the eruption of this rock.

In the weathering of antimony deposits, the stibnite usually alters to form insoluble white or yellowish oxides, which are sometimes called "antimony ocher." These tend to accumulate in the oxide zone through the removal of the more soluble accompanying minerals. Secondary sulphide enrichment of antimony deposits, if it occurs at all, is negligible.

ARSENIC ORES

ECONOMIC FEATURES

About two-thirds of the arsenic consumed in recent years has been used in agriculture, where various arsenic compounds—arsenic trioxide or "white arsenic," Paris green, lead arsenate, etc.—are used as insecticides and weed killers. Arsenic compounds are also used in "cattle-dips" for killing vermin. The only other large use of arsenic is in the glass industry, arsenic trioxide being added to the molten glass to purify and decolorize the product. Small quantities of arsenic compounds are used in the preparation of drugs and dyeing materials, and metallic arsenic is used for hardening lead in shot-making.

The principal arsenic-producing countries are the United States, Germany, France, Great Britain, Canada, and Mexico. Spain, Portugal, Japan, and China are also producers, and recent trouble with the "prickly-pear" pest in Queensland, Australia, has led to local development of arsenic mining in that country. For the most part, European production has been used in Europe and American production in the United States.

Arsenic is recovered almost wholly as a by-product of smelting ores for the metals. The potential supply is ample in most countries where smelting is conducted, but owing to the elaborate plant required to recover the arsenic, apparatus is not usually installed much in advance of the demand for production. Rapid expansion is not possible.

Before the war the arsenic needs of the United States (chiefly agricultural) were supplied by a few recovery plants in the United States, Mexico, and Canada. Several large smelters had not found it profitable to install recovery plants, as the market might have been oversupplied and prices were low. During the war, with the extensive demand for insecticides for gardening, there was a considerable deficiency of arsenic supplies. With rising prices production was stimulated, but was still unable to meet the increased demand. This situation resulted in regulation of the prices of white arsenic by the Food Administration.

Production of arsenic in the United States comes chiefly from smelters in Colorado, Washington, Utah, Montana, and New Jersey. Small amounts are produced by arsenic mines in Virginia and New York. A Mexican plant at Mapimi has been shipping important quantities to the United States. The plant at Anaconda, Montana, is expected to produce an ample supply in the future.

The United States is entirely independent in arsenic supplies and will probably soon have an exportable surplus. Export trade, after the reconstruction period, will probably meet competition from France and Germany where production was formerly large.

GEOLOGIC FEATURES

Arsenic-bearing minerals are numerous and rather widely distributed, but only a few of them are mined primarily for their content of arsenic. Arsenopyrite or "mispickle" (iron-arsenic sulphide) has been used intermittently as a source of white arsenic in various places,—notably at Brinton, Virginia, and near Carmel, New York. The former deposits contain arsenopyrite and copper-bearing pyrite impregnating a mica-quartz-schist, adjacent to and in apparent genetic relation with aplite or pegmatite intrusives. In the latter locality arsenopyrite is found associated with pyrite in a gangue of quartz, forming a series of parallel stringers in gneiss close to a basic dike.

The orange-red sulphides of arsenic, orpiment and realgar, are formed both as primary minerals of igneous source and as secondary products of weathering. They are rather characteristic of the oxide zones of certain arsenical metallic ores, and are believed in many cases to have formed from arsenopyrite. They are mined on a commercial scale in China.

The great bulk of the world's arsenic, as previously stated, is obtained as a by-product of smelting operations. The enargite of the Butte copper ores (pp. 201-203) contains a considerable amount of arsenic, a large part of which will be recovered from the smelter fumes by new processes which are being installed. The gold-silver ores of the Tintic district (pp. 235) also yield important amounts, the arsenic-bearing minerals being enargite and tennantite (copper-arsenic sulphides) and others. The silver ores of the Cobalt district of Ontario (pp. 234-235), containing nickel and cobalt arsenides, produce considerable arsenic. Many other metallic ores contain notable amounts of arsenic, which are at present allowed to escape through smelter flues, but which could be recovered under market conditions which would repay the cost of installing the necessary apparatus.

BISMUTH ORES

ECONOMIC FEATURES

Bismuth metal is used in alloys, to which it gives low fusibility combined with hardness and sharp definition. Bismuth alloys are employed in automatic fire sprinklers, in safety plugs for boilers, in electric fuses, in solders and dental amalgams, and in some type and bearing metals. Bismuth salts find a considerable application for pharmaceutical purposes, especially in connection with intestinal disorders, and the best grades of bismuth materials are used for this purpose. The salts are also used in porcelain painting and enameling and in staining glass.

Bolivia is the most important producer of bismuth ore. The output is controlled entirely by British smelting interests. An important deposit exists in Peru, the output of which is limited by the same British syndicate. Considerable bismuth is produced in Australia, Tasmania, and New Zealand, all of which likewise goes to England. Germany before the war had three smelters which produced bismuth from native ores in Saxony; bismuth was one of the few metals of which Germany had an adequate domestic supply. Recently southern China is reported to be mining increasing amounts of bismuth.

The United States produces the larger part of its bismuth requirements, chiefly from plants installed at two lead refineries. A further installation would make this country entirely independent of foreign supplies if occasion required. Imports, from England and South America, have been steadily declining, but during the war were somewhat increased. The United States does not export bismuth so far as known.

Geologic Features

The principal minerals of bismuth are bismuthinite (bismuth sulphide), bismutite (hydrated carbonate), bismite or bismuth ocher (hydrated oxide), and native bismuth.

The native metal and the sulphide are believed to be formed mainly as primary minerals of igneous origin. In the deposits of New South Wales they are found associated with molybdenite in quartz gangue, in pipe-like deposits in granite. The oxide and the carbonate are probably products of surface weathering. The Bolivian deposits contain the native metal, the oxide, and the carbonate, associated with gold, silver, and tin minerals, in one locality in slates and in another locality in porphyry. The origin is not well known.

In the United States, the sulphide, bismuthinite, is found in the siliceous ores of Goldfield, Nevada (p. 230), and in minor amounts in a great number of the sulphide ores of the Cordilleran region. The ores of the Leadville and Tintic districts (pp. 219 and 235) yield the larger part of the United States production, the bismuth being recovered as by-product from the electrolytic refining of the lead bullion. Large amounts of bismuth pass out of the stacks of smelters treating other western ores, and while it would not be cheap nor easy to save the bismuth thus lost, it could probably be done in case of necessity.

CADMIUM ORES

Economic Features

Cadmium is used in low melting-point alloys—as, for example, those employed in automatic fire-extinguishers and electric fuses,—in the manufacture of silverware, and in dental amalgams. During the war the critical scarcity of tin led to experiments in the substitution of cadmium for tin in solders and anti-friction metals. Results of some of these experiments were promising, but the war ceased and demands for tin decreased before the cadmium materials became widely used. Future developments in this direction seem not unlikely. Cadmium compounds are used as pigments, particularly as the sulphide "cadmium yellow," and to give color and luster to glass and porcelain. Cadmium salts are also variously used in the arts, in medicine, and in electroplating.

Practically the entire cadmium output of the world comes from Germany and the United States. In addition, England produces a very small quantity. Before the war Germany produced about two-thirds of the world's total, and supplied the European as well as a considerable part of the United States consumption. During the war the United States production increased three to four fold, imports ceased, and considerable quantities were exported to the allied nations in Europe and to Japan. At present the United States is entirely independent as regards cadmium supplies. Production is sufficient to supply all the home demand and to permit exports of one-third of the total output. A considerable number of possible cadmium sources are not being used, and the production is capable of extension should the need arise.

Geologic Features

Nearly the only cadmium mineral known is the sulphide, greenockite, but no deposits of this mineral have been found of sufficient volume to be called cadmium ores. Sphalerite almost always contains a little cadmium, probably as the sulphide; and in zinc deposits crystals of sphalerite in cavities are frequently covered with a greenish-yellow film or coating of greenockite. These coatings have probably been formed by the decomposition of cadmium-bearing zinc sulphide in the oxide zone, the carrying down of the cadmium in solution, and its precipitation as secondary cadmium sulphide. The zinc oxide minerals in the surficial zone also are sometimes colored yellow by small amounts of greenockite. In the zinc ores of the Joplin district of Missouri, cadmium is present in amounts ranging from a trace to 1 per cent and averaging 0.3 per cent.

Germany's cadmium is produced by fractional distillation of the Silesian zinc ores, which contain at most 0.3 per cent cadmium. In the United States there are large potential sources in the zinc ores of the Mississippi valley, and considerable cadmium is recovered in roasting them. Much of the American cadmium is also obtained from bag-house dusts at lead smelters.

The general geologic conditions of the cadmium-bearing ores are indicated in the discussion of lead and zinc deposits in an earlier chapter.

COBALT ORES

Economic Features

Cobalt finds its largest use in the form of cobalt salts, employed in coloring pottery and glass and in insect poisons. Cobalt is also used in some of the best high-speed tool steels. "Stellite," which is used to a limited extent in non-rusting tools of various sorts, and in considerable quantity to replace high-speed tool steels, is an alloy of cobalt, chromium, and small quantities of other metals. Considerable experimental work has been done on the properties and uses of cobalt alloys, and their consumption is rapidly on the increase.

Cobalt is an item of commerce of insignificant tonnage. There are only two countries, Canada (Ontario) and the Belgian Congo, which produce noteworthy amounts. The Katanga district in the Congo produces blister copper that contains as much as 4 per cent of cobalt, though usually less than 2 per cent. This product formerly went to Germany, and now goes entirely to Great Britain. Just how much cobalt is saved is unknown, but probably several hundred tons annually. It is probable that most of the cobalt in these ores will be lost on the installation of a leaching process for recovery of the copper. Canada exports most of its product to the United States, though the amount is small. Domestic production in this country has been too small to record. The United States has been dependent on imports from Canada.

Geologic Features

The principal cobalt minerals are smaltite (cobalt arsenide), cobaltite (cobalt-arsenic sulphide), and linnæite (cobalt-nickel sulphide). Under weathering conditions these minerals oxidize readily to form asbolite, a mixture of cobalt and manganese oxides, and the pink arsenate, erythrite or "cobalt bloom."

Cobalt minerals are found principally in small quantities disseminated through ores of silver, nickel, and copper. The production of Canada is obtained mainly as a by-product of the silver ores of the Cobalt district (described on pp. 234-235), and smaller amounts are recovered from the Sudbury nickel ores (pp. 180-182). The cobalt of Belgian Congo is obtained from rich oxidized copper ores which impregnate folded sediments (p. 205).

MERCURY (QUICKSILVER) ORES

Economic Features

Uses of mercury are characterized by their wide variety and their application to very many different phases of modern industry; they will be named here in general order of decreasing importance. About one-third of the mercury consumed in this country goes into the manufacture of drugs and chemicals, such as corrosive sublimate, calomel, and glacial acetic acid. Mercury fulminate is used as a detonator for high explosives and to some extent for small-arms ammunition—a use which was exceedingly important during the war, but is probably of minor consequence in normal times. Mercuric sulphide forms the brilliant red pigment, vermilion, and mercuric oxide is becoming increasingly important in anti-fouling marine paint for ship-bottoms. Either as the metal or the oxide, mercury is employed in the manufacture of electrical apparatus (batteries, electrolyzers, rectifiers, etc.), and in the making of thermostats, gas governors, automatic sprinklers, and other mechanical appliances. Mercuric nitrate is used in the fabrication of felt hats from rabbits' fur. In the extraction of gold and silver from their ores by amalgamation, large amounts of metallic mercury have been utilized, but of late years the wide application of the cyanide process has decreased this use. Minor uses include the making of certain compounds for preventing boiler-scale, of cosmetics, and of dental amalgam.

The ores of mercury vary greatly in grade. Spanish ores yield an average in the neighborhood of 7 per cent, Italian ores 0.9 per cent, and Austrian ores 0.65 per cent of metallic mercury. In the United States the ores of California yield about 0.4 per cent and those of Texas range from about 0.5 to 4 per cent. In almost all cases the ores are treated in the immediate vicinity of the mines, and fairly pure metal is obtained by a process of sublimation and condensation. This is usually marketed in iron bottles or flasks containing 75 pounds each.

The large producers of mercury are, in order of normal importance, Spain, Italy, Austria, and United States. Mexico, Russia, and all other countries produce somewhat less than 5 per cent of the world's total.

The largest quicksilver mines of the world are those of Almaden in central Spain, which are owned and operated by the Spanish government. This government, after reserving a small amount for domestic use, sells all the balance of the production through the Rothschilds of London. In addition British capital controls some smaller mines in northern Spain. England thus largely controls the European commercial situation in this commodity, and London is the world's great quicksilver market, where prices are fixed and whence supplies go to all corners of the globe. Reserves of the Almaden ore bodies are very large. Sufficient ore is reported to have been developed to insure a future production of at least 40,000 metric tons—an amount equivalent to the entire world requirements at pre-war rates of consumption for 100 years.

The mercury deposits of the Monte Amiata district of central Italy were in large part dominated by German capital, but during the war were seized by the Italian government. The mines of Idria, Austria-Hungary, were owned by the Austrian government and their ultimate control is at present uncertain. Reserves are very large, being estimated at about one-half those of Almaden. Although England has had a considerable control over the prices and the market for mercury, the Italian and Austrian deposits have provided a sufficient amount to prevent any absolute monopoly. English interests have now secured control of the Italian production, and it is expected that they will also control the Austrian production—thus giving England control of something over three-fourths of the world's mercury.

In the United States about two-thirds of the mercury is produced in the Coast Range district of California, and most of the remainder in the Terlingua district of Texas. Smaller quantities come from Nevada, Oregon, and a few other states. The output before the war was normally slightly in excess of domestic demand and some mercury was exported to various countries. Due to the exhaustion of the richer and more easily worked deposits, however, production was declining. During the war, with increased demands and higher prices, production was stimulated, the United States became the largest mercury-producing country in the world, and large quantities were exported to help meet the military needs of England and France.

With the end of war prices and with high costs of labor and supplies, production in the United States has again declined. Many of the mines have passed their greatest yield, and though discovery of new ore bodies might revive the industry, production is probably on the down grade. Future needs of this country will probably in some part be met by imports from Spain, Italy, and Austria, where the deposits are richer and labor is cheaper. This situation has caused much agitation for a tariff on imports. The present tariff of 10 per cent is not sufficient to keep out foreign mercury.

Outside of the United States large changes in distribution of production of quicksilver are not expected for some time. The reserves of the European producers are all large and are ample to sustain present output for a considerable number of years. It is reported that there will be a resumption of mining in the once very productive Huancavelica District of Peru and in Asia Minor, and with restoration of political order there may be an increase in output from Mexico and Russia,—but these districts will be subordinate factors in the world situation. On geologic grounds, new areas of mercury ores may be looked for in regions of recent volcanic activity, such as the east coast of Asia, some islands of Oceania, the shores of the Mediterranean, and the Cordilleras of North and South America,—but no such areas which are likely to be producers on a large scale are now known.

Geologic Features

The chief mineral of mercury, from which probably over 95 per cent of the world's mercury comes, is the brilliant red sulphide, cinnabar. Minor sources include the black or gray sulphide, metacinnabar, the native metal, and the white mercurous chloride, calomel. The ores are commonly associated with more or less iron sulphide, and frequently with the sulphides of antimony and arsenic, in a gangue consisting largely of quartz and carbonates (of calcium, magnesium, and iron). The precious metals and the sulphides of the base metals are rare.

Mercury deposits are in general related to igneous rocks, and have associations which indicate a particular type of igneous activity. They are not found in magmatic segregations, in pegmatites, nor in veins which have been formed at great depths and under very high temperatures. On the contrary, the occurrence of many deposits in recent flows which have not been eroded, their general shallow depth (large numbers extending down only a few hundred feet), and the association of some deposits with active hot springs now carrying mercury in solution, suggest an origin through the work of ascending hot waters near the surface. The mercury minerals are believed to have been carried in alkaline sulphide solutions. Precipitation from such solutions may be effected by oxidation, by dilution, by cooling, or by the presence of organic matter. Being near the surface, it is a natural assumption that the waters doing the work were not intensely hot. At Sulphur Bank Springs, in the California quicksilver belt, deposition of cinnabar by moderately hot waters is actually taking place at present; also these waters are bleaching the rock in a manner often observed about mercury deposits.

The Coast Ranges of California contain a great number of mercury deposits extending over a belt about 400 miles long. The ore bodies are in fissured zones in serpentine and Jurassic sediments, and are related in general to recent volcanic flows. A considerable amount of bituminous matter is found in the ores, and is believed to have been an agent in their precipitation.

The Terlingua ores of Texas are found in similar fractured zones in Cretaceous shales and limestones associated with surface igneous flows. The occurrence of a few ore bodies in vertical shoots in limestone, apparently terminating upward at the base of an impervious shale, furnishes an additional argument for their formation by ascending waters.

In the few deposits (*e. g.*, those of Almaden, Spain, and of the deep mines of New Almaden and New Idria, California,) where there is no such clear relation to volcanic rocks as generally observed, but where the ores contain the same characteristic set of minerals, it is concluded that practically the same processes outlined above have been active in their formation; and that the volcanic source of the hot solutions either failed to reach the surface or has been removed by erosion. The same line of reasoning is carried a step further, and in many gold-quartz veins in volcanic rocks, where cinnabar and its associated minerals are present, it is believed that waters of a hot-spring nature have again been effective. Thus cinnabar, when taken with its customary associations, is regarded as a sort of geologic thermometer.

In the weathering of mercury deposits, cinnabar behaves somewhat like the corresponding silver sulphide, argentite. In the oxide zone, native mercury and the chloride, calomel, are formed. In the Texas deposits a red oxide and a number of oxychlorides are also present. The carrying down of the mercury and its precipitation as secondary sulphide may have taken place in some deposits, but this process is unimportant in forming values.

TIN ORES

Economic Features

The largest use of tin is in the manufacture of tin-plate, which is employed in containers for food, oil, and other materials. Next in importance is its use in the making of solder and of babbitt or bearing metal. Tin is also a constituent of certain kinds of brass, bronze, and other alloys, such as white metal and type metal. Minor uses include the making of tinfoil, collapsible tubes, wire, rubber, and various chemicals. Tin oxide is used to some extent in white enameling of metal surfaces. Roughly a third of the tin consumed within the United States goes into tin-plate, a third into solder and babbitt metal, and a third into miscellaneous uses.

The ores of tin in general contain only small quantities of the metal. Tin has sufficient value to warrant the working of certain placers containing only a half-pound to the cubic yard, although the usual run is somewhat higher. The tin content of the vein deposits ranges from about 1 per cent to 40 per cent, and the average grade is much closer to the lower figure.

Great Britain has long controlled the world's tin ores, producing about half of the total and controlling additional supplies in other countries. The production is in small part in Cornwall, but largely in several British colonies—the Malay States, central and south Africa, Australia, and others. The Malay States furnish about a third of the world's total. Another third is produced in immediately adjacent districts of the Dutch East Indies, Siam (British control), and China, and some of the concentrates of these countries are handled by British smelters, especially at Singapore.

Tin is easily reduced from its ores and most of the tin is smelted close to the sources of production. Considerable quantities, however, have gone to England for treatment. London has been the chief tin market of the world, and before the war the larger portion of the tin entering international trade went through this port. During the war a good deal of the export tin from Straits Settlements was shipped direct to consumers rather than via London, but it is not certain how future shipments may be made.

Significant features of the tin situation in recent years have been a decline of production in the Malay States, and a large and growing production in Bolivia. Malayan output has decreased because of the exhaustion of some of the richer and more accessible deposits; certain governmental measures have also had a restrictive effect. Bolivian production now amounts to over a fifth of the world's total and bids fair to increase. About half the output is controlled by Chilean, and small amounts by American, French, and German interests. A large portion of the Bolivian concentrates formerly went to Germany for smelting, but during the war American smelters were developed to handle part of this material; large quantities are also smelted in England.

The United States produces a small fraction of 1 per cent of the world's tin, and consumes a third to a half of the total. The production is mainly from the Seward Peninsula of northwestern Alaska. For American tin smelters, Bolivia is about the only available source of supplies; metallic tin can be obtained from British possessions, but no ore, except by paying a 33-⅓ per cent export tax. The United States exports tin-plate in large amounts, and in this trade has met strong competition from English and German tin-plate makers.

A world shortage of tin during the war required a division of available supplies through a central international committee. Somewhat later, with the removal of certain restrictions on the distribution of tin, considerable quantities which had accumulated in the Orient found their way into Europe and precipitated a sensational slump in the tin market.

Geologic Features

The principal mineral of tin is cassiterite (tin oxide). Stannite, a sulphide of copper, iron, and tin, is found in some of the Bolivian deposits but is rare elsewhere.

About two-thirds of the world's tin is obtained from placers and one-third from vein or "lode" deposits. Over 90 per cent of the tin of southeastern Asia and Oceania is obtained from placers. Tin placers, like placers of gold, platinum, and tungsten, represent concentrations in stream beds and ocean beaches of heavy, insoluble minerals—in this case chiefly cassiterite—which were present in the parent rocks in much smaller quantities, but which have been sorted out by the classifying action of running water.

The original home of cassiterite is in veins closely related to granitic rocks. It is occasionally found in pegmatites, as in certain small deposits of the Southern Appalachians and the Black Hills of South Dakota, or is present in a typical contact-metamorphic silicated zone in limestone, as in some of the deposits of the Seward Peninsula of Alaska. In general, however, it is found in well-defined fissure veins in the outer parts of granitic intrusions and extending out into the surrounding rocks. With the cassiterite are often found minerals of tungsten, molybdenum, and bismuth, as well as sulphides of iron, copper, lead, and zinc, and in some cases there is evidence of a rough zonal arrangement. The deposits of Cornwall and of Saxony show transitions from cassiterite veins close to the intrusions into lead-silver veins at a greater distance. The gangue is usually quartz, containing smaller amounts of a number of less common minerals—including lithium mica, fluorite, topaz, tourmaline, and apatite. The wall rocks are usually strongly altered and in part are replaced by some of the above minerals, forming coarse-grained rocks which are called "greisen."

The origin of cassiterite veins, in view of their universal association with granitic rocks, is evidently related to igneous intrusions. The occurrence of the veins in distinct fissures in the granite and in the surrounding contact-metamorphic zone indicates that the granite had consolidated before their formation, and that they represent a late stage in the cooling. The association with minerals containing fluorine and boron, and the intense alteration of the wall rocks, indicate that the temperature must have been very high. It is probable that the temperature was so high as to cause the solutions to be gaseous rather than liquid, and that what have been called "pneumatolytic" conditions prevailed; but evidence to decide this question is not at present available.

The most important deposits of tin in veins are those of Bolivia, some of which are exceptionally rich. These are found in granitic rocks forming the core of the high Cordillera Real and in the adjacent intruded sediments, in narrow fissure veins and broader brecciated zones containing the typical ore and gangue minerals described above, and also, in many cases, silver-bearing sulphides (chiefly tetrahedrite). There appear to be all gradations in type from silver-free tin ores to tin-free silver ores, although the extremes are now believed to be rare. In the main the tin ores, with abundant tourmaline, appear to be more closely related to the coarse-grained granites, and to indicate intense conditions of heat and pressure, while the more argentiferous ores, with very little or no tourmaline, are found in relation to finer-grained quartz porphyries and even rhyolites, and seem to indicate less intense conditions at the time of deposition. The ores of the whole area, which is a few hundred miles long, have been supposed to represent a single genetic unit, and the sundry variations are believed to be local facies of a general mineralization. Processes of secondary enrichment have in places yielded large quantities of oxidized silver minerals and wood tin near the surface, with accumulations of ruby silver ores at greater depths.

The only other vein deposits which are at present of consequence are those of Cornwall. Here batholiths of granite have been intruded into Paleozoic slates and sandstones, and tin ores occur in fissures and stockworks in the marginal zones. With the exhaustion of the more easily mined placers, the lode deposits will doubtless be of increasing importance.

Cassiterite is practically insoluble and is very resistant to decomposition by weathering. Oxide zones of tin deposits are therefore enriched by removal of the more soluble minerals. Stannite probably alters to "wood tin," a fibrous variety of cassiterite. Secondary enrichment of tin deposits by redeposition of tin minerals is negligible.

URANIUM AND RADIUM ORES

Economic Features

Radium salts are used in various medical treatments—especially for cancer, internal tumors, lupus, and birth marks—and in luminous paints. During the latter part of the war it is estimated that over nine-tenths of the radium produced was used in luminous paints for the dials of watches and other instruments. In addition part of the material owned by physicians was devoted to this purpose, and it is probable that the accumulated stocks held by the medical profession were in this way reduced by one-half. The greatly extended use of radium, together with the distinctly limited character of the world's known radium supplies, has led to some concern; and considerable investigation has been made of the possibilities of mesóthorium as a substitute for radium in luminous paints. Low-grade radium residues are used to some extent as fertilizers.

Uranium has been used as a steel alloy, but has not as yet gained wide favor. Uranium salts have a limited use as yellow coloring agents in pottery and glass. The principal use of uranium, however, is as a source of radium, with which it is always associated.

European countries first developed the processes of reduction of radium salts from their ores. Most of the European ores are obtained from Austria, where the mines are owned and operated by the Austrian government, and small quantities are mined in Cornwall, England, and in Germany. Production is decreasing. The European hospitals and municipalities have acquired nearly all of the production.

The United States has the largest reserves of radium ore in the world, and the American market has in recent years been supplied from domestic plants. Before the war, radium ores were shipped to Europe for treatment in Germany, France, and England, and radium salts were imported from these countries. There are now radium plants in the United States capable of producing annually from domestic ores an amount several times as large as the entire production of the rest of the world. Practically all the production has come from Colorado and Utah. Known reserves are not believed to be sufficient for more than a comparatively few years' production, but it is not unlikely that additional deposits will be found in the same area.

Geologic Features

Uranium is one of the rarer metals. Radium is found only in uranium ores and only in exceedingly small quantities. The maximum amount which can be present in a state of equilibrium is about one part of radium in 3,000,000 parts of uranium. The principal sources of uranium and radium are the minerals carnotite (hydrous potassium-uranium vanadate) and pitchblende or uraninite (uranium oxide).

The deposits of Joachimsthal, Bohemia, contain pitchblende, along with silver, nickel, and cobalt minerals and other metallic sulphides, in veins associated with igneous intrusions.

The important commercial deposits of Colorado and Utah contain carnotite, together with roscoelite (a vanadium mica) and small amounts of chromium, copper, and molybdenum minerals, as impregnations of flat-lying Jurassic sandstones. The ores carry up to 35 per cent uranium oxide (though largely below 2 per cent), and from one-third as much to an equal amount of vanadium oxide. The ore minerals are supposed to have been derived from a thick series of clays and impure sandstones a few hundred feet above, containing uranium and vanadium minerals widely disseminated, and to have been carried downward by surface waters containing sulphates. The ore bodies vary from very small pockets to deposits yielding a thousand tons or so, and are found irregularly throughout certain particular beds without any special relation to present topography or to faults. The association of many of the deposits with fossil wood and other carbonaceous material suggests that organic matter was an agent in their precipitation, but the exact nature of the process is not clear. In a few places in Utah the beds dip at steep angles, and the carnotite appears in spots along the outcrops and generally disappears as the outcrops are followed into the hillsides; this suggests that the carnotite may be locally redissolved and carried to the surface by capillary action, forming rich efflorescences. Because of the nature of the deposits no large amount of ore is developed in advance of actual mining; but estimates based on past experience indicate great potentialities of this region for future production.

In eastern Wyoming is a unique deposit of uranium ore in a quartzite which lies between mica-schist and granite. The principal ore mineral is uranophane, a hydrated calcium-uranium silicate, which is believed to be an oxidation product of pitchblende. Some of the ore runs as high as 4 per cent uranium oxide, and the ore carries appreciable amounts of copper but very little vanadium.

Very recently radium ores have been discovered in the White Signal mining district of New Mexico, which was formerly worked for gold, silver, copper, and lead. The radium-bearing minerals are torbernite and autunite (hydrous copper-uranium and calcium-uranium phosphates), and are found in dark felsite dikes near their intersections with east-west gold-silver-quartz veins. The possibilities of this district have not yet been determined.

Pitchblende has been found in gold-bearing veins in Gilpin County, eastern Colorado, and in pegmatite dikes in the Appalachians, but these deposits are of no commercial importance. Pitchblende is grayish-black, opaque, and so lacking in distinctive characteristics that it may readily be overlooked; hence future discoveries in various regions would not be surprising.

CHAPTER XIII

MISCELLANEOUS NON-METALLIC MINERALS

NATURAL ABRASIVES

Economic Features

Natural abrasives are less important commercially in the United States than artificial abrasives, but a considerable industry is based on the natural abrasives.

Silica or quartz in its various crystalline forms constitutes over three-fourths of the tonnage of natural abrasives used in the United States. It is the chief ingredient of sand, sandstone, quartzite, chert, diatomaceous earth, and tripoli. From the sand and sandstone are made millstones, buhrstones, grindstones, pulpstones, hones, oilstones, and whetstones. Sand, sandstone, and quartzite are also ground up and used in sand-blasts, sandpaper, and for other abrasive purposes. Chert or flint constitutes grinding pebbles and tube-mill linings, and is also ground up for abrasives. Diatomaceous (infusorial) earth is used as a polishing agent and also as a filtering medium, an absorbent, and for heat insulation. Tripoli (and rottenstone) are used in polishing powders and scouring soaps as well as for filter blocks and many other purposes.

Other important abrasives are emery and corundum, garnet, pumice, diamond dust and bort, and feldspar.

Imports of abrasive materials into the United States have about one-third of the value of those locally produced. While all of the various abrasives are represented in these imports, the United States is dependent on foreign sources for important parts of its needs only of emery and corundum, garnet, pumice, diamond dust and bort, and grinding pebbles.

Emery and corundum are used in various forms for the grinding and polishing of hard materials—steel, glass, stone, etc. The principal foreign sources of emery have been Turkey (Smyrna) and Greece (Naxos) where reserves are large and production cheap. Production of corundum has come from Canada, South Africa, Madagascar, and India. The domestic production of emery is mainly from New York and Virginia, and corundum comes from North Carolina. Domestic supplies are insufficient to meet requirements, and cannot be substituted for the foreign material for the polishing of fine glass and other special purposes. Curtailment of imports during the war greatly stimulated the development of artificial abrasives and their substitution for emery and corundum.

Garnet is used chiefly in the form of garnet paper for working leather, wood, and brass. Garnet is produced mainly in the United States and Spain. The United States is the only country using large amounts of this mineral and imports most of the Spanish output. The domestic supply comes mainly from New York, New Hampshire, and North Carolina.

Pumice is used in fine finishing and polishing of varnished and enameled surfaces, and in cleaning powders. The world's principal source for pumice is the Lipari Islands, Italy. There is a large domestic supply of somewhat lower-grade material (volcanic ash) in the Great Plains region, and there are high-grade materials in California and Arizona. Under war conditions these supplies were drawn on, but normally the high-quality Italian pumice can be placed in American markets more cheaply.

Diamond dust is used for cutting gem stones and other very hard materials, and borts or carbonadoes (black diamonds) for diamond-drilling in exploration. Most of the black diamonds come from Brazil, and diamond dust comes from South Africa, Brazil, Borneo, and India.

Chert or flint pebbles for tube-mills are supplied mainly from the extensive deposits on the French and Danish coasts. The domestic production has been small, consisting principally of flint pebbles from the California beaches, and artificial pebbles made from rhyolite in Nevada and quartzite in Iowa. War experience demonstrated the possibility of using the domestic supply in larger proportion, but the grade is such that in normal times this supply will not compete with importations.

Feldspar as an abrasive is used mainly in scouring soaps and window-wash. Domestic supplies are ample. The principal use of feldspar is in the ceramic industry and the mineral is discussed at greater length in the chapter on common rocks (p. 86).

For the large number of abrasives produced from silica, outside of flint pebbles, domestic sources of production are ample. Siliceous rocks are available almost everywhere. For particular purposes, however, rocks possessing the exact combinations of qualities which make them most suitable are in many cases distinctly localized. *Millstones and buhrstones*, used for grinding cereals, paint ores, cement rock, fertilizers, etc., are produced chiefly in New York and Virginia; partly because of trade prejudice and tradition, about a third of the American requirements are imported from France, Belgium, and Germany. *Grindstones and pulpstones*, used for sharpening tools, grinding wood-pulp, etc., come mainly from Ohio and to a lesser extent from Michigan and West Virginia; about 5 per cent of the consumption is imported from Canada and Great Britain. *Hones*, *oilstones*, and *whetstones* are produced largely from a rock called "novaculite" in Arkansas, and also in Indiana, Ohio, and New England; imports are negligible. *Flint linings* for tube-mills were formerly imported from Belgium, but American products, developed during the war in Pennsylvania, Tennessee, and Iowa, appear to be wholly satisfactory substitutes. *Diatomaceous earth* is produced in California, Nevada, Connecticut, and Maryland, and *tripoli and rottenstone* in Illinois, Missouri, and Oklahoma; domestic sources are sufficient for all needs, but due to questions of back-haul and cost of rail transportation there has been some importation from England and Germany.

GEOLOGIC FEATURES

The geologic features of silica (quartz), feldspar, and diamonds are sufficiently indicated elsewhere (Chapter II; pp. 84, 196, 86, 291-292).

Diatomaceous earth is made up of remains of minute aquatic plants. It may be loose and powdery, or coherent like chalk. It is of sedimentary origin, accumulated originally at the bottoms of ponds, lakes, and in the sea.

Tripoli and rottenstone are light, porous, siliceous rocks which have resulted from the leaching of calcareous materials from various siliceous limestones or calcareous cherts in the process of weathering.

Grinding pebbles are derived from the erosion of limestone or chalk formations which contain concretions of extremely fine-grained and dense chert. Under stream and wave action they are rounded and polished. The principal sources are ocean beaches.

Corundum as an abrasive is the mineral of this name—made up of anhydrous aluminum oxide. Emery is an intimate mechanical mixture of corundum, magnetite, and sometimes spinel. Corundum is a product of contact metamorphism and also a result of direct crystallization from molten magma. Canadian corundum occurs as a constituent of syenite and nepheline-syenite in Lower Ontario. In North Carolina and Georgia, the corundum occurs in vein-like bodies at the contact of peridotite with gneisses and schists, and also in part in the peridotite itself. In New York the emery deposits are segregations of aluminum and iron oxides in norite (a basic igneous rock). The emery of Greece and Turkey occurs as lenses or pockets in crystalline limestones, and is the result of contact metamorphism by intrusive granites.

Garnets result mainly from contact metamorphism, and commonly occur either in schists and gneisses or in marble. The principal American occurrences are of this type. Being heavy and resistant to weathering, they are also concentrated in placers. The Spanish garnets are reported to be obtained by washing the sands of certain streams.

Pumice is solidified rock froth formed by escape of gases from molten igneous rocks at the surface. It is often closely associated with volcanic ash, which is also used for abrasive purposes.

In general, the geologic processes entering into the formation of abrasives cover almost the full range from primary igneous processes to surface alterations and sedimentation.

ASBESTOS

Economic Features

The principal uses of asbestos are in high-pressure packing in heat engines, in thermal and electrical insulation, in fire-proofing, and in brake-band linings.

The largest producers of asbestos are Canada (Quebec) and, to a considerably less extent, Russia. United States interests have financial control of about a fourth of the Canadian production, and practically the entire export trade of Canada goes to the United States. Russia exports nearly all her product to Germany, Austria, United Kingdom, Belgium, and the Netherlands. Previous to the war the output was largely controlled by a German syndicate. There is a considerable recent production in South Africa, which is taken by England and the United States, and small amounts are produced in Italy, Cyprus, and Australia.

The United States has been a large importer of asbestos, from Canada and some other sources. Domestic production is relatively insignificant, and exports depend chiefly on an excess of import. Georgia is the principal local source. Arizona and California are also producers, their product being of a higher grade. The United States is the largest manufacturer of asbestos goods, and exports go to nearly all parts of the world.

So long as the abundant Canadian material is accessible on reasonable conditions, the United States is about as well situated as if independent. Some Canadian proposals of restriction during the war led to a study of other supplies and showed that several deposits, such as those in Russia and Africa, might compete with the Canadian asbestos.

Geologic Features

Asbestos consists mostly of magnesium silicate minerals—chrysotile, anthophyllite, and crocidolite. The term asbestos covers all fibrous minerals with some tensile strength which are poor conductors and can be used for heat-protection. Like talc, they are derived principally from the alteration of olivine, pyroxene, and amphibole,—or more commonly from serpentine, which itself results from the alteration of these minerals. Chrysotile is the most common, and because of the length, fineness, and flexibility of its fibers, enabling it to be spun into asbestos ropes and fabrics, it is the most valuable. Anthophyllite fibers, on the other hand, are short, coarse, and brittle, and can be used only for lower-grade purposes. Crocidolite or blue asbestos is similar to chrysotile but somewhat inferior in fire-resisting qualities.

Asbestos deposits occur chiefly as veinlets in serpentine rock, which is itself the alteration of some earlier rock like peridotite. They are clearly formed in cracks and fissures through the agency of water, but whether the waters are hot or cold is not apparent. The veinlets have sometimes been interpreted as fillings of contraction cracks, but more probably are due to recrystallization of the serpentine, proceeding inward from the cracks. In Quebec the chrysotile asbestos (which is partly of spinning and partly of non-spinning grade) forms irregular veins of this nature in serpentine, the fiber making up 2 to 6 per cent of the rock.

In Georgia the asbestos, which is anthophyllite, occurs in lenticular masses in peridotite associated with gneiss. It is supposed to have formed by the alteration of olivine and pyroxene in the igneous rocks. In Arizona chrysotile is found in veins in cherty limestone, associated with diabase intrusives. Here it is believed to be an alteration product of diopside (lime-magnesia pyroxene) in a contact-metamorphic silicated zone.

Crocidolite is mined on a commercial scale only in Cape Colony, South Africa. The deposits occur in thin sedimentary layers interbedded with jaspers and ironstones. Their origin has not been worked out in detail.

The deposits of Russia, the Transvaal, Rhodesia, and Australia are of high-grade chrysotile, probably similar in origin to the Quebec deposits. The asbestos of Italy and Cyprus is anthophyllite, more like the Georgia material.

BARITE (BARYTES)

Economic Features

Barite is used chiefly as a material for paints. For this purpose it is employed both in the ground form and in the manufacture of lithopone, a widely used white paint consisting of barium sulphate and zinc sulphide. Ground barite is also used in certain kinds of rubber goods and in the making of heavy glazed paper. Lesser amounts go into the manufacture of barium chemicals, which are used in the preparation of hydrogen peroxide, in softening water, in tanning leather, and in a wide variety of other applications.

Germany is the world's principal producer of barite and has large reserves of high grade. Great Britain also has extensive deposits and produces perhaps one-fourth as much as Germany. France, Italy, Belgium, Austria-Hungary, and Spain produce smaller but significant amounts.

Before the war the United States imported from Germany nearly half the barite consumed in this country, and produced the remainder. Under the necessities of war times, adequate domestic supplies were developed and took care of nearly all the greatly increased demands. Production has come from fourteen states, the large producers being Georgia, Missouri, and Tennessee. During the war, also, an important movement of barite-consuming industries to the middle west took place, in order to utilize more readily and cheaply the domestic product. For this reason it is not expected that German barite will play as important a part as formerly in American markets,—although it can undoubtedly be put down on the Atlantic seaboard much more cheaply than domestic barite, which requires long rail hauls from southern and middle-western states.

Geologic Features

The mineral barite is a heavy white sulphate of barium, frequently called "barytes" or "heavy spar." Witherite, the barium carbonate, is a much rarer mineral but is found with barite in some veins.

All igneous rocks contain at least a trace of barium, which is probably present in the silicates, and these small quantities are the ultimate source of the more concentrated deposits. Barite itself is not found as an original constituent of igneous rocks or pegmatites, but is apparently always formed by deposition from aqueous solutions. It is a common gangue mineral in many deposits of metallic sulphides, both those formed in relation to igneous activity and those which are independent of such activity, but in these occurrences it is of little or no commercial importance.

The principal deposits of barite are found in sedimentary rocks, and especially in limestones and dolomites. In these rocks it occurs in veins and lenses very similar in nature to the lead and zinc deposits of the Mississippi valley (p. 211 *et seq.*), and, like them, probably deposited by cold solutions which gathered together small quantities of material from the overlying or surrounding rocks. The Missouri deposits are found in limestones in a region not far from the great southeastern Missouri lead district, and vary from the lead deposits in relative proportions rather than in kind of minerals; the veins consist chiefly of barite, with minor quantities of silica, iron sulphide, galena, and sphalerite. The deposits of the southern Appalachians occur as lenses in limestones and schists.

Barite is little affected by surface weathering, and tends to remain behind while the more soluble minerals of the associated rock are dissolved out and carried away. A limited amount of solution and redeposition of the barite takes place, however, resulting in its segregation into nodules in the residual clays. Most of the barite actually mined comes from these residual deposits, which owe their present positions and values to katamorphic processes. The accompanying clay and iron oxide are removed by washing and mechanical concentration.

Certain investigators of the deposits of the Mississippi valley are extremely reluctant to accept the idea that the ores are formed by surface waters of ordinary temperatures, and are inclined to appeal to heated waters from a hypothetical underlying magmatic source. The fact that barite is a characteristic mineral of many igneous veins, and the fact that in this same general region it is found in the Kentucky-Illinois fluorspar deposits,—where a magmatic source is generally accepted,—together with doubts as to the theoretical efficacy of meteoric waters to transport the minerals found in the barite deposits, have led certain writers to ascribe to these barite deposits a magmatic origin. The magmatic theory has not been disproved; but on the whole the balance of evidence seems strongly to indicate that the barite deposits as well as the lead and zinc ores, which are essentially the same in nature though differing in mineral proportions, have been concentrated from the adjacent sediments by ordinary surface waters.

BORAX

ECONOMIC FEATURES

Borax-bearing minerals are used almost entirely in the manufacture of borax and boric acid. Fully a third of the borax consumed in the United States is used in the manufacture of enamels or porcelain-like coatings for such objects as bathtubs, kitchen sinks, and cooking utensils. Other uses of borax or of boric acid are as a flux in the melting and purification of the precious metals, in decomposing chromite, in making glass, as a preservative, as an antiseptic, and as a cleansing agent. Recent developments indicate that the metal, boron, may play an important part in the metallurgy of various metals. It has been used in making very pure copper castings for electrical purposes, in aluminum bronzes, and in hardening aluminum castings; and an alloy, ferroboron, has been shown experimentally to act on steel somewhat like ferrovanadium.

The bulk of the world's borax comes from the Western Hemisphere, the United States and Chile being the two principal producers. There are additional large deposits in northern Argentina, southern Peru, and southern Bolivia, which have thus far been little drawn on because of their inaccessibility. English financial interests control most of these South American deposits.

The only large European producer of borax is Turkey. Italy and Germany produce small amounts. There has also been small production of borax in Thibet, brought out from the mountains on sheep-back.

The United States supplies of borax are sufficient for all domestic requirements and probably for export. Small quantities of boric acid are imported, but no borax in recent years. The domestic production comes entirely from California, though in the past deposits in Nevada and Oregon have also been worked.

GEOLOGIC FEATURES

The element boron is present in various complex boro-silicates, such as datolite and tourmaline, the latter of which is used as a precious stone (pp. 290, 293). None of these are commercial sources of borax. The principal boron minerals are borax or "tincal" (hydrated sodium borate), colemanite (hydrated calcium borate), ulexite (hydrated calcium-sodium borate), and boracite (magnesium chloro-borate). Commercially the term borax is sometimes applied to all these materials. These minerals appear in nature under rather widely differing modes of origin.

The borax production of Italy is obtained from the famous "soffioni" or "fumaroles" of Tuscany. These are volcanic exhalations, in which jets of steam carrying boric acid and various borates, together with ammonium compounds, emerge from vents in the ground. The boric acid material is recovered by a process of condensation.

Borates, principally in the form of borax, occur in hot springs and in lakes of volcanic regions. The Thibet deposits, and those formerly worked at Borax Lake, California, are of this type. Certain of the hot-spring waters of the California coast ranges and of Nevada carry considerable quantities of boron, together with ammoniacal salts, and in some places they deposit borax along with sulphur and cinnabar. It seems probable (see p. 40) that these waters may come from an igneous source not far beneath.

Most of the borax deposits of California, Nevada, and Oregon, though not at present the largely producing ones, and probably most of the Chilean and adjacent South American deposits, are formed by the evaporation of desert lakes. They are products of desiccation, and in Chile are associated with the great nitrate deposits (pp. 102-104), which are of similar origin. The salts contained in these deposits are mainly borax, ulexite, and colemanite. The sources of these materials are perhaps deposits of the type mentioned in the last paragraph, or, in California, certain Tertiary borate deposits described below. Whatever their source, the borates are carried in solution by the waters of occasional rains to shallow basins, which become covered with temporary thin sheets of water or "playa lakes." Evaporation of these lakes leaves broad flats covered with the white salts. These may subsequently be covered with drifting sands and capillary action may cause the borates to work up through the sands, becoming mixed with them and efflorescing at the surface. One of the largest of the California deposits of this general class is that at Searles Lake, from which it has been proposed to recover borax along with the potash (pp. 113-114).

The deposits which at present constitute the principal source of domestic borax are not the playa deposits just described, but are masses of colemanite in Tertiary clays and limestones with interbedded basaltic flows. The principal deposits are in Death Valley and adjacent parts of California. The colemanite occurs in irregular milky-white layers or nodules, mingled with more or less gypsum. The deposits are believed to be of the replacement type, rather than ones formed contemporaneously with the sediments. Whether they are due to magmatic solutions carrying boric acid from the associated flows, or to surface waters carrying materials leached from other sediments, is not clear. The crude colemanite as mined carries an average of about 25 per cent B_2O_3; it is treated with soda in the manufacture of borax, or with sulphuric acid in making boric acid.

Boron is present in minute quantities in sea water. When such water evaporates, it becomes concentrated, along with the magnesium and potassium salts, in the "mother liquor"; and upon complete evaporation, it crystallizes out as boracite and other rarer minerals. Thus the Stassfurt salts of Germany (p. 113) contain borates of this type in the carnallite zone of the upper part of the deposits. This is the only important case known of borate deposits of marine origin.

BROMINE

Economic Features

Bromine finds a considerable use in chemistry as an oxidizing agent, in separating gold from other metals, and in manufacturing disinfectants, bromine salts, and aniline colors. The best known and most widely used bromine salts are the silver bromide, used in photography, and the potassium bromide, used in medicine to depress the nervous system. During the war, large quantities of bromine were used in asphyxiating and lachrymating gases.

The chief center of the bromine industry in Europe prior to 1914 was Stassfurt, Germany. No other important commercial source in foreign countries is known, though small quantities have been obtained from the mother liquors of Chile saltpeter and from the seaweed, kelp, in various countries. India has been mentioned as a possible large producer in the future.

The United States is independent of foreign sources for bromine. The entire domestic tonnage is produced from brines pumped in Michigan, Ohio, West Virginia, and Pennsylvania. A large part of the output is not actually marketed as bromine, but in the form of potassium and sodium bromides and other salts. During the war considerable quantities of bromine materials were exported to Great Britain, France, and Italy.

GEOLOGIC FEATURES

Bromine is very similar chemically to chlorine, and is found under much the same conditions, though usually in smaller quantities. The natural silver bromide (bromyrite) and the combined silver chloride and bromide (embolite) are fairly common in the oxide zones of silver ores, but are not commercial sources of bromine.

Bromine occurs in sea water in appreciable amounts, as well as in some spring waters and many natural brines. When natural salt waters evaporate, bromine is one of the last materials to be precipitated, and the residual "mother liquors" or bitterns frequently show a considerable concentration of the bromine. Where complete evaporation takes place, as in the case of the Stassfurt salt deposits (p. 113), the bromine salts are crystallized out in the final stages along with the salts of sodium, magnesium, and potassium. The larger part of the world's bromine has come from the mother liquor resulting from the solution and fractional evaporation of these Stassfurt salts.

The bromine obtained from salt deposits in the eastern United States is doubtless of a similar origin. It is produced as a by-product of the salt industry, the natural or artificial brines being pumped from the rocks (p. 295), and the bromides being extracted either from the mother liquors or directly from the unconcentrated brines.

FULLER'S EARTH

Economic Features

Fuller's earth is used chiefly for bleaching, clarifying, or filtering mineral and vegetable oils, fats, and greases. The petroleum industry is the largest consumer. Minor uses are in the manufacture of pigments for printing wall papers, in detecting coloring matters in certain food-products, and as a substitute for talcum powder.

Fuller's earths are in general rather widely distributed. The principal producers are the United States, England, and the other large consuming countries of Europe. The only important international trade in this commodity consists of exports from the United States to various countries for treating mineral oils, and exports from England for treating vegetable oils.

There is a large surplus production in the United States of fuller's earth of a grade suitable for refining mineral oils, but an inadequate production of material for use in refining edible oils, at least by methods and equipment now in most general use. However, the imports needed from England are more than offset by our exports to Europe of domestic earth particularly adapted to the petroleum industry. Production in the United States comes almost entirely from the southern states; Florida produces over three-fourths of the total and other considerable producers are Texas, Georgia, California, and Arkansas. Imports from England are normally equivalent to about a third of the domestic production.

Geologic Features

Fuller's earth is essentially a variety of clay having a high absorptive power which makes it useful for decolorizing and purifying purposes. Fuller's earths are in general higher in water content and have less plasticity than most clays, but they vary widely in physical and chemical properties. Chemical analyses are of little value in determining whether a given clay will serve as fuller's earth, and an actual test is the only trustworthy criterion.

Deposits of fuller's earth may occur under the same variety of conditions as deposits of other clays. The deposits of Florida and Georgia consist of beds in slightly consolidated flat-lying Tertiary sediments, which are worked by open cuts. The Arkansas deposits are residual clays derived from the weathering of basic igneous rocks, and are worked through shafts.

GRAPHITE (PLUMBAGO)

ECONOMIC FEATURES

Crystalline graphite is used principally in the manufacture of crucibles for the melting of brass, bronze, crucible steel, and aluminum. About 45 per cent of the quantity and 70 per cent of the value of all the graphite consumed in the United States is employed in this manner. Both *crystalline* and *amorphous graphite* are used in lubricants, pencils, foundry facings, boiler mixtures, stove-polishes and paint, electrodes, and fillers or adulterants for fertilizers. The most important use of amorphous graphite is for foundry facings, this application accounting for about 25 per cent of the total United States consumption of graphite of all kinds. *Artificial graphite* is not suitable for crucibles or pencils but is adapted to meet other uses to which natural graphite is put. It is particularly adapted to the manufacture of electrodes.

The grade of graphite deposits varies widely, their utilization being largely dependent on the size of the grains and the ease of concentration. Some of the richest deposits, those of Madagascar, contain 20 per cent or more of graphite. The United States deposits contain only 3 to 10 per cent. The graphite situation is complicated by the differences in the quality of different supplies. Crucibles require coarsely crystalline graphite, but pencils, lubricants, and foundry facings may use amorphous and finely crystalline material.

The largest production of high-grade crucible graphite has come from Ceylon, under British control, and about two-thirds of the output has come to the United States. The mines are now worked down to water-level and costs are increasing.

In later years a rival supply has come from the French island of Madagascar, where conditions are more favorable to cheap production, and where reserves are very large. French, British, and Belgian interests are concerned in the development of these deposits. The quality of graphite is different from the Ceylon product; it has not found favor in the United States but is apparently satisfactory to crucible makers in Europe. Most of the output is exported to Great Britain and France, and smaller amounts to Germany and Belgium.

Less satisfactory supplies of crystalline graphite are available in many countries, including Bavaria, Canada, and Japan. Large deposits of crystalline material have been reported in Greenland, Brazil, and Roumania, but as yet have assumed no commercial importance.

Amorphous graphite is widely distributed, being produced in about twenty countries,—chiefly in Austria, Italy, Korea, and Mexico. Certain deposits have been found to be best for special uses, but most countries could get along with nearby supplies.

A large part of the world's needs of crucible graphite will probably continue to be met from Ceylon and Madagascar, while a large part of the amorphous graphite will come from the four sources mentioned.

The United States has been largely dependent upon importations from Ceylon for crucible graphite. Domestic supplies are large and capable of further development, but for the most part the flake is of such quality that it is not desired for crucible manufacture without large admixture of the Ceylon material. Restrictions during the war required crucible makers to use at least 20 per cent of domestic or Canadian graphite in their mixtures, with 80 per cent of foreign graphite. This created a demand for domestic graphite which caused an increased domestic output. Most of the production in the United States comes from the Appalachians, particularly from Alabama, New York, and Pennsylvania, and smaller amounts are obtained from California, Montana, and Texas. One of the permanently beneficial effects of the war was the improvement of concentrating practice and the standardization of output, to enable the domestic product to compete more effectively with the well-standardized imported grades. Whether the domestic production will hold its own with foreign competition under peace conditions remains to be seen. Domestic reserves are large but of low grade.

The Madagascar graphite, in the shape and size of the flakes, is more like the American domestic graphite than the Ceylon product. Small amounts have been used in this country, but American consumers appear in general to prefer the Ceylon graphite in spite of its greater cost. The Madagascar product can be produced and supplied to eastern United States markets much more cheaply than any other large supply; and, in view of the possible exhaustion of the Ceylon deposits, it may be desirable for American users to adapt crucible manufacture to the use of Madagascar material as has already apparently been done in Europe.

Expansion of the American graphite industry during the war, and its subsequent collapse, have resulted in agitation for a duty on imports of foreign graphite.

Amorphous graphite is produced from some deposits in the United States (Colorado, Nevada, and Rhode Island), but the high quality of Mexican graphite, which is controlled by a company in the United States, makes it likely that imports from this source will continue. Since the war the Mexican material has practically replaced the Austrian graphite in American markets. The output of Korea is divided between the United States and England.

Artificial graphite, in amounts about equal to the domestic production of amorphous graphite, is produced from anthracite or petroleum coke at Niagara Falls.

GEOLOGIC FEATURES

The mineral graphite is a soft, steel-gray, crystalline form of carbon.

Ceylon graphite occurs in veins and lenses cutting gneisses and limestones. Usually the veins consist almost entirely of graphite, but sometimes other minerals occur in important amounts, especially pyrite and quartz. The association of graphite with these minerals, and also with feldspar, pyroxene, apatite, and other minerals, suggests that the veins are of igneous origin, like some of the pegmatite veins in the Adirondacks of New York. The graphite is mined from open pits and shafts, and sorted by hand and mechanically. The product consists of angular lumps or chips with a relatively small amount of surface in proportion to their volume.

In Madagascar the graphite is mainly disseminated in a graphitic schist, though to some extent it is present in the form of veins and in gneiss. Most of the graphite is mined from a weathered zone near the surface, and the material is therefore soft and easily concentrated. The product is made up of flakes or scales, and in the making of crucibles requires the use of larger amounts of clay binder than the Ceylon graphite.

The flake graphite of the United States, principally in the Appalachian region, occurs in crystalline graphitic schists, resulting from the anamorphism of sedimentary rocks containing organic matter. Certain beds or zones of comparatively narrow width carry from 3 to 10 per cent of disseminated graphite. The graphite is recovered by mechanical processes of sorting. The graphite is believed to be of organic origin, the change from organic carbon to graphite having been effected by heat and pressure accompanying mountain-building stresses. Some of the graphite also occurs in pegmatite intrusives and adjacent wall rocks. This graphite is considered to be of inorganic origin, formed by the breaking up of gaseous oxides of carbon in the original magma of the pegmatites. The Montana graphite is similar in origin. This inorganic graphite in pegmatite veins resembles Ceylon graphite, in breaking into large lumps and chips, but supplies are very limited.

Amorphous graphite is formed in many places where coal and other carbonaceous materials have undergone extreme metamorphism. It represents simply a continuation in the processes by which high grade coals are formed from plant matter (pp. 123-127). The Mexican deposits are of this type, and occur in beds up to 24 feet in thickness interbedded with metamorphosed sandstones.

In general, graphite is primarily concentrated both by igneous processes in dikes, and by sedimentary processes in beds. In the latter case anamorphism is necessary to recrystallize the carbon into the form of graphite.

GYPSUM

ECONOMIC FEATURES

The principal use of gypsum is in structural materials. About two-thirds of the gypsum produced in the United States is used in the manufacture of various plasters—wall plaster, plaster of Paris, and Keene's cement (for statuary and decorative purposes),—and about a fifth is used as a retarder in Portland cement. Another important structural use is in the manufacture of plaster boards, blocks, and tile for interior construction. Gypsum is used as a fertilizer under the name of "land plaster," and with the growing recognition of the lack of sulphur in various soils an extension of its application is not unlikely. Minor uses are in the polishing of plate glass, in the manufacture of dental plaster, in white pigments, in steampipe coverings, and as a filler in cotton goods.

The world's gypsum deposits are widely distributed. Of foreign countries, France, Canada, and the United Kingdom are the principal producers. Germany, Algeria, and India produce comparatively meager amounts. The United States is the largest producer of gypsum in the world. In spite of its large production, the United States normally imports quantities equivalent to between one-fifteenth and one-tenth of the domestic production, mainly in the crude form from Nova Scotia and New Brunswick for consumption by the mills in the vicinity of New York. This material is of a better grade than the eastern domestic supply, and is cheaper than the western supply for eastern consumption. During the war this importation was practically stopped because of governmental requisition of the carrying barges for the coal-carrying trade, but with the return of normal conditions it was resumed. There is no prospect of importation of any considerable amount from any other sources. The domestic supply is ample for all demands.

Production of gypsum in the United States comes from eighteen states. Four-fifths of the total comes from New York, Iowa, Michigan, Ohio, Texas, and Oklahoma. There are extensive deposits in some of the western states, the known reserves in Wyoming alone being sufficient for the entire world demands for many decades.

The United States exports a small amount of crude gypsum to Canada, principally for use in Portland cement manufacture. This exportation is due to geographic location. The United States is the largest manufacturer of plaster boards, insulating materials, and tile, and exports large quantities of these products to Cuba, Australia, Japan, and South America.

GEOLOGIC FEATURES

Gypsum is a hydrated calcium sulphate. It is frequently associated with minor quantities of anhydrite, which is calcium sulphate without water, and under the proper natural conditions either of these materials may be changed into the other.

Common impurities in gypsum deposits include clay and lime carbonate, and also magnesia, silica, and iron oxide. In the material as extracted, impurities may range from a trace to about 25 per cent. *Gypsite*, or gypsum dirt, is an impure mixture of gypsum with clay or sand found in Kansas and some of the western states; it is believed to have been produced in the soil or in shallow lakes, by spring waters carrying calcium sulphate which was leached from gypsum deposits or from other rocks.

Gypsum deposits, like deposits of common salt, occur in beds which are the result of evaporation of salt water. Calcium makes up a small percentage of the dissolved material in the sea, and when sea waters are about 37 per cent evaporated it begins to be precipitated as calcium sulphate. Conditions for precipitation are especially favorable in arid climates, in arms of the sea or in enclosed basins which may or may not once have been connected with the sea. Simultaneously with the deposition of gypsum, there may be occasional inwashings of clay and sand, and with slight changes of conditions organic materials of a limey nature may be deposited. Further evaporation of the waters may result in the deposition of common salt. Thus gypsum beds are found interbedded with shales, sandstones, and limestones, and frequently, but not always, they are associated with salt beds. The nature of these processes is further discussed under the heading of salt (pp. 295-298).

The anhydrite found in gypsum deposits is formed both by direct precipitation from salt water and by subsequent alteration of the gypsum. The latter process involves a reduction of volume, and consequently a shrinkage and settling of the sediments. The hydration of anhydrite to form gypsum, on the other hand, involves an increase of volume and may result in the doming up and shattering of the overlying sediments.

Gypsum is fairly soluble in ground-water, and sink-holes and solution cavities are often developed in gypsum deposits. These may allow the inwash of surface dirt and also may interfere with the mining.

All the important commercial gypsum deposits are believed to have been formed by evaporation of salt water in the manner indicated. Small quantities of gypsum are formed also when pyrite and other sulphides oxidize to sulphuric acid and this acid acts on limestone. Thus gypsum is found in the oxide zones of some ore bodies. These occurrences are of no commercial significance.

MICA

Economic Features

The principal use of sheet mica is for insulating purposes in the manufacture of a large variety of electrical equipment. The highest grades are employed particularly in making condensers for magnetos of automobile and airplane engines and for radio equipment, and in the manufacture of spark plugs for high tension gas engines. Sheet mica is also used in considerable amounts for glazing, for heat insulation, and as phonograph diaphragms. Ground mica is used in pipe and boiler coverings, as an insulator, in patent roofing, and for lubricating and decorative purposes.

India, Canada, and the United States are the important sheet mica-producing countries, before the war accounting for 98 per cent of the world's total. India has long dominated the sheet mica markets of the world, and will probably continue to supply the standard of quality for many years. The bulk of the Indian mica is consumed in the United States, Great Britain, and Germany. The mica of India and the United States is chiefly muscovite. Canada is the chief source of amber mica (phlogopite), though other deposits of potential importance are known in Ceylon and South Africa. Canadian mica is produced chiefly in Quebec and Ontario, and is exported principally to the United States.

Important deposits of mica (principally muscovite) are also known in Brazil, Argentina, and German East Africa. Large shipments were made from the two former countries during the war, both to Europe and the United States, and Brazil particularly should become of increasing importance as a producer of mica. The deposits in German East Africa were being quite extensively developed immediately before the war and large shipments were made to Germany in 1913.

The United States is the largest consumer of sheet mica and mica splittings, absorbing normally nearly one-half of the world's production. Approximately three-fifths of this consumption is in the form of mica splittings, most of which are made from muscovite in India and part from amber mica in Canada. Due to the cheapness of labor in India and the amenability of Indian mica to the splitting process, India splittings should continue to dominate the market in this country. Amber mica is a variety peculiarly adapted to certain electrical uses. There are no known commercial deposits of this mica in the United States, but American interests own the largest producing mines in Canada. Shipments of Brazilian mica are not of such uniformly high quality as the Indian material, but promise to become of increasing importance in American markets.

Of the sheet mica consumed annually, the United States normally produces about one-third. War conditions, although stimulating the production of domestic mica very considerably, did not materially change the situation in this country as regards the dependence of the United States on foreign supplies for sheet mica.

About 70 per cent of the domestic mica comes from North Carolina and 25 per cent from New Hampshire. The deposits are small and irregular, and mining operations are small and scattered. These conditions are largely responsible for the heterogeneous nature of the American product. It is hardly possible for any one mine to standardize and classify its product, although progress was made in this direction during the war by the organization of associations of mica producers. This lack of standardization and classification is a serious handicap in competition with the standard grades and sizes which are available in any desired amounts from foreign sources.

For ground mica, the domestic production exceeds in tonnage the total world production of sheet mica, and is adequate for all demands.

Geologic Features

Mica is a common rock mineral, but is available for commerce only in igneous dikes of a pegmatite nature, where the crystallization is so coarse that the mica crystals are exceptionally large. Muscovite mica occurs principally in the granitic pegmatite dikes. The phlogopite mica of Canada occurs in pyroxenite dikes. The distribution of mica within the dikes is very erratic, making predictions as to reserves hazardous. The associated minerals, mainly quartz and feldspar, are ordinarily present in amounts greater than the mica. Also, individual deposits are likely to be small. For these reasons mining operations cannot be organized on a large scale, but are ordinarily hand-to-mouth operations near the surface. A large amount of hand labor is involved, and the Indian deposits are favored by the cheapness of native labor. The output of a district is from many small mines rather than from any single large one.

Pegmatites which have been subjected to dynamic metamorphism are often not available as a source of mica, because of the distortion of the mica sheets.

The mining of a mica is facilitated by weathering, which softens the associated feldspar, making it an easier task to take out the mica blocks. On the other hand, iron staining by surface solutions during weathering may make the mica unfit for electrical and certain other uses.

Scrap or ground mica is obtained as a by-product of sheet mica and from deposits where the crystals are not so well developed. Black mica (biotite) and chlorite minerals, which are soft and flexible but not elastic and are found extensively developed in certain schists, have been used to a limited extent for the same purposes.

MONAZITE (THORIUM AND CERIUM ORES)

ECONOMIC FEATURES

The mineral monazite is the source of the thorium and cerium compounds which, glowing intensely when heated, form the light-giving material of incandescent gas mantles. Welsbach mantles consist of about 99 per cent thorium oxide and 1 per cent cerium oxide. Cerium metal, alloyed with iron and other metals, forms the spark-producing alloys used in various forms of gas lighters and for lighting cigars, cigarettes, etc. Mesothorium, a by-product of the manufacture of thorium nitrate for gas mantles, is used as a substitute for radium in luminous paints and for therapeutic purposes. The alloy ferrocerium is used to a small extent in iron and steel.

The world's supply of monazite is obtained mainly from Brazilian and Indian properties. Before the war German commercial interests controlled most of the production, as well as the manufacture of the thorium products. During the war German control was broken up.

The United States has a supply of domestic monazite of lower grade than the imports, but is dependent under normal conditions on supplies from Brazil and India. The American deposits are chiefly in North and South Carolina, and have been worked only during periods of abnormally high prices or of restriction of imports. Known reserves are small and the deposits will probably never be important producers. During the war, however, the United States became the largest manufacturer of thorium nitrate and gas mantles and exported these products in considerable quantity. An effort is now being made to secure protective legislation against German thorium products.

GEOLOGIC FEATURES

Monazite is a mineral consisting of phosphates of cerium, lanthanum, thorium, and other rare earths in varying proportions. The content of thorium oxide varies from a trace up to 30 per cent, and commercial monazite sands are usually mixed so as to bring the grade up to at least 5 per cent.

Yellowish-brown crystals of monazite have been found scattered through granites, gneisses, and pegmatites, but in quantities ordinarily too small to warrant mining. In general the mineral is recovered on a commercial scale only from placers, where it has been concentrated along with other dense, insoluble minerals such as zircon, garnet, ilmenite, and sometimes gold. The Indian and Brazilian monazite is obtained principally from the sands of ocean beaches, in the same localities from which zircon is recovered (p. 189). The North and South Carolina monazite has been obtained chiefly from stream beds, and to a slight extent by mining and washing the rotted underlying rock, which is a pegmatized gneiss. Monazite, together with a small amount of gold, is also known in the stream gravels of the Boise Basin, Idaho, where a large granitic batholith evidently carries the mineral sparsely distributed throughout. These deposits have not been worked.

PRECIOUS STONES

Economic Features

Precious stones range high in the world's annual production of mineral values. A hundred or more minerals are used to some degree as precious stones; but those most prized, representing upwards of 90 per cent of the total production value, are diamond, pearl, ruby, sapphire, and emerald. In total value the diamonds have an overwhelming dominance. Over a ton of diamonds is mined annually.

Diamonds come mainly from South Africa, which produces over 99 per cent of the total. Pearls come chiefly from the Indian and Pacific oceans. Burma is the principal source of fine rubies. Siam is the principal producer of sapphires. Colombia is the principal source of fine emeralds.

The United States produces small amounts of sapphires (in Montana) and pearls (from fresh-water molluscs). Diamonds, rubies, and emeralds are practically absent on a commercial scale. Of other precious and semi-precious gem stones produced in the United States, the principal ones are quartz, tourmaline, and turquoise.

On the other hand, the United States absorbs by purchase over half of the world's production of precious stones. It is estimated roughly that there are now in the United States nearly one billion dollars' worth of diamonds, or over one-half of the world's accumulated stock, and probably the proportions for the other stones are not far different.

Value attaches to a precious stone because of its qualities of beauty, coupled with endurance and rarity, or because of some combination of these features which has caught the popular fancy. No one of these qualities is sufficient to make a stone highly prized; neither does the possession of all of them insure value. Some beautiful and enduring stones are so rare that they are known only to collectors and have no standard market value. Others fail to catch the popular fancy for reasons not obvious to the layman. While the intrinsic qualities go far in determining the desirability of a stone, it is clear that whim and chance have been no small factors in determining the demand or lack of demand for some stones. As in other minerals, value has both its intrinsic and extrinsic elements.

For the leading precious stones above named, the values are more nearly standard throughout the world than for any other minerals, with the exception of gold and possibly platinum. Highly prized everywhere and easily transported, the price levels show comparatively little variation over the world when allowance is made for exchange and taxes. The valuation of precious stones is a highly specialized art, involving the appraisal not only of intrinsic qualities, but of the appeal which the stone will make to the buying public. In marking a sale price for some exceptional stone not commonly handled in the trade, experts in different parts of the world often reach an almost uncanny uniformity of opinion.

It is estimated that the world stock of precious stones approximates three billion dollars, or a third of the world's monetary gold reserve. Because of small bulk and standard value, this wealth may be easily secreted, carried, and exchanged. When the economic fabric of civilization is disturbed by war or other conditions, precious stones become a medium of transfer and exchange of wealth of no inconsiderable importance.

The beauty of a stone may arise from its color or lack of color, from its translucency or opaqueness, from its high refraction of light, and from the manner of cutting and polishing to bring out these qualities. Hardness and durability are desirable qualities.

The diamond is the hardest known mineral and the sapphire, ruby, and emerald rank high in this regard. On the other hand the pearl is soft and fragile and yet highly prized.

GEOLOGIC FEATURES

The principal precious stones above named are of simple composition. Diamond is made of carbon; the pearl is calcium carbonate; ruby and sapphire are aluminum oxide—varieties of the mineral corundum; the emerald is silica and alumina, with a minor amount of beryllia. Minute percentages of chromite, iron, manganese, and other substances are often responsible for the colors in these stones. Carbon also constitutes graphite and is the principal element in coal. Lime carbonate is the principal constituent of limestone and marble. Alumina is the principal constituent of bauxite, the ore of aluminum, and of the natural abrasives, emery and corundum. Silica, the substance of common quartz, also constitutes gem quartz, amethyst, opal, agate, onyx, etc.

Most of the world's diamonds come from the Kimberley and Transvaal fields of South Africa, where they are found in a much decomposed volcanic rock called "blue ground." This is a rock of dull, greasy appearance consisting largely of serpentine. It was originally peridotite, occurring in necks or plugs of old volcanoes penetrating carbonaceous sediments. When the rock is mined and spread at the surface, it decomposes in the course of six months or a year, allowing it to be washed and mechanically sorted for its diamond content. The amount of ground treated in one of the large mines is about equal to that handled in operating the huge porphyry copper deposit of Bingham, Utah; the annual production of diamonds from the same mine could be carried in a large suit-case.

The diamonds were clearly formed at high temperatures and pressures within the igneous rocks. It has been suggested that the igneous magma may have secured the carbon by the melting of carbonaceous sediments through which it penetrated, but proof of this is difficult to obtain. Artificial diamonds of small size have been made in the electric furnace under high-pressure conditions not unlike those assumed to have been present in nature.

Weathering and transportation of rocks containing diamonds have resulted in the development of diamond-bearing placers.

The South African diamonds were first found in stream placers, leading to a search for their source and its ultimate discovery under a blanket of soil which completely covered the parent rock. The proportion of diamonds now mined from placers is very small.

The diamonds of Brazil come from placer deposits. This is the principal source of the black diamond so largely used in diamond-drilling.

The United States produces no diamonds on a commercial scale. Small diamonds have been found in peridotite masses in Pike County, Arkansas, but these are of very little commercial value. A few diamonds have been found in the glacial drift of Wisconsin and adjacent states, indicating a possible diamond-bearing source somewhere to the north which has not yet been located (p. 317).

Pearls are concretions of lime carbonate of organic origin, and are found in the shells of certain species of molluscs. Their color or luster is given by organic material or by the interior shell surface against which the pearl is formed. The principal supply comes from the Indian and Pacific Oceans, but some are found in the fresh water mussels of North America, in the Caribbean, and on the western coast of Mexico and Central America.

From the beginning of history the principal source of rubies has been upper Burma, where the stones are found in limestone or marble near the contact with igneous rocks, associated with high-temperature minerals. The weathering of the rock has developed placers from which most of the rubies are recovered. Siam is also an important producer. In the United States rubies have been found in pegmatites in North Carolina, but these gems are of little commercial importance.

Sapphires are of the same composition as rubies and are found in much the same localities. Most of the sapphires of the best quality come from Siam, where they are found in sandy clay of placer origin. In the United States sapphires are recovered from alluvial deposits along the Missouri River near Helena, Montana, where they are supposed to have been derived from dikes of andesite rocks. In Fergus County, Montana, they are mined from decomposed dikes of lamprophyre (a basic igneous rock). In North Carolina sapphire has been found in pegmatite dikes.

The principal source of fine emeralds is in the Andes in Colombia. Their occurrence here is in calcite veins in a bituminous limestone, but little seems to be known of their origin. The only other emerald locality of commercial importance is in the Ural Mountains of Siberia. Emeralds have been found in pegmatite dikes in North Carolina and New England, but the production is insignificant.

Tourmaline is a complex hydrous silicate of aluminum and boron, with varying amounts of magnesium, iron, and alkalies. It is a rather common mineral in silicated zones in limestones near igneous contacts, but gem tourmalines are found principally in pegmatite dikes. They have a wide variety of colors, the red and green gems being the most prized. Maine, California, and Connecticut are the principal American producers.

Turquoise is a hydrated copper-aluminum phosphate. It is found in veinlets near the surface in altered granites and other igneous rocks. It is usually associated with kaolin and frequently with quartz, and is believed to have been formed by surface alterations. In the United States it is produced chiefly in Nevada, Arizona, and Colorado.

In general the principal gem minerals, except pearl and turquoise, occur as original constituents in igneous intrusives, usually of a pegmatite or peridotite nature. Sapphire, ruby, emerald, and tourmaline result also from contact metamorphism of sediments in the vicinity of igneous rocks. Weathering softens the primary rocks, making it possible to separate the gem stones from the matrix. When eroded and transported the gems are concentrated in placers.

SALT

ECONOMIC FEATURES

The principal uses of salt are in the preserving and seasoning of foods and in chemical industries. Chemical industries require salt for the manufacture of many sodium compounds, and also as a source of hydrochloric acid and chlorine. A minor use of salt is in the making of glazes and enamel on pottery and hardware.

Because of the wide distribution of salt in continental deposits and because of the availability of ocean and salt-lake brines as other sources, most countries of the world either possess domestic supplies of salt adequate for the bulk of their needs, or are able to obtain supplies from nearby foreign countries. Certain sea salts preferred by fish packers and other users are, however, shipped to distant points. About a fifth of all the salt consumed in the world annually is produced in the United States, and other large producers are Great Britain, Germany, Russia, China, India, and France.

The United States produces almost its entire consumption of salt, which is increasing at a very rapid rate. Salt is produced in fourteen states, but over 85 per cent of the total output comes from Michigan, New York, Ohio, and Kansas. Reserves are practically inexhaustible.

Exports and imports of salt form a very minor part of the United States industry, each being equivalent to less than 5 per cent of the domestic production. A large part of the imported material is coarse solar-evaporated sea salt, which is believed by fish and pork packers to be almost essential to their industry. Imports of this salt come from Spain, Italy, Portugal, and the British and Dutch West Indies; during the war, on account of ship shortage, they were confined chiefly to the West Indies. A considerable tonnage of specially prepared kiln-dried salt, desired by butter-makers, is imported from Liverpool, England. There are also some small imports from Canada, probably because of geographic location. Exports of domestic salt go chiefly to Canada, Cuba, and New Zealand, with smaller amounts to practically all parts of the world.

Salt is recovered from salt beds in two ways. About a fourth of the salt produced in the United States is mined through shafts in the same manner as coal, the lumps of salt being broken and sized just as coal is prepared for the market. The larger part of the United States production, however, is derived by pumping water down to the beds to dissolve the salt, and pumping the resulting brine to the surface where it is then evaporated. A considerable amount of salt, also, is recovered from natural brines—which represent the solution of rock salt by ground-waters—and from the waters of salt lakes and the ocean.

Common salt constitutes the mineral halite, the composition of which is sodium chloride. It is rarely found perfectly pure in nature, but is commonly mixed with other saline materials, such as gypsum and anhydrite, and occasionally with salts of potassium and magnesium. The general grade of rock-salt deposits, where not admixed with clay, is perhaps 96 to 99 per cent of sodium chloride.

The ultimate source of salt deposits is the sodium and chlorine of igneous rocks. In the weathering of these rocks the soda, being one of the more soluble materials, is leached out and carried off by ground-waters, and in the end a large part of it reaches the sea. The chlorine follows a similar course; however, the amount of chlorine in ordinary igneous rocks is so extremely small that, in order to explain the amount of chlorine present in the sea, it has been thought necessary to appeal to volcanic emanations or to some similar agency. Ocean water contains about 3.5 per cent by weight of dissolved matter, over three-fourths of which consists of the constituents of common salt. Chief among the other dissolved materials are magnesium, calcium, potassium, and SO_4 (the sulphuric acid radical).

When sea water evaporates it becomes saturated with various salts, according to the amounts of these salts present and their relative solubilities. In a general way, after 37 per cent of the water has evaporated gypsum begins to separate out, and after 93 per cent has evaporated common salt begins to be deposited. After a large part of the common salt has been precipitated, the residual liquid, called a "bittern" or "mother liquid," contains chiefly a concentration of the salts of magnesium and potassium. Still further evaporation will result in their deposition, mainly as complex salts like those found in the Stassfurt deposit (p. 113).

The actual processes of concentration and precipitation in sea water or other salt waters are much more complex than is indicated by the above simple outline. The solubility of each of the various salts present, and consequently the rate at which each will crystallize out as evaporation proceeds, depends upon the kinds and concentrations of all the other salts in the solution. Temperature, pressure, mass-action, and the crystallization of double salts are all factors which influence the nature and rate of the processes and add to their complexity. During a large part of the general process, several different salts may be crystallizing out simultaneously. It is evident that gypsum may be precipitated in some quantity, and that external conditions may then change, so that evaporation ceases or so that the waters are freshened, before any common salt is crystallized out. This fact may explain in part why gypsum beds are more widely distributed than beds of common salt. At the same time the much greater amount of sodium chloride than of calcium sulphate in sea water may explain the greater thickness of many individual salt beds.

The evaporation of salt waters, either from the ocean or from other bodies of water, is believed to have been responsible for nearly all of the important deposits of common salt. This process has been going on from Cambrian time down through all the intervening geologic ages, and can be observed to be actually operative today in various localities. The beds of salt so formed are found interstratified with shales, sandstones, and limestones, and are frequently associated with gypsum. On a broad scale, they are always lens-shaped, though they vary greatly in extent and thickness.

The necessary conditions for the formation of extensive salt beds include arid climate and bodies of water which are essentially enclosed—either as lakes, as lagoons, or as arms of the sea with restricted outlets,—where evaporation exceeds the contributions of fresh water from rivers, and where circulation from the sea is insufficient to dilute the water and keep it at the same composition as the sea water. Under such conditions the dissolved salts in the enclosed body become concentrated, and precipitation may occur. A change of conditions so that mud or sand is washed in or so that calcareous materials are deposited, followed by a recurrence of salt-precipitation, results in the interstratification of salt beds with shales, sandstones, and limestones.

For the formation of very thick beds of salt, and especially of thick beds of fairly pure composition, however, this simple explanation of conditions is insufficient. The deposits of Michigan and New York occur in beds as much as 21 feet in thickness, with a considerable number of separate beds in a section a few hundred feet thick. Beneath the potash salt deposits of Stassfurt, beds of common salt 300 to 500 feet in thickness are found, and beds even thicker are known in other localities. When we come to investigate the volume of salts deposited from a given volume of sea water, we find it to be so small that for the formation of 500 feet of salt over a given area, an equivalent area of water 25,000 feet deep would be required. It has therefore been one of the puzzling problems of geology to determine the exact physical conditions under which deposition of these beds took place.

One of the most prominent theories, the "bar" theory, suggests that deposition may have taken place in a bay separated from the sea by a bar. Sea water is supposed to have been able to flow in over the bar or through a narrow channel, so that evaporation in the bay was about balanced by inflow of sea water. Thus the salts of a very large quantity of sea water may have accumulated in a small bay. As the process went on, the salts would become progressively more concentrated, and would be precipitated in great thickness. A final complete separation of the basin from the sea, for instance by the relative elevation of the land, might result in complete desiccation, and deposition of potassium-magnesium salts such as those found at Stassfurt (p. 113).

Another suggestion to explain the thickness of some salt beds is that the salts in a very large basin of water may, as the water evaporated and the basin shrank, have been deposited in great thickness in a few small depressions of the basin.

Other writers believe that certain thick salt deposits were formed in desert basins (with no necessary connection with the sea), through the extensive leaching of small quantities of salt from previous sediments, and its transportation by water to desert lakes, where it was precipitated as the lakes evaporated. Over a long period of time large amounts of salt could accumulate in the lakes, and thick deposits could result. Such hypotheses also explain those cases where common salt beds are unaccompanied by gypsum, since land streams can easily be conceived to have been carrying sodium chloride without appreciable calcium sulphate; in ocean waters, on the other hand, so far as known both calcium sulphate and sodium chloride are always present, and gypsum would be expected to accompany the common salt.

A partial explanation of some great thicknesses found in salt beds is that these beds, especially when soaked with water, are highly plastic and incompetent under pressure. In the deformation of the enclosing rocks, the salt beds will flow somewhat like viscous liquids, and will become thinned on the limbs of the folds and correspondingly thickened on the crests and troughs.

The salt deposits of the Gulf Coast of Texas and Louisiana should be referred to because of their exceptional features. They occur in low domes in Tertiary and more recent sands, limestones, and clays. Vertical thicknesses of a few thousand feet of salt have been found, but the structure is known only from drilling. In some of these domes are also found petroleum, gypsum, and sulphur (p. 110). No igneous rocks are known in the vicinity. It has been thought by some that the deposits were formed by hot waters ascending along fissures from underlying igneous rocks, and the upbowing of the rocks has been variously explained as due to the expanding force of growing crystals, to hydrostatic pressure of the solutions, and to laccolithic intrusions. On the other hand, the uniform association of other salt and gypsum deposits with sedimentary rocks, and the absence of igneous rocks, suggest that these deposits may have had essentially a sedimentary origin, and that they have been modified by subsequent deformation and alteration. The origin is still uncertain.

Other mineral deposits formed under much the same conditions as salt are gypsum, potash, borax, nitrates, and minerals of bromine; and in a study of the origin of salt deposits these minerals should also be considered.

TALC AND SOAPSTONE

ECONOMIC FEATURES

Soapstone is a rock composed mainly of the mineral talc. Popularly the terms *talc* and *soapstone* are often used synonymously. The softness, greasy feel, ease of shaping, and resistance to heat and acids of this material make it useful for many purposes. Soapstone is cut into slabs for laundry tubs, laboratory table tops, and other structural purposes. Finer grades are cut into slate pencils and acetylene burners. Ground talc or soapstone is used as a filler for paper, paint, and rubber goods, and in electrical insulation. Fine grades are used for toilet powder.

Pyrophyllite (hydrated aluminum silicate) resembles talc in some of its properties and is used in much the same way. Fine English clays (p. 85) are sometimes used interchangeably with talc as paper filler.

The United States produces nearly two-thirds of the world's talc. The other large producers are France, Italy, Austria, and Canada (Ontario).

The United States is independent of foreign markets for the bulk of its talc consumption, but some carefully prepared talc of high quality is imported from Canada, Italy, and France. Italy is our chief source of talc for pharmaceutical purposes, though recently these needs have been largely supplied by high-grade talc from California. In the United States, Vermont and New York are the leading producers of talc and Virginia of soapstone slabs. Reserves are large.

Geologic Features

Talc is hydrated magnesium silicate, as is also serpentine, a mineral with which talc is closely associated. Both are common alteration products of magnesian silicate minerals such as olivine, pyroxene, and amphibole. Talc is also derived from the recrystallization of magnesian carbonates.

Talc deposits consist of lenses and bands in metamorphic limestones, schists, and gneisses of ancient age. The talc itself is usually schistose like the wall rocks, and is largely a product of mechanical mashing. In some cases, also, talc results from the alteration of igneous rocks without mashing—as in the case of the large talc and soapstone deposits of Virginia, which are the result of rather complete alteration of basic igneous rocks such as peridotites and pyroxenites.

Talc is known to result from the weathering of magnesian silicates under surface conditions, but the common occurrence of the principal deposits, in highly crystalline rocks which have undergone extensive deep-seated metamorphism, is an indication that processes other than weathering have been effective. It has been suggested that hot ascending solutions have been responsible for the work, but without much proof. A more plausible explanation for many deposits is that the talc results from the dynamic metamorphism or shearing of impure magnesian carbonates (as in highly magnesian limestones), the process resulting in elimination of the carbon dioxide and recrystallization of the residue. Certain talc deposits, such as those of Ontario, show clearly traces of the original bedding planes of limestone crossing the cleavage of the talc, and the rock bears all the evidence of having formed in the same manner as a common slate. Talc and slate are almost the only mineral products which owe their value principally to dynamic metamorphism.

CHAPTER XIV

EXPLORATION AND DEVELOPMENT

THE GENERAL RELATIONS OF THE GEOLOGIST TO EXPLORATION AND DEVELOPMENT

The economic geologist is more vitally concerned with exploration and development than with any other phase of his work. This comes closest to being his special field. Here is a fascinating element of adventure and chance. Here is the opportunity to converge all his knowledge of geology and economics to a practical end. The outcome is likely to be definite one way or the other, thus giving a quantitative measure of the accuracy of scientific thinking which puts a keen edge on his efforts. It is not enough merely to present plausible generalizations; scientific conclusions are followed swiftly either by proof or disproof. With this check always in mind, the scientist feels the necessity for the most rigid verification of his data, methods, and principles.

The general success of the application of geology to exploration and development is indicated by the rapid increase in demand for such service in recent years, and by the large part it plays in nearly all systematic and large-scale operations. The argument is sometimes made that many mineral deposits have been found without geologic assistance, and that therefore the geologist is superfluous. The answer to this argument is that there are often hundreds of "practical" explorers in the field to one geologist, and that in proportion to numbers the story is quite a different one. The very fact that many large mining organizations, as a result of their experience, now leave these matters of exploration and development largely in the hands of geologists, is a tribute to the usefulness of the science. Also, it is to be remembered that not all applications of geology are made by geologists. It is hard to find a prospector or explorer who has not absorbed empirically some of the elements of geology, and locally this may be enough. Very often men who take pride in the title of "practical prospectors" are the ones with the largest stock of self-made geological theories.

During a prospecting boom it is not uncommon for speculators and promoters to attempt to discount geologic considerations where these run counter to their plans. The catching phrase "bet against the geologist" has a broad appeal to an instinctive preference for the practical as opposed to the theoretical. If the public would stop to note the character of the support behind the geologist, including as it does the larger and more successful operators, it would not be so ready to accept this implication.

Another aspect of this question might be mentioned. There is scarcely an oil field or mining camp in the world without a cherished tradition to the effect that, prior to discovery, the mineral possibilities had been reported on unfavorably by the geologists,—again implying that success has been due to the hard common sense of the horny-handed prospector. These traditions persist in the face of favorable geological reports published before discovery; they are natural expressions of the instinctive distrust of any knowledge which is beyond the field of empirical experience. In many cases the discoveries were made long before geologists appeared on the scene. In others, possibly one or two geologic reports were unfavorable, while many were favorable. In the aggregate, there can be no question that, in proportion to the scale of its use, geological advice has had more than its proportion of success.

Even under the most favorable conditions, the chances against the success of an individual drill hole or underground development are likely to be greater than the chances for it. The geologist may not change this major balance; but if he can reduce the adverse chances by only a few per cent, his employment is justified on purely commercial grounds.

The above comments refer to sound geological work by competent scientists. The geologic profession, like many others, is handicapped by numbers of ill-trained men and by many who have assumed the title of geologist without any real claim whatever,—who may do much to discredit the profession. The very newness of the field makes it difficult to draw a sharp line between qualified and unqualified men. With the further development of the profession this condition is likely to be improved (see pp. 427-428).

So new is the large-scale application of geology to exploration and development, and so diverse are the scientific methods of approach, that it is difficult to lay out a specific course for a student which will prepare him for all the opportunities he may have later. In the writer's experience, both in teaching and practice, the only safe course for the student is to prepare broadly on purely scientific lines. With this background he will be able later to adapt himself to most of the special conditions met in field practice.

Partly Explored Versus Virgin Territories

In selecting an area to work, the geologic explorer will naturally consider various factors mentioned in succeeding paragraphs; but the natural first impulse is to start for some place where no one else has been, and to keep away from the older principal mining camps,—on the assumption that such grounds have been thoroughly explored and that their geological conditions bearing on exploration are fully understood. It is safe to say that very few mineral districts are thoroughly understood and explored. Numerous important discoveries of recent years have been in the extensions of old mines and old districts; and when one considers the scale of even the most extensive mine openings in comparison with the vast body of rock available for exploration, it is clear that this will continue to be the situation far into the future. It is the writer's belief that the economic geologist stands at least as good a chance of success in exploration in the older districts as he does in new fields. Nature is exceedingly erratic and economical in providing places favorable for mineral production; in a producing district the geologic conditions have been proved to be right, and the explorer starts here with this general pragmatic advantage. The explorer here has another great advantage, that much essential information has been gathered which can be built into his plan of operations. He can start, scientifically and practically, where the other man left off. One of the best-known economic geologists has maintained that the more previous work done, the better, because it furnished him more tools to work with. There is no such thing as "skimming the cream" from a geologic problem; there is no end in sight in the search for more knowledge.

This attitude toward the problem of exploration has also proved advantageous on the business or financial side. A successful backer of mineral enterprises once remarked that his best prospecting was done from the rear platform of a private car,—meaning that this mode of transportation had carried him to the center of important mining activities, where the chances for large financial success showed a better percentage than in more general and miscellaneous exploration.

The Use of All Available Information

Effective scientific exploration requires the use of all available information applying to the specific area. This might seem to be too obvious to require mention, yet observance of the methods of explorers seems to call for warning against the rather common tendency to go into a field unprepared with a thorough knowledge of preceding work. It is easy to forget or overlook some investigation made many years previously; or to assume that such work is out of date, and of no special consequence in the application of new thought and method which is the basis of the faith and confidence of each new geologic explorer. A study of the reports on an old camp shows how often the younger generations have ignored the results of the older. Many of the same elementary truths are rediscovered by successive generations, after large efforts which could have been saved by means of proper care and investigation of the previous literature and mapping.

In outlying parts of the world, the existing information bearing on exploration may be at a minimum. In many of the older mining camps and throughout most civilized countries, however, careful investigation will usually disclose a considerable range of useful information bearing on the territory to be explored. In the United States the natural course to be pursued is to hunt carefully through the reports of the U. S. Geological Survey, the Bureau of Mines, various state surveys, universities, and private organizations (so far as these reports are available), and through the technical journals and the reports of technical societies, for something bearing on the district to be explored. Even if no specific report or map is to be found, it is usually possible to locate general maps or accounts which are likely to be of use.

Coöperation in Exploration

Competition in exploration often develops an atmosphere of suspicion and furtiveness which is highly unfavorable to coöperative efforts. Individuals and companies may handicap themselves greatly by a desire to play a lone hand, and by failure to take advantage of an exchange of information. This action may be based, particularly on the part of strong mining companies, on the assumption that they know all that is necessary about the problem, and that an outsider has nothing to contribute. Financial and other conditions may require this attitude; but in large part it is a result of temperament, as clearly indicated by the difference in methods followed by different groups and in different mining districts. From the scientific point of view this attitude can hardly be justified, in view of the extremely narrow limits of human knowledge as compared with the scientific field to be explored. The sum total of knowledge from all sources is only a small fraction of that necessary for the most effective results. The mutual exchange of information and discussion is usually justified on the basis of self-interest alone, to say nothing of the larger interest to the mineral district, to the country, or to science.

National and state survey organizations exercise considerable effort to secure records of drilling. In some cases they have the legal power to command this information, particularly in relation to appraisals for taxation and "blue sky" laws. In a larger number of cases drill records are secured through voluntary coöperation with explorers. A considerable number of records are nevertheless not filed with public agencies and some of these are permanently lost. Even where the records are turned in to a public organization, they are in most cases not directly available to explorers.

Public registration of all drilling records is a highly desirable procedure in the interests of the development of the mineral industry as a whole. A vast amount of unnecessary duplication can thus be avoided. The record of a drill hole, even though barren, may be of vast significance in the interpretation of future developments and should be recorded as carefully as an abstract of land title. The property right of the explorer in such information can be and usually is protected by withholding the record from public inspection until sufficient time has elapsed to give him full opportunity to use the information to his own best advantage.

The opportunities for coöperation with specialists of public organizations are almost unlimited. These organizations are likely to have an accumulation of data and experience extending through long periods and over large areas, which the private explorer ordinarily cannot hope to duplicate. With proper restrictions this information may be available for public use. A good illustration of current coöperative effort of this kind is in the deep exploration for oil in the Trenton limestone of Illinois. Outcrops and other specific indications are not sufficient to localize this drilling; but the information along broad geologic and structural lines which has been collected previously by the Illinois Survey is sufficient so that, with a comparatively small amount of shallow drilling, the locus of the more favorable structural conditions may be determined. In this case the Survey is directing the initial exploration, which is financed by private capital.

ECONOMIC FACTORS IN EXPLORATION

The approach to the problem of exploration is very often determined by local requirements and conditions; but if one were to come at the problem from a distance and to keep matters in broad perspective, the first step would be a consideration of what might be called the *economic factors*. Let us suppose that the geologist is free to choose his field of exploration. An obvious preliminary step is to eliminate from consideration mineral commodities which are not in steady or large demand and are much at the mercy of market conditions, or which are otherwise not well situated commercially. The underlying factors are many and complex. They include the present nature and future possibilities of foreign competition, the domestic competition, the grades necessary to meet competition, the cost of transportation, the cost of mining under local conditions—including considerations of labor and climatic and topographic conditions,—the probability of increase or decrease in demand for the product, the possible changes in metallurgical or concentrating practice (such as those which made possible the mining of low-grade porphyry copper ores), the size of already available reserves, and the mining laws in relation to ownership and regulation. Most of these factors are discussed at some length on other pages. After looking into the economic conditions limiting the chromite, nickel, or tin developments in the United States, the explorer might hesitate to proceed in these directions,—for he would find that past experience shows little promise of quantities and grades equivalent to those available in other countries, and that there is little likelihood of tariffs or other artificial measures to improve the domestic situation. Before and during the war, commercial conditions might have shown the desirability of hunting for pyrite, but more recent developments in the situation cast some doubt on this procedure. To go ahead blindly in such a case, on the assumption that the pyrite market would in some fashion readjust itself, would not be reasoned exploration. Again, in considering exploration for copper, account should be taken in this country of the already large reserves developed far in advance of probable demand, which require that any new discoveries be very favorably situated for competition. In oil, on the other hand, a very brief survey of the economic factors of the situation indicates the desirability of exploration. The comparative shortage of lead supplies at the present time suggests another favorable field for exploration.

In short, before actual field exploration is begun, intelligent consideration of the economic factors may go far toward narrowing the field and toward converging efforts along profitable lines. Looked at broadly, this result is usually accomplished by the natural working of general laws of supply and demand; but there are many individual cases of misdirected effort, under the spell of provincial conditions, which might easily be avoided by a broader approach to the problem.

GEOLOGIC FACTORS IN EXPLORATION

Coming to the geological aspects of exploration, the procedure in its early stages is again one of elimination. Oil and coal, for instance, are found in certain sediments of certain ages, and one would not look for them in an area of granite. For every mineral resource there are broad geologic conditions of this sort, particularly the genetic, structural, and metamorphic conditions, which make it possible to eliminate vast areas from consideration and to concentrate on relatively small areas.

After the elimination of unfavorable areas, there comes the hunt for positively favorable geologic conditions—for a definite kind of sediment or igneous rock, for a definite structure, for the right kind of mineralogic and metamorphic conditions, or for the right combination of these and other geologic elements. The geologic considerations used in exploration for the various mineral deposits are so many and so diverse, and they require so much adjustment and interpretation in their local application, that one would be rash indeed to attempt anything in the nature of an exhaustive discussion. It is hardly practicable to do more than to outline, for illustrative purposes, a few of the geologic factors most commonly used in exploration.

Mineral Provinces and Epochs

Mineral deposits may be similar in their mineralogic and geologic characters and relations over a considerable area. They may give evidence of having developed under the same general conditions of origin; perhaps they may even be of the same geologic age. The gold-silver deposits of Goldfield, of Tonopah, the Comstock Lode of Virginia City, and many other deposits through the Great Basin area of the southwestern United States and Mexico have group characteristics which have led geologists to refer to this area as a "metallogenic" or "metallographic" province. The gold-silver ores on the west slope of the Sierra Nevadas, for nearly the entire length of California, likewise constitute a metallogenic province. The Lake Superior copper ores on the south shore of Lake Superior, the silver ores on the north shore, miscellaneous small deposits of copper, silver, and gold ores to the east of Lake Superior, the nickel ores of Sudbury, and the silver-nickel-cobalt ores of the Cobalt district are all characterized by similar groups of minerals (though in highly differing proportions), by similar geologic associations, by similar age, and probably by similar conditions of origin. This area is a metallogenic province. The lead and zinc ores of the Mississippi Valley constitute another such province. The oil pools of the principal fields are characterized by common geologic conditions over great areas (p. 149), which may likewise be considered as forming mineral provinces; for them the term "petroliferous provinces" has been used. The list might be extended indefinitely. Knowledge of such group distributions of minerals is a valuable asset to the explorer, in that it tends to localize and direct search for certain classes of ores in certain provinces; also, within a province, it tells the explorer what is to be normally expected as regards kinds and occurrences of mineral deposits. In searching for minerals of sedimentary origin, the explorer will use stratigraphic methods in following definite sedimentary horizons. In searching for ores related to igneous intrusions he will naturally hunt for the intrusions, and then follow the periphery of the intrusions for evidences of mineralization, taking into account possible features of zonal arrangement of minerals about the intrusives (see pp. 42-44), and the preference of the ores for certain easily replaced horizons like limestones, or for certain planes or zones of fracturing.

Just as minerals may be grouped by provinces, they may be grouped by geologic ages. Such groupings are especially useful in the case of minerals which are closely related to certain stratigraphic horizons, such as coal, oil, and iron. The greater number of the productive coal deposits of the United States are of Carboniferous age, and the distribution of sediments of this age is pretty well understood from general geologic mapping. The Clinton iron ores all follow one general horizon in the lower-middle Paleozoic. The Lake Superior iron ores are pre-Cambrian, and over three-fourths of them occur at one horizon in the pre-Cambrian. Gold deposits of the United States were formed mainly in the pre-Cambrian, the early Cretaceous, and the Tertiary. Copper deposits of the United States were formed chiefly in pre-Cambrian, Cretaceous, and Tertiary time. While there are many exceptions and modifications to general classifications of this sort, they seem to express essential geologic facts which can be made very useful in localizing exploration.

Classification of Mineral Lands

In recent years there has been considerable development of the practice of classifying mineral lands in given areas for purposes of exploration and valuation, or for purposes of formulation and administration of government laws. This has been done both by private interests and by the government. These classifications take into account all of the geologic and economic factors ascertainable. The classes of mineral land designated vary with the mineral, the district, and the purpose for which the classification is made.

Common procedure for commercial exploration purposes is to divide the lands of a given territory into three groups—(1) lands which are definitely promising for mineral exploration, (2) lands of doubtful possibilities, and (3) lands in which the mineral possibilities are so slight that they may be excluded from practical consideration. Each of these classes may be subdivided for special purposes. Another commonly used classification is, (1) proved mineral lands, (2) probable mineral lands, usually adjacent to producing mines, (3) possible mineral lands, and (4) commercially unpromising mineral lands.

The classification of the public mineral lands by government agencies is fully discussed by George Otis Smith and others in a bulletin of the United States Geological Survey.[37] The purposes, methods, and results of this classification should be familiar to every explorer. Nowhere else is there available such a vast body of information of practical value. Quoting from this report:

A study of the land laws shows the absolute necessity of some form of segregation of the lands into classes as a prerequisite to their disposition. Agricultural entry may not be made on lands containing valuable minerals, nor coal entry on lands containing gold, silver, or copper; lands included in desert entries or selected under the Carey Act must be desert lands; enlarged-homestead lands must not be susceptible of successful irrigation; placer claims must not be taken for their timber value or their control of watercourses; and lands included in building-stone, petroleum, or salt placers must be more valuable for those minerals than for any other purpose. So through the whole scheme of American land laws runs the necessity for determining the use for which each tract is best fitted.

For this purpose the Geological Survey has made extensive classification of coal lands, oil and gas lands, phosphate lands, lands bearing potash and related salines, metalliferous mineral lands, miscellaneous non-metalliferous mineral lands, and water resources. The scope of the work may be indicated by the factors considered. For instance coal is investigated in relation to its character and heat-giving qualities (whence comes its value), quantity, thickness, depth, and other conditions that effect the cost of its extraction. Metalliferous mineral lands are considered in relation to general geology, country rock, intrusions and metamorphism, structure, outcrops and float of lodes, prospects and mines, samples, and history of the region.

Classifications of this kind have often proved useful to large holders of land as a basis for intelligent handling of problems of sale, taxation, and the granting of rights to explorers. Because of the lack of this elementary information, there has been in some quarters timidity about dealing with large holdings, for fear of parting with possible future mineral wealth,—with the result that such tracts are carried at large expense and practically removed from the field of exploration. To the same cause may be attributed some of the long delays on the part of the government in opening lands for mineral entry or in issuing patents on land grants.

Outcrops of Mineral Deposits

Many mineral deposits have been found because they outcrop at the surface; the discoveries may have been by accident or they may have been aided by consideration of geologic factors. There are still vast unexplored areas in which mineral deposits are likely to be found standing out at the surface. For much of the world, however, the surface has been so thoroughly examined that the easy surface discoveries have been made, and the future is likely to see a larger application of scientific methods to ground where the outcrops do not tell an obvious story. Mineral deposits may fail to outcrop because of covering by weathered rock or soil, by glacial deposits, or by younger formations (surface igneous flows or sediments), or the outcrop of a deposit may be so altered by weathering as to give little clue to the uninitiated as to what is beneath. Mineral deposits formed in older geologic periods have in most cases been deeply covered by later sediments and igneous rocks. Such deposits are in reach of exploration from the surface only in places where erosion has partly or wholly removed the later covering. An illustration of this condition is furnished in the Great Basin district of Nevada, where ore bodies have been covered by later lava flows. The ore-bearing districts are merely islands exposed by erosion in a vast sea of lava and surface sediments. Beyond reasonable doubt many more deposits are so covered than are exposed, and it is no exaggeration to say that by far the greater part of the mineral wealth of the earth may never be found. Where a mineral-bearing horizon is exposed by erosion at the surface, underground operations may follow this horizon a long way below the capping rocks; but, after all, such operations are geographically small as compared with the vast areas over which the covering rocks give no clue as to what is beneath. One of the principal problems of economic geology for the future is to develop means for exploration in territories of this sort. A beginning has been made in various districts by the use of reconnaissance drilling, combined with interpretation of all the geologic and structural features. The discovery of one of the largest nickel deposits in the Sudbury district of Canada was made by reconnaissance drilling to ascertain the general geologic features, in an area so deeply covered as to give little suggestion as to the proper location for attack.

Some Illustrative Cases

The use of outcrops in oil exploration has been noted on other pages (pp. 146-147).

Outcrops of coal seams may be found in folded or deeply eroded areas. For the most part, however, and especially in areas of flat-lying rocks, the presence of coal is inferred from stratigraphic evidence and from the general nature of the geologic section—which has been determined by outcrops of associated rocks or by information available at some distant point. The structural mapping of coal beds on the basis of outcrops and drill holes has been referred to (pp. 126-127).

Iron ores are very resistant to solution. Where hard and compact they tend to form conspicuous outcrops, and where soft they may be pretty well covered by clay and soil. In glaciated areas, like the Lake Superior region, outcrops of iron ore are much less numerous because of the drift covering. Certain of the harder iron ores of the Marquette, Gogebic and Menominee districts of Michigan and of the Vermilion district of Minnesota project in places through the glacial drift, and these ores were the first and most easily found. Much the greater number of iron ore deposits of Lake Superior, including the great soft deposits of the Mesabi range of Minnesota, fail to outcrop. On the other hand the *iron formation*, or mother rock of the ore, is hard and resistant and outcrops are numerous. The hematite ores of Brazil have many features in common with the Lake Superior ores in age and occurrence, but they have not been covered with glacial deposits. Outcrops of the iron ore are large and conspicuous, and the surface in this territory gives one some idea of what the Lake Superior region may have looked like before the glaciers came along. Certain of the soft iron ores of the lateritic type, as in Cuba, outcrop over great areas where their topographic situation is such that erosion has not swept them off. On erosion slopes they are seldom found. The Clinton iron ores of the southeastern United States outcrop freely.

Some of the lead and zinc deposits of the Mississippi Valley outcrop at the grass roots as varying mixtures of iron oxide, galena, chert, and clay, though they seldom project above the general surface. The old lead ranges of Wisconsin and Illinois, found at the surface a century ago by the early explorers and traders, have served as starting points for deeper exploration which has located the zinc deposits. Erosion channels have freely exposed these ore bodies, and in the Wisconsin-Illinois deposits most of the ores thus far found are confined to the vicinity of these channels. The greater number of the lead and zinc deposits of the Mississippi Valley, however, are covered with weathered material or with outliers of overlying sediments, with the result that underground exploration is necessary to locate them.

Sulphide deposits in general, including those carrying gold, silver, copper, lead, zinc, and other metals, have many common features of outcrop. The iron sulphide commonly present in these ore bodies is oxidized to limonite at the surface, with the result that prospectors look for iron-stained rocks. These iron-stained rocks are variously called the "gossan," the "iron capping," the "colorado," or the "eiserner Hut" (iron hat). The gossan is likely to resist erosion and to be conspicuous at the surface,—though this depends largely on the relative resistance of the wall rocks, and on whether the gangue is a hard material like quartz, or some material which weathers more rapidly like limestone or igneous rock. The gossan does not often carry much value, though it may show traces of minerals which suggest what may be found below. Gold, silver, and lead are not easily leached out of the surface outcrops. Copper and zinc are much more readily leached, and in the outcrop may disclose their existence only by traces of staining. It happens not infrequently, therefore, that copper and zinc deposits are found through the downward exploitation of oxidized gold, silver, and lead ores. The veins at Butte were first worked for silver, and the ore bodies at Bingham, Utah, and Jerome, Arizona, were first mined for gold. Exceptionally, copper ore in enriched, oxidized form outcrops, as at Bisbee, Arizona.

It is not always true that valuable sulphide deposits have an iron-stained outcrop, for in some of them iron sulphide or pyrite is so scarce that the surface outcrops may be light-colored clayey and siliceous rocks.

Silver is often represented in the outcrop by silver chloride or cerargyrite, which may be easily identified. The prospecting for such surface ores is sometimes called "chloriding."

The presence in the outcrop of dark manganese oxides associated with vein quartz sometimes indicates the presence below of copper and zinc and other minerals, as at Butte.

Extensive alterations of the country rock in the way of silicification and sericitization, and the presence of minerals like garnet, tourmaline, diopside, and others, known to be commonly deposited by the same hot solutions which make many ore deposits, may furnish a clue for exploration below. These characteristics of the country rock, however, are likely to be masked at the outcrop by later weathering, which superposes a kaolinic or clayey alteration.

Topography and Climate As Aids in Searching for Mineral Outcrops

The topographic expression of a mineral deposit depends upon its hardness and resistance to erosion as compared with the adjacent rocks. If more resistant it will stand out at the surface; if less resistant, it will form a depression. The conditions determining resistance are exceedingly variable, and no broad generalization can be made; but within a local province a given group of mineral deposits may characteristically form depressions or ridges, and thus topographic criteria may be very useful in exploration. Even with such limitations, the variations of the topographic factor may be so great as to require much care in its use. Sulphide ores in quartzites are likely to develop depressions under erosion. In limestones they are more likely to stand out in relief, because of the softer character of the limestone, though this does not always work out. Crystalline magnetite and hematite are more resistant to erosion than almost any other type of rock, and stand out at the surface with proportional frequency.

Climatic conditions may determine the locus of search for certain surface minerals. Bauxite and lateritic iron ores, for instance, are known to favor tropical climates. In exploration for these minerals, the climatic factor must be applied in connection with the topographic considerations already mentioned, and both, in turn, in connection with the character of the country rock as determined by general geologic surveys. A combination of climatic, topographic, and other physiographic conditions may be used also in exploration for certain types of residual clays.

Size and Depth of Ore Bodies As Determined From Outcrop

Where the ore body is harder than the surrounding rock, it stands out in conspicuous outcrops and is likely to show a narrowing below. Where it is softer than the surrounding rocks, and outcrops in a topographic depression, it is perhaps more likely to show widening below. These features are due to the general facts that, where the ore body is hard and resistant, the downward progress of erosion is likely to be arrested where the adjacent rocks occupy the larger part of the surface, that is, where the ore body is narrower. This principle is often vaguely recognized in the assumption that an exceptionally large outcrop of an ore vein may be "too good to last." Again, such a generalization must be applied to a specific case with much caution.

Attempts to forecast the depth of veins from their extent at the surface meet with only partial success. In a very general way great persistence horizontally suggests persistence in depth, on the ground that the section exposed on the surface is as likely to be a section of average dimensions as one along vertical lines.

Faith is the first article of the prospector's creed, and it is hard to shake his conviction that every ore outcrop must widen and improve below. As expressed by the French-Canadian prospector in the Cobalt district, the "vein calcite can't go up, she must go down." While the scientist may have grounds to doubt this reasoning, he is not often in a position to offer definite negative evidence.

The Use of Placers in Tracing Mineral Outcrops

Outcrops of ore-bearing rock may occasionally be located by tracing a placer deposit back to its source, or by following up ore fragments in the "wash" on mountain sides to the place of origin, or by noting ore fragments in glacial deposits. The presence of an ore mineral in a placer naturally raises the question as to whence it came. If it is a recent placer, it may be comparatively easy to follow up the stream channels to the head-water territory which is delivering the main mass of sediment, and there to locate a vein in place. The problem is complicated by multiplicity of tributaries and by large size of the drainage areas. In such cases careful panning and testing of the gravels at frequent intervals may show which of several tributaries are contributing most of the values, and thus may further localize the area of search. Many important mining districts, including Butte, Bisbee, the Mother Lode region of southern California, the diamond fields of Africa, and others, have been found by tracing up placers in this manner. In the case of an older placer deposit, where the topography and drainage have been much altered since its formation, or where the deposit has been covered by later sediments, the problem is of course much more difficult.

Much less than a commercially valuable placer deposit in unconsolidated surface rocks may start a search for the mother lode. A single fragment of ore in the "wash" naturally directs attention up the slope, and the repetition of fragments in a certain direction may lead unerringly to the source. The fragments may not even in themselves carry value, but may consist of detrital material from the leached outcrop—such as iron or manganese oxides, which, because of their red or black color, stand out conspicuously in the rock débris.

In the Lake Superior region large angular fragments of iron ore or iron formation in the glacial drift immediately raise question as to source. If the fragments are rounded and small, they usually indicate a very distant source. The general direction of glacial movement is known in most places, and by tracing up the fragments in this direction the outcrop may be found; or the chain of fragments may be traced to a point where they stop, which point may serve to locate the parent bedrock carrying the ore body, even though it does not outcrop.

An interesting suggestion was made some years ago with reference to the diamonds found sporadically in the terminal moraines in Wisconsin and other mid-west states. The diamonds are of such size and quality as to indicate surely the existence of a real diamond field somewhere to the north. The locations of these diamond finds were platted on a glacial map, and lines were projected in a general northerly direction along the known lines of the glacial movement. It was found that these lines converged at a point near Hudson's Bay. The data were too meager and the base line too short for this long projection, and the indicated source of the diamonds can be regarded as the merest speculation. However, with the finding of additional diamonds in the drift, as seems very likely, the refinement of this method might conceivably bring results in time.

THE USE OF MAGNETIC SURVEYS IN TRACING MINERAL LEDGES

Magnetic surveys are often useful in tracing iron-bearing rocks beneath the surface, in the discovery of outcrops of such rocks, and in working out their lines of connection. This method is in general use for the crystalline iron ores in the Lake Superior region, Canada, the Adirondacks, and elsewhere in the glaciated portions of the United States. It is not so useful for the brown ores and the Clinton ores of the southeastern United States, which are only slightly magnetic and can be commonly located by other methods.

Where the ore is strongly magnetic, and is associated with other rocks which are non-magnetic, the nature of the magnetic field determined by a surface survey with vertical and horizontal needles may tell something about the shape and size of the ore body. Commonly, however, magnetic ores are associated with leaner magnetic rocks,—with the result that the magnetic survey, unless it happens to lead to an outcrop of ore, indicates only the general area through which underground exploration might be warranted. In the hematitic iron ores of Lake Superior, magnetism is less pronounced than in the magnetites; and in the soft hydrous hematites, like those of the Mesabi district, it may cause only slight disturbance of the magnetic needle. This disturbance is usually sufficient to locate the position of the iron-bearing formation, though not the position of the ore.

Where the iron formation has been highly metamorphosed, and rendered resistant to weathering and erosion so that it will not concentrate into ore, it is likely to have higher magnetic attraction than the richer ores. For this reason an area of strong magnetic attraction is ordinarily regarded as not particularly favorable to the finding of important hematite deposits. However, this attraction may be very useful in tracing out the formation to a place where it is less metamorphic, less resistant to erosion, less likely to outcrop, and yet more promising for the discovery of iron ore. For instance, on the east end of the Mesabi and on the east and west ends of the Gogebic district, magnetic surveys trace the iron formation with great ease to points where the attraction is low and the conditions for exploration more favorable.

The magnetic needle has also been used in the search for nickel ore in the Sudbury district of Ontario, but without great success, because of the variety of rocks other than nickel which are more or less magnetic, and because of the slight magnetic properties of the nickel ore itself. In a large-scale exploration of this type, conducted some years ago, a favorable magnetic belt was discovered, and a pit was sunk to water level but not to bedrock. Years later, the extension of this pit by only a few feet disclosed one of the great ore bodies of the district.

Experimental work on the use of the magnetic needle on copper deposits has yielded some interesting and suggestive results, but this investigation is still under way and the results have not been published.

The Use of Electrical Conductivity and Other Qualities Of Rocks in Exploration

In addition to magnetism, rocks and ores have other properties susceptible to observations made at a distance, such as electrical conductivity, transparency to X-rays, specific induction, elasticity, and density. All these qualities have been of interest to geologists in some connection or another, but none of them have yet been used effectively in exploration for mineral resources. The only one of these properties that has thus far seemed to promise practical results is electrical conductivity. The results yet obtained are slight, and this kind of investigation has rested under something of a cloud, due to extravagant claims of inventors. Nevertheless, there has been a considerable amount of scientific work by physicists, geologists, and engineers, supplemented by special war-time investigations of rock and earth conductivity in connection with ground telephones and the tapping of enemy conversations, which seems to indicate a distinct possibility of practical results in the future,—perhaps not so much in locating specific ore bodies as in locating general types of formation and structures,—which may serve to supplement other methods of search.[38]

The transmission and reflection of sound waves in rocks have also been more or less investigated with reference to their possible military use. It seems not impossible that these phenomena may be of some geologic aid in the future, but experimental work is yet in a very early stage.

The Use of Structure and Metamorphism in Exploration

The necessity for careful use of structural data in exploration scarcely requires discussion. References have been made to structural features in connection with coal, oil, iron ore, and other minerals. This phase of study can scarcely be too intensively followed. The tracing of a folded or faulted vein, in a particularly complex system of veins, requires application of all of the methods and principles of structural geology.

Similarly, the importance of applying the principles of metamorphism, embodied in the *metamorphic cycle* (pp. 27-28) is almost self-evident. Certain kinds of metamorphism are suggestive of the nature of the mineral deposits with which they are associated. One would not look for minerals known to be caused mainly by surficial processes in rocks which have been altered mainly by deep-seated processes. The presence of metamorphism indicating high temperatures and pressures to some extent limits the kinds of minerals which one may expect to find. On the other hand, minerals known to be primarily formed at great depths, providing they are resistant to surface weathering, may be found in deposits which are the result of surficial alterations or katamorphic processes; that is, they may become concentrated as residual materials in weathered zones or as placers.

Drilling in Exploration

In the absence of distinctive outcrops, as well as when outcrops are found, drilling is a widely used method of underground exploration in advance of the sinking of shafts or the driving of tunnels. Drilling is more useful in the locating and proving of mineral deposits of large bulk, like deposits of coal, iron, and oil, than mineral deposits of small bulk and high value, like gold and silver deposits. However, it is not always used in the exploration of the first class of deposits and is not always eliminated in the exploration of the second class. With the development of better mechanical devices, better methods of controlling and ascertaining the direction of the drill hole, and more skillful interpretation of drill samples, the use of drilling is rapidly extending into mineral fields where it was formerly thought not applicable.

The geologist takes an active part in drilling operations by locating the drill holes, by determining the angle of the holes, by identifying and interpreting the samples, by studying bedding, cleavage, and other structures as shown in the samples, and determining the attitude of these structures in the ground, by determining when the horizon is reached which is most promising for mineral, and by determining when the hole shall be stopped. With a given set of surface conditions, the problem of locating and directing a drill hole to secure the maximum possible results for the amount expended requires the careful consideration of many geologic factors,—and, what is more important, their arrangement in proper perspective and relationship. Faulty reasoning from any one of the principal factors, or over-emphasis on any one of them, or failure to develop an accurate three-dimensional conception of the underground structural conditions, may lead to failure or extra expense. Success or failure is swiftly and definitely determined. The geologist is usually employed by the company financing the drilling; but in recognition of the importance of his work, some of the large contracting drill companies now employ their own geologists. The technique of the geologic interpretation and direction of drilling has become rather complicated and formidable, and has resulted in the introduction of special college courses in these subjects.

The desirability of public registration of drilling records is discussed on another page (pp. 305-306).

Quantitative Aspects of Geologic Exploration

In recent years there has been a tendency to reduce the geologic factors in exploration to some kind of a quantitative basis. While these factors may be very variable and very complex, their net effect frequently may be expressed in terms of quantitative averages. In various mines and mining districts where operations are of wide extent, local quantitative factors have been worked out which are useful in predicting results from proposed explorations in undeveloped portions. Figures of this sort may be useful and practical guides in planning any given exploration, its cost, and its probable outcome.

Quantitative methods are illustrated in the general account of Lake Superior iron ore exploration in a later section.

Curves of production from oil wells and from oil districts have been found to have certain characteristic features in common which are often used in predicting the future output and life of a given well, property, or district. Where associated with coal, the percentage of fixed carbon in the coal may be a guide to the presence and nature of the oil (see Chapter VIII).

The geological staff of the Netherlands East Indies estimated the tin reserves of one of these islands by the use of a factor or coefficient, based on the experience of another island.

In the Cobalt district of Canada a factor for future discoveries and output, based on past experience, was similarly developed.

Hoover[39] made a statistical study of several hundred metal mines in various parts of the world, and found that not 6 per cent of the mines that yielded profits ever made them from ore mined below 2,000 feet; and that of the mines that paid dividends, 80 per cent did not yield profit below 1,500 feet, and most of them died above 500 feet.

Attempt has been made by a Swedish geologist to estimate the iron ore resources of continents by the use of an iron coefficient. This coefficient was obtained by dividing the known iron ore resources of the comparatively well-investigated portions of the world by the number of square miles in which they occurred, and was then multiplied into the area of the continents whose resources were to be determined.

The application of quantitative methods of this kind has not yet become very general, nor is it possible to use them in some cases; where applied many of them have been very crude and others have been partly disproved by experience. With increasing knowledge and experience, such methods are becoming more accurate and useful, and are likely to have wider use in the future.

Origin of Mineral Deposits As a Factor in Exploration

In exploration, the geologist is keen to ascertain the origin of the mineral deposit. This is often a source of wonder to the layman or "practical" man, and the geologist may be charged with having let his fondness for theory run away with him. A widespread fatalistic conception is expressed in the Cornishman's dictum on ore, "Where it is, there it is." Yet an understanding of the origin of any particular ore, the "why" of it, is coming to be recognized as the most effective means of reaching sound practical conclusions. By ascertaining the approximate origin of the ore, it may be possible at once to infer a whole group of practical considerations based on experience with ores of like origin in other localities. The origin of the ore is the geologist's primary interest, and it is this which gives him his most effective and distinctive tool in exploration. Many other phases of exploration work may be picked up empirically by any one familiar with the local conditions; but when the man without sound geologic training attempts to go into this particular field, his lack of background and perspective often leads to fantastic hypotheses which may vitiate the inferences on which he plans his exploration.

The scientific investigator, while not accepting the fanciful theories of the local observer, will make a mistake if he fails to recognize the residuum of solid fact on which they are built. Many practical explorers are shrewd observers of empirical facts, even though their explanations may show a lack of comprehension of the processes involved. Any assumption of superiority, intolerance, or lack of sympathy, on the part of the geologist, toward the inadequate explanations and descriptions given him by the practical man, is likely to indicate a weakness or limitation in his own mental processes. The geologist's business is to sift out the fact from the inference, and not to throw over the whole structure because some of the inferences are faulty.

Lake Superior Iron Ore Exploration As an Illustration

To illustrate the application of some of the methods of exploration of the kinds described in this chapter, the writer selects an example from his own experience in the Lake Superior iron fields.[40]

In this region, consideration of the economic aspects of the problem may eliminate from the best explorable field certain Canadian portions which are far from water transportation, because the conditions in these sections would prevent the use of anything but an exceptionally large and rich deposit. Economic conditions determine in advance also that it is not worth while looking for ores of certain grades, either because they are not usable on account of deleterious constituents or low content of iron, or because these particular grades have already been developed in excess of requirements. Having determined what ore is desired, whether Bessemer or non-Bessemer, whether open-hearth or foundry, further elimination of area is possible on the basis of past experience.

Coming to the geologic phases of the problem, the first step is to eliminate great areas of rock which are known never to contain iron ore, like the granite areas and the quartzite and limestone areas. Within the remaining areas, by examination of the surface outcrops and with the aid of magnetic surveys, iron formations are found which are the mother rock of the ores. In Michigan, it has been possible to use certain percentage expectations in the areal location of iron formations within certain series of rocks extending over wide areas. Such percentage coefficients have been useful, not only in exploration, but also in the valuation of lands which are so covered with drift that no one knows whether they carry an iron formation or not.

Examination of the iron formations results in elimination of large parts of them, because their metamorphic condition is not favorable to ore concentration. In the remaining areas more intensive methods are followed. It is scarcely possible to summarize briefly all of the structural and stratigraphic methods used in locating the ore bodies. These have often been described in print.[41] Comparatively recent advances in this phase of exploration work have been in the more detailed application of stratigraphic methods to the iron formation. The group characteristics of the iron formation are fairly uniform and distinctive as compared with all other rocks; yet within the iron formation there are so many different kinds of layers represented that it is possible to use these variations with great effectiveness, in correlating favorable horizons for ore deposition, in interpreting drill records, and in other ways. Another method of approach, employed chiefly on the Mesabi Range, relates to the slumping of the ore layers which results from the leaching of silica during the concentration of the ore. This slumping can be measured quantitatively, and has been used to much advantage in exploration, in correlation of ore horizons, in preparation of sections and ore estimates, etc.

Early geologic explorations in the Lake Superior country were based on the assumption that the ores were concentrated by waters working down from the present erosion surface; but recognition of the fact that the waters which did the work were related to a far older and different erosion surface, under conditions which allowed of a far deeper penetration, has modified exploration plans for certain of the districts like the Marquette and Gogebic.

Notwithstanding the complexity of the geologic factors involved, their net result has been to concentrate iron ores in a surprisingly uniform ratio to the mass of the formation in different parts of the region,—with the result that on an average it may be predicted for any district, in an exploration of sufficient magnitude, how much ore is likely to be cut in either vertical or horizontal dimension. Thirteen per cent of the productive area of the Mesabi iron formation is iron ore. For the remainder of the Lake Superior region five or six per cent is the factor. These figures mean that, if a person could explore a broad enough area of iron formation, any miscellaneous group of drill holes or underground openings would tend to yield these percentage results. Such percentages are amply sufficient to pay a large profit on the exploration. The question may be raised why the application of geology is required, if such average results can be secured from miscellaneous undirected work. The answer is that seldom is it possible to conduct an exploration on a sufficiently large scale to be sure of approximating this average, and that geologic study has made it possible in many cases to secure a better percentage result. If the geologist is able to raise the percentage ever so little, the expenditure is amply justified. He is not expected to have 100 per cent success; but he is expected to better the average returns, and in this on the whole he has not failed.

Applying this method specifically to the Gogebic Range, it appears that up to January 1, 1918, exploration and development had covered 3,650 acres of iron formation, measured along the dip in the plane of the footwall, within the limits of the area in which the formation is in such condition as to allow concentration of the ore. The total area of the footwall to a depth of 3,000 feet is approximately 9,650 acres. The range, therefore, was 38 per cent developed to this depth. In the developed area, 160,000,000 tons of ore had been found, or approximately one ton per square foot of footwall area, or 43,800 per acre of footwall explored. The total area of ore measured on the footwall was 785 acres. The ratio of ore area to total explored area, measured in the plane of the footwall, was 21-½ per cent. This may be taken in a rough way to indicate the average exploring possibilities in new ground, where local conditions to the contrary do not exist. This means that over the whole range about one drill hole or cross-cut in five will strike ore on an average. Or, looked at in another way, about 200 feet of drifting in every 1,000 on the footwall will be in ore. Applying this factor to the unexplored area, amounting to 6,000 acres, the range had an expectation on January 1, 1918, to a depth of 3,000 feet, over and above ores already discovered, of approximately 262,800,000 tons. This was sufficient to extend the life of the range by about forty-four years. Knowing the average cost of development of ore per foot in the past, and knowing the annual output and its rate of acceleration, it is possible to figure with some accuracy how much expenditure should be planned for annually in the future in order to maintain a safe margin of reserves against output.

Such quantitative considerations in the Lake Superior region serve not only to guide the general conduct of the exploration and development work, but in some cases as a basis for valuation both for commercial and taxation purposes.

Development and Exploitation of Mineral Deposits

The search for new ore bodies is closely related to the development, extension, and mining of ore bodies already found. In this field the geologist finds wide application of his science. Here he may not be so much concerned with the economic factors or with the broader methods of geologic elimination; his study is more likely to be based mainly on the local geologic conditions.

Some of the larger and more successful mining companies, perhaps the greater number of them these days, have geologists whose business it is to follow closely the underground operations, with a view to advising on the conduct of the development work. This requires the most precise and intensive study. For instance, the Anaconda Copper Mining Company has a staff of several geologists, who follow the underground work in the utmost detail and whose approval must be obtained by the operating department in the formulation of any development plan. The complexity and fault relations of the veins in this company's mines are such that the application of these methods has abundantly justified itself on the cost sheet.

Too often mining companies leave the planning and execution of the underground development work to the local management, commonly to the underground mining captain, without geologic consultation. This procedure does not eliminate the economic geologist; for when the development fails at any point, or new and unexpected conditions are met, the geologist is likely to be called in. In such cases the practice of a geologist is like that of the ordinary medical practitioner; he is called in only when his patients are in trouble. The use of adequate geologic advice in the planning stages is about as little advanced in some localities as the practice of preventive medicine.

The work of the economic geologist may not be ended by the finding and development of the ore; for the moment this is accomplished, he should again consider the economic phases of the problem—the grade of his ore, its probable amount, and other features, in relation to the general economic setting. In his enthusiasm for physical results, he may be carried into expenditures not justified by the economic factors in the problem. Some one else may and usually does look out for the economic elements, but the prudent geologist will at least see to it that someone is on the job.

FOOTNOTES:

[37] Smith, George Otis, and others, The classification of the public lands: *Bull. 537, U.S. Geol. Survey*, 1913.

[38] Schlumberger, C., *Study of underground electrical prospecting*: Translated from the French by Sherwin F. Kelly, Paris, 1920.

Bergstrom, Gunnar, and Bergholm, Carl, "Teknisk Tidskrift, Kemi och Bergvetenskap," 1918, Book 12.

[39] Hoover, Herbert C., *Principles of mining*: McGraw-Hill Book Co., New York, 1909, p. 32.

[40] Leith, C. K., Use of geology in iron ore exploration: *Econ. Geol.*, vol. 7, 1912, pp. 662-675.

[41] Van Hise, C. R., and Leith, C. K., Geology of the Lake Superior region: *Mon. 52, U.S. Geol. Survey*, 1911.

CHAPTER XV

VALUATION AND TAXATION OF MINERAL RESOURCES

POPULAR CONCEPTION OF MINERAL VALUATION

The total returns from mining may not in the aggregate be far above the expenditure for exploration, development, and extraction; yet the total mineral wealth of the United States, on the basis of earning power and aside from the industries based on it, cannot be far from sixty billions of dollars, and this wealth has virtually come into existence since the 1849 gold rush to California. The mining industry supports a large population. These facts are the solid basis for the widespread popular interest in mineral investment—and mineral speculation. But there are other reasons for this interest,—the gambler's chance for quick returns, the "lure of gold," the possibility of "getting something for nothing," the mushroom nature of certain branches of the industry, the element of mystery related to nature's secrets, and the conception of minerals as bonanzas with ready-made value, merely awaiting discovery and requiring no effort to make them valuable. In the United States a factor contributing to the popular interest is the large freedom allowed by the laws to discover and acquire minerals on the public domain. Perhaps no other field of industry comes so near being common ground for all classes of people. The mineral industry is a field in which it is easy to capitalize not only honest and skillful endeavor, but hopes, guesses, and greed. It is not to be wondered at, therefore, that in the popular mind the valuation of a mineral resource is little more than a guess, and sometimes not even an honest one.

Nevertheless, the mineral industry has become second only to agriculture in its capital value and in its earning capacity. In this industry it is hardly possible to arrive at valuations as securely based as in many other industries, but the elements of hazard are not so hopeless of measurement as might be supposed. The great mineral and financial organizations do not depend on mere guesses, but use well-tried methods. If the general investor were to give more attention to these methods he would doubtless save himself money, and the mineral industry would be rid of a great incumbrance of parasites who live on the credulity of the public. To anyone familiar with the mineral field, it is often surprising to see the rashness with which a conservative business man, who would not think of entering another industrial field without close study of all the factors in the situation, will invest in minerals without using ordinary methods of analysis of values.

In the following account of valuation of minerals in the ground, and the closely related subject, taxation of such minerals, the attempt is made to state some of the principles briefly and simply with a view to making them intelligible to the layman. Values beyond the mine are concerned with so many factors of a non-geologic nature that they are not here discussed.

VALUATION AND TAXATION OF MINES

INTRINSIC AND EXTRINSIC FACTORS IN VALUATION

It is essential to recognize at the outset that the value of a mineral deposit, like the value of any other commercial material, comprises two main elements; an intrinsic element based on the qualities of the material itself, and an extrinsic element based on its availability and the nature of the demands for it. The two elements may not be sharply separated, and neither exists without the other. A mineral deposit in easy reach of a populous community, which has sufficiently advanced methods and requirements to use it, may have high value; an exactly similar deposit, if far removed from points of consumption, handicapped by transportation, or available only to people without developed methods for its use, may have little or no value. Intrinsically the deposits are alike; but extrinsically they are far different, and their values are correspondingly unlike. Even two adjacent properties, differently managed and controlled, and with different relations to markets, may have somewhat different values depending on the use made of them. The value of a deposit may vary from year to year with changes in demand for its output, or with changes in metallurgical and other processes which make its use possible. Minerals of small bulk and high value, as for instance gold, platinum, and diamonds, have a nearly standard value related to their intrinsic properties, because they can be transported so easily to any part of the world. On the other hand, materials of large bulk and low unit value, such as coal, iron ore, and clay, may have highly varying values independently of their physical characteristics, because of their relative immobility. But the values even of gold and precious stones represent a combination of intrinsic qualities and of demand. A diamond is made of carbon but is more valuable than coal or graphite because it appeals to the esthetic taste. It is only because man introduces an element of demand that the diamond takes on value. In short, man is the multiplier and the mineral substance is the multiplicand in the product known as value.

Recognition of the two elements of value is vital to a clear understanding of the methods and problems of valuation of minerals. It is too often assumed that the physical properties constitute the sole factor.

Looked at in a large way, the returns from the mineral industry are commensurate with the effort put into discovery and development of mineral resources, even though the returns to lucky individuals have been excessive. In respect to the importance of the human energy element, the mining of minerals is not unlike the cropping of soils. Some interesting economic studies have been made of mining districts to ascertain whether the total return has been equal to the total investments by both successful and unsuccessful participants. The results show that, even in some of the most successful districts, there is not a large "social surplus,"—that is, a surplus of receipts over total expenditures. It is difficult to generalize from such studies with any degree of accuracy; but it seems likely that if we could measure the vast amount of fruitless effort which has been expended in non-productive territories, the result would tend to bear out the general conclusion that the social surplus for the mineral industry as a whole is a modest one, if it exists at all. Of course, it is to be remembered that the total benefits from mineral resources are not to be measured in terms of gain to the producers,—but that their measurement must take into account the satisfying of all the complex demands of modern civilization.

Values of Mineral Deposits Not Often Established By Market Transfers

While minerals as extracted and used may have standard market values, mineral deposits in the ground are not bought and sold on the open market with sufficient frequency to establish standard market values. A sale may establish a criterion of value for the particular deposit, but not for the class of deposits,—for no two mineral deposits are exactly alike. Stock quotations may establish a certain kind of market value, but these are often vitiated by extraneous considerations. For these reasons the valuation of a mineral deposit is in each case a special problem.

THE AD VALOREM METHOD OF VALUATION

The ordinary commercial method of valuing mineral deposits recognizes the two main elements of value above discussed. This method is sometimes called the *rational* or *ad valorem* method. The profit per ton (or per other unit) of the product is established, on the basis either of past performance of the property or of experience with other similar properties. This profit is multiplied by the total tonnage estimated in the deposit, the estimate including known reserves, probable reserves, and in some cases possible and prospective reserves. The product of the profit per ton and the total tonnage gives the total net amount which will be received; it does not, however, give the present value, because the commodity cannot all be taken out and sold at once, but must be mined and absorbed by the market through a considerable period of years. The returns receivable some years in the future have obviously a lower proportionate present worth than amounts to be received at once. The interest rate comes into play, making it necessary to discount each annual payment for the number of years which will elapse before it is received. It is evident, therefore, that an estimate of the *life* of the property is necessary, involving not only knowledge of the reserves, but also a forecast of the annual extraction or *rate of depletion*.

As a simple case of *ad valorem* valuation for illustrative purposes, a deposit containing 1,000,000 tons in reserve has an estimated output of 100,000 tons a year for ten years, on which the profit per ton has in the past averaged $1 and is expected to average $1 in the future. Ten annual instalments or dividends of $100,000 are to be received. The present value of the total of these instalments is figured by an annuity method. It is the value upon which the series of dividends will pay interest at a predetermined rate, in addition to paying to a sinking fund annual instalments which, safely invested each year at a low rate of interest (usually 4%), will repay the present value at the end of the ten years. In our hypothetical case, if an interest rate of 8% be taken, the present value of $1,000,000, to be received through ten years in ten equal instalments, is $612,000. In other words, the sum of $612,000 will be replaced by the sinking fund at the end of ten years, and will pay 8% interest during this period,—this requiring total receipts of $1,000,000 in ten equal annual instalments. If the deposit here cited as an illustration were to be worked out in three years, thus yielding three annual instalments of $333,000, its value would be $833,000.

Each of the factors entering into this method of valuation covers a wide range of variables, any one of which may be difficult to determine.

The profit per ton for a given deposit may have been extremely variable in the past, making it difficult to determine whether the highest or lowest figure should be projected into the future or whether some average should be taken; and if an average, whether the time covered by the average should be long or short. For a small, short-lived deposit obviously the most recent conditions would be taken into account in estimating future profits. For a long-lived property there would be more tendency to consider the long-time average vicissitudes, as reflected in the average profits of the past. For some mineral commodities there are cycles of prices, costs, and profits, of more or less definite length, established during the long past history of the industry; and in such cases it is desirable in calculating averages to use a period covering one or more of these cycles, rather than some shorter or longer period. For many minerals, however, these cycles have been too irregular to afford a sound basis for future estimates. If the experience of the property itself is too short to afford a sufficient foundation for forecasting profits, or if there has been no previous work on the property, then it is necessary to use averages based on other properties or other districts; or if there are none strictly comparable, to build up a hypothetical figure from various estimated costs of labor, supplies, and transportation, selling prices, etc. In the estimate of the profit factor, the geologist is not primarily concerned.

In estimating the total reserves in a mine, geological considerations nearly always play a large part. An ore body may in some few cases be completely blocked out by underground work or drilling, eliminating the necessity for inferring conditions beyond those actually seen; but in the huge majority of mineral deposits the reserves are not so definitely known, and it becomes necessary for the geologist, through knowledge of similar occurrences, through study of the structural features of the deposit, its origin, and its history, to arrive at some sort of an estimate of reserves.

In estimating the life of a mineral deposit it is necessary to start with the figure of total reserves, and from a study of conditions of mining and of markets to estimate the number of years necessary to exhaust the deposit. This is a more nearly commercial phase of the problem, in which the geologist takes only part of the responsibility. Perhaps more estimates of value have gone wrong because of misjudgment of this factor than for any other cause. If the physical conditions are satisfactory, it is easy to assume a rate of extraction and life based on hope, which experience will not substantiate.

The choice of the interest rate to be used in discounting future receipts to present worth likewise is a financial and not a geologic matter. Again, however, the geologist must give consideration to this factor, in view of the fact that the interest rate must be varied to cover the different degrees of hazard and doubt in the geologic factors. For instance, to the extent to which the estimate of ore reserves is doubtful, it is necessary to use a high rate of interest to allow for this hazard. In a large, well-developed mineral deposit, with the geological factors all well known and the demand and market well established, it is reasonable to use a lower rate of interest. In general, the mineral industry is regarded in financial circles as being more hazardous than many other industrial lines; and money is put into the industry with the expectation of a high rate of interest, no matter how safe the investment may be. In actual practice interest rates used in making valuations vary from 6 to 15 or 20 per cent.

It is clear that, where a property has long life, the interest will very materially reduce the present value of the ores to be mined far in the future. Reserves to be mined more than thirty years hence have relatively little or no present value. Beyond a certain point, therefore, the acquirement and holding of reserves for future use by private companies has little commercial justification.

This is a matter which is too often not sufficiently well considered. Man's natural acquisitiveness often leads him into investments which, because of the time and interest factor, have little chance of successful outcome. Of course a large corporation, anticipating an indefinitely long life, or perhaps aiming at monopoly, may afford to hold reserves as a matter of general insurance longer than a small company,—even though, because of the interest rate, these reserves have no present value on their books. It is likewise true that governments, looking forward to the future of the nation, and without the necessity of paying so much attention to interest and taxes, are not so limited by this consideration.

An illustration of the limiting effect of the interest rate on the acquirement of long-lived coal deposits by private interests is discussed in Chapter XVII on Conservation. Investments made many years ago have so augmented, even at low interest rates, as to make it practically impossible to count on a return of capital and interest; or if the return were to be exacted from the public it would mean excessive charges, which are not possible in competition with other mines not so burdened.

In the commercial valuation of oil wells and pools, much the same method is used as has been described for mineral resources in the solid form, but the estimate of reserves or life is based on consideration of curves of production of the sort mentioned on pages 134-136.

The essence of the *ad valorem* method of valuation above described is income-producing capacity. This method recognizes the fact that the value of the mineral deposit depends, not only on its physical constitution, but also on what performance can be expected from it.

Stock quotations on mineral properties in the standard markets are based substantially on estimates of income capacity, more or less on the *ad valorem* basis. However, the quotations also reflect the hopes and fears of the public, often resulting in valuations quite different from those based on studies of the objective conditions.

The war introduced new considerations into the problems of *ad valorem* valuation. Under peace conditions there is a tendency toward the establishment of normal costs, selling prices, and markets, which can be taken more or less for granted by anyone attempting to value mineral deposits. Under war and post-war conditions, few of these elements can be taken for granted; it becomes necessary to consider the entire world situation in regard to a mineral commodity, the effects of the Peace Treaty (which greatly concerns minerals), future international relations, tariffs, and other matters of a similar sort. If a person were today valuing a manganese deposit according to the method above outlined, and were to confine himself solely to a narrow consideration of past markets and profits on individual properties, he would be very likely to go wrong,—for the world manganese situation has an immediate and practical bearing on each local problem (see pp. 173-176).

OTHER METHODS OF MINERAL VALUATION AND TAXATION

We have discussed the *ad valorem* method of valuation at some length because it is the one in widest commercial use, and also because the principles involved underlie practically all other methods of mineral valuation. The *ad valorem* method is used in appraisals for taxation in some districts and for some commodities, as, for instance, the iron mines of Michigan and Wisconsin. Its application, however, requires skill and judgment if equitable results are to be secured. For taxation purposes, therefore, it is not uncommon to adopt purely arbitrary or empirical methods which eliminate the element of judgment, and which often result in valuations quite different from those used commercially.

The state of Minnesota divides its iron ore deposits into a series of classes, on each of which a more or less arbitrary flat value per ton is placed, based on the spread between cost and selling price. The adjustments of flat values on the several classes through a series of years, however, as well as the assigning of specific ores to the different classes, have been based on the same factors as are used in *ad valorem* valuations.

The state of Wisconsin uses a so-called "equated income" method of valuation and taxation for the lead and zinc deposits of the southwestern part of the state. Under this method the state puts such a tax on the mine incomes for the preceding year as will yield approximately the same total return as under the *ad valorem* method,—the whole being based on the assumption that each deposit has about the average life figured for the mines of the entire district. So far as individual ore deposits vary from this average life, the value fixed departs from the true or *ad valorem* value.

Several states impose specific taxes based on the operations of the mines for the preceding year or for some combination of preceding years, as expressed in tonnage output or net profits or net proceeds, regardless of life or reserves. So far as output or net proceeds for a year are proportional to the real value of the property, a rough approximation to equitable taxation as between mines is accomplished. Often, however, the valuation thus obtained has little relation to the true value, because it does not take into account the great differences between properties in reserves, in life, and in capacity for future profit.

Income taxes, national and state, are of course based on the profits of the preceding year; but in the collection of these taxes from mineral operations, it is recognized that mineral deposits are wasting assets, and therefore a considerable part of the income may under the law be regarded as a distribution of capital assets, and be deducted from taxable income. The amount to be deducted obviously depends on the size of the reserves and the life,—with the result that progressive adjustment of income tax valuations tends to take into consideration exactly the same factors as are used in the *ad valorem* method. It is obviously unjust, for instance, to collect the same proportion of tax from the annual income of a mine which has a life of only two years as from a mine which has a life of fifty years. Under the federal income tax a capital value is placed on the mineral deposit as of March 1, 1913, which total capital value may be increased with subsequent discoveries. As the ore is taken out of the ground and sold, income tax is paid only on the difference between the assigned capital value per unit and the selling profit. If, for instance, the capital value as of March 1, 1913, is placed at 50c. per ton of mineral in the ground, and ten years later a ton is sold for a profit of $1, income tax is paid on 50c. The figure of 50c. per ton as value in the ground is actually obtained by estimating a profit, when the ore is ultimately mined and sold, of $1 per ton, and discounting this dollar to present worth as of March 1, 1913. Therefore the total amounts on which taxes are paid during the life of the mine should represent approximately the total accruals of interest from March 1, 1913. In this manner the proportion of annual income to be taxed becomes larger with the length of the life period. With a deposit having a life of thirty years the net result is that about half of the aggregate income is taxed, though this figure of course varies somewhat with the interest rate used.

In the collection of income taxes from coal mines in England, and in the collection of certain state income taxes in the United States, a considerably smaller allowance is made for the retirement of capital value (or for *depletion*, as this is commonly called). In these cases the deduction allowed is a small fixed percentage of the capital value, regardless of the actual life of the property.

The treatment of mineral resources as wasting assets in the United States income tax law meets one considerable practical difficulty—namely, that the law really requires physical or *ad valorem* valuation of every mineral property by the government, as a check on the claims for depletion allowance. This immense and expensive task is too much for the tax collection agencies as now organized, and it may be questionable whether it will ever be desirable to expand these agencies to the extent required for such a purpose. This is the principal argument for the use of arbitrary depletion factors such as those sometimes used abroad.

There are many advocates of the straight tonnage tax on mineral deposits, on the ground that it is simple, definite, and easily applied. The present tendency is to extend the application of this form of tax. It is clear, however, that to assume the same value per ton for taxing purposes on a property making a large profit, and on another property which, because of physical conditions, is barely able to operate at a profit, imposes a relative injustice. To meet this difficulty, it is sometimes proposed that the tonnage tax should be graded in such a manner as to allow for differences in physical conditions and in profit at different mines. When one attempts to apply a graded tonnage tax, however, it soon becomes apparent that, in order to make such a valuation equitable as between properties, it is necessary to use all of the factors of the *ad valorem* method for each of the properties. The wide appeal of arguments for a flat tonnage tax is based partly on popular misconception of the complexity of elements entering into mineral valuations.

There are many forms of more or less indirect tax which are substituted in different parts of the world for direct taxes. For instance, certain states in South America do not tax ores in the ground, but collect the revenue in the form of mining licenses or export taxes.

GENERAL COMMENTS ON TAXATION OF MINERAL RESOURCES

There has been a noticeable tendency in recent years to regard mineral resources as a heritage of the people, to be held in trust, rather than as property to be acquired and managed solely for private interest. This tendency has been indicated by the adoption in various parts of the world of laws affecting rights to explore and acquire minerals on the public domain; laws relating to the right of eminent domain over minerals already alienated from the government; laws regulating the exploitation of minerals in the interests of conservation; laws relating to tariffs and other restrictions on the export of mineral commodities; and laws relating to taxation.

The feeling that mineral resources really do not belong in private hands has undoubtedly been an underlying factor in the imposition of heavy taxes. Contributing to this action also are the popular belief in the intrinsic bonanza values in mineral resources, the failure to recognize the large element of value which is put into such resources by human efforts, and the failure to realize that the social surplus in the aggregate is small. To some tax officials an ore is an ore, more or less regardless of situation, of conditions of mining, of the demand for the product, and of the time when the demand will allow the ore to be mined,—in short, more or less regardless of what the ore may be made to yield as a going business. In this way heavy taxes are sometimes imposed on mineral reserves, which are based on unwarrantably high appraisals of future possibilities, and which cannot be paid out of earnings.

Ultimately, a tax must be adjusted to the capacity of the mine to pay out of its earnings, and this capacity in turn is determined both by the physical characters of the ore and by the success with which it may be made available for consumption. This view of valuation for taxing purposes is sometimes opposed by mining men on the grounds that it taxes brains, skill, and initiative, and that it puts a premium on shiftless management. The same argument might be applied to the valuing of any business or profession. To the writer the argument is not sound, in that it fails to recognize the element of human energy in resource values. If value were to be confined solely to the intrinsic character of the ore itself, there would be required an almost impossible degree of discrimination on the part of taxing officials to dissociate this value from other considerations; and there would be required further the differentiation between efficient and inefficient management, which involves so many considerations that the conclusion would be worthless.

In the application of income taxes to mining operations, there is sometimes another tendency toward over-taxation in that the income is regarded as more or less permanent, and insufficient allowance is made for exhaustion of the mineral deposit. Under the United States income tax, mineral deposits are definitely recognized as wasting assets and this factor is allowed for; but in state income taxes and in England and other parts of the world, allowances for this purpose are small.

There is wide belief that heavy taxation of mineral resources, particularly on the *ad valorem* basis, retards exploration and prevents the development of the reserves which are necessary to stabilize the mineral industry. High taxes have undoubtedly had this effect in some cases, especially where taxes have been imposed on resources long prior to development; but, in the writer's view, this tendency in general has not yet passed the danger point, and is not likely to do so until taxes become positively confiscatory of the industry. To argue that increase of taxes may even have certain beneficial results on the mineral industry may lead to suspicion of one's mental soundness; but it is hard to escape the conclusion that the incidence of high taxes has led to a much more careful study of the question of reserves, has eliminated in some cases the expenditure of money for development of excessive reserves to be used far in the future, and has tended to prevent over-production.

Where mineral reserves are developed too far ahead of demand, the interest on the investment piles up an economic loss to be charged against the industry. It may be assumed that the urge for exploration will continue as long as there is demand for mineral resources; and that, to keep the industry on a sound basis, a certain amount should be set aside and charged to cost for the purpose of keeping up reserves in a proper ratio to production. Much remains to be learned about the most desirable ratio between reserves and production. In many camps, before the incidence of high taxes, this ratio was not properly determined; and there was a tendency, due to natural acquisitiveness and in the absence of anything to hinder it, to build up reserves indefinitely. The first effect of high taxes in such camps has frequently been the curtailment of exploration and development. Later, as production has begun to approach the end of the reserves, exploration has been resumed, but only on a scale necessary to insure production for a limited period in advance.

The argument that high taxes inhibit exploration is good only beyond the point where the industry itself becomes no longer profitable. If there is sufficient demand for the resource, it is obvious that such a condition cannot long continue; for, as production and the development of reserves fall off, the resulting increase in the price received for the product is likely to offset any effect of taxes, and to restimulate production and exploration.

Nevertheless, in this period of high taxes following the war, there is much discouragement in the matter of exploration, suggesting that the danger point is being approached. Some relief has been afforded by recent special provisions of the federal income tax law, recognizing mineral resources as wasting assets, allowing recent discoveries to be included with total assets for depletion purposes, and recognizing special and peculiar circumstances with reference to each mine. Also a certain amount of exploration goes on through the momentum gained from past conditions, without sufficiently full recognition of the effect of present high taxes. This is not surprising when it is remembered that the people actively engaged in field exploration often do not think sufficiently fully of the tax situation, until after a discovery or development has brought them face to face with it.

Because of the vital importance of the reserve factor in mineral valuation, geologic aid and advice are extensively sought by both public and private organizations. Mining geologists are playing an important part in the application of the national income tax. A larger number are acting for private companies in appraisals required by this tax. Many geologists are used in making valuations for state taxes, and in two cases the state geological surveys have complete charge of appraisals. These appraisals include not only examinations of specific properties, but general surveys of large regions, to ascertain possible values of undeveloped lands and to establish broad principles of valuation based on a consideration of all the physical factors in the situation.

CHAPTER XVI

LAWS RELATING TO MINERAL RESOURCES

This heading is likely to suggest mining law and the vast literature devoted to it. Mining law has mainly to do with questions of the ownership and leasing of mineral deposits. In addition, there are laws relating to the extraction of mineral products, including those having to do with methods of mining and with safety and welfare measures. There are laws affecting the distribution of mineral products, such as those relating to tariffs, duties, international trade agreements, and many other matters. There are laws relating to underground water, to shore lines, and to various geologic engineering fields.

In the formulation of these laws, as well as in the litigation growing out of their infraction, basic geologic principles are involved; and thus it is that the geologist finds much practice in the application of his science to legal questions. It will be convenient to consider some of the laws relating to mineral resources under three headings: first, ownership and control; second, extraction; and third, distribution.

I

LAWS RELATING TO OWNERSHIP AND CONTROL OF MINERAL RESOURCES

Large use of mineral resources is of comparatively recent date. Some of the mineral industries are not more than a decade or two old and the greater number of them are scarcely a century old. In the United States the mineral industry dates mainly from the gold rush to California in 1849. The formulation of laws relating to the ownership of minerals has on the whole followed rather than preceded the development of the mineral industries; and hence mining laws relating to ownership are not of great age, although historical precedent may be traced far back.

ON ALIENATED LANDS

Where lands came into private ownership, or were "alienated" from the governments before the formulation of mining laws, varied procedures have been followed in different countries.

In England and the United States, under the old régime in Russia, and to a slight extent in other parts of the world, mineral titles remain with the owner of the land and the government does not exercise the right of eminent domain. But even in England, where private property rights have been held peculiarly sacred, the discovery of oil during the later years of the war led to an attempt to expropriate the oil rights for the government. Because of the objection of landowners this attempt has not reached the statute books, but the movement is today an extremely live political question in England. A somewhat similar question is involved also in the movement to nationalize the coal resources of England, now being so vigorously urged by the labor party. In the United States, no serious attempt has yet been made to take over mineral resources from private ownership.

Other countries have gone farther in retroactive measures in regard to alienated lands. Under the leadership of France, most of the countries of western Europe have appropriated to their governments the undiscovered mineral resources on private ground, particularly those beneath the surface, except where previously they had been specifically conveyed to the private owners, or with the exception of certain designated areas and minerals which had been conveyed to private ownership prior to certain dates. Some minerals occurring at the surface, variously specified and defined in different countries, are allowed to remain with the private owners, although often subject to government regulation in regard to their development and use.

In varying degree this treatment of mineral resources on alienated lands is followed in the British colonial laws—in South Africa, Australia, New Zealand, and Canada—and in the Latin-American laws. The laws are usually based on specified classifications of minerals. Those occurring at or near the surface, and called "quarries," "placer deposits," "non-mines," or "surface deposits," usually remain with the surface owners. Those beneath the surface, called "sub-surface deposits" or "mines," in general belong to the government. In some of the countries of South America the state exercises eminent domain even over the surface deposits; and in others even sub-surface minerals remain in private ownership, where specifically granted, or where the transfer of property took place prior to certain dates.

Where the government has acquired mineral ownership of lands previously alienated, the resources are open for development either by the owners of the surface or by others, on a rental, lease, specific tax, labor, or concession basis. The government holds the title, exacts tribute, and more or less directs and controls the operation. Exceptionally, as in Ontario, British Columbia, Quebec, and Newfoundland, the government grants patents, that is, it disposes of its rights to purchasers.

On the Public Domain

Where the development of mineral resources began before the lands had passed from governmental ownership, special mining laws were enacted. Looked at broadly, these laws may be regarded as based on two partly conflicting considerations.

(1) The assumption that mineral resources, which are wasting assets, accumulated through long geologic periods, are peculiarly public property,—not to be allowed to go into private ownership, but to be treated as a heritage for the people as a whole and to be transferred to posterity in the best possible condition. Some of the early minerals to be developed were used either for money or for war purposes, leading naturally to the acceptance of the idea that these belonged to the government or to the sovereign.

(2) The assumption that the discovery and development of mineral resources requires a free field for individual initiative, and that the fewest possible obstacles are to be put in the way of private ownership. Governments have not as a rule been greatly interested nor particularly successful in exploration. Therefore, in framing laws of ownership, concessions have been made to encourage private initiative in exploration and development. In the case of the United States this idea was coupled with the broad doctrine that the government held public lands only in the interest of the people, and that its people were entitled to secure these lands for private ownership with the least possible restriction.

A survey of the mining laws affecting the public domain or non-alienated lands of different parts of the world, as well as the history of changes in the mining laws, indicates a wide range of relative emphasis on these two underlying considerations. In the United States, at one extreme, the laws have been such as to give the maximum possible freedom to private initiative, and to allow easy acquirement of mineral resources from the government. At the other extreme, in South Africa, Australia, and South America, it is impossible for the individual to secure title in fee simple from the government; he must develop the mineral resources on what amounts to a lease or rental basis, the ownership remaining in the government.

The trend of events in mineral laws is toward the latter procedure. This is evidenced in the United States by the withdrawal of large areas of public lands from entry, and by the recent enactment substituting leasing privileges for specified minerals for the outright ownership which was allowed under the federal law before the lands were withdrawn from entry. The withdrawal of oil lands from public entry in other parts of the world is another illustration (see pp. 131-132).

NATIONALIZATION OF MINERAL RESOURCES

Nationalization, as this term is popularly understood, means financial control and management of mineral resources by the government, either through actual ownership or through measures of public control designed to eliminate private interest from the active direction of the resources. In a broader sense, it may be used to include a considerable variety of restrictive and coercive measures adopted by the government in the proposed interests of public welfare,—as illustrated by the war-time measures instituted by the United States and other governments relating to the mining and distribution of coal, and to coal prices. In this broader sense various aspects of nationalization are indicated under other headings in this and other chapters.

It is clear that other countries of the world have gone farther in the direction of nationalization of mineral resources than the United States. The tendency was manifest before the war, and has been strongly emphasized during and since the war. In the United States, notwithstanding war-time measures, the subject has not yet come prominently forward, at least by name. On the other hand, there has been growing recognition of the dependence of public welfare on the proper handling of mineral resources—particularly of the energy resources, coal and oil,—as evidenced by a variety of proposals and measures under consideration in legislative and administrative branches of our national and state governments. Even taxation, both local and national, has in effect reached a stage where private interest has become considerably minimized by the increasing burdens laid on the industry by government requirements. The immediate purpose of taxation is to raise money for the needs of the government; but in the formulation of tax measures there is clearly to be discerned a growth of underlying sentiment that natural resources belong in some fashion to the public, and that private control is to be regarded not as a sacred property right but as a trust held on sufferance of the public.

In view of the obvious trend toward nationalization in other parts of the world and the significant tendencies in the United States, it seems likely that the subject of nationalization of mineral resources will come prominently to the front in this country in the comparatively near future. If so, it is time that students of mineral resources should recognize the comprehensiveness of this problem, and should attempt to develop basic principles to serve as a guide in the direction and formulation of the numerous and complex measures which are sure to be proposed. At present there is no government or technical organization related to the industry which is studying the problem in its broader aspects and is in a position to advise wisely with public officials interested in this problem.

It is beyond the scope of this book to discuss the pro and con of an economic question of this magnitude. The writer would, however, record his belief, which is implied also in discussions in other chapters, that the discovery and intelligent management of mineral resources by their very nature and infinite variety require private initiative, and that the history of government efforts in this field in this and other countries does not promise that nationalization can supply sufficient advantages to counterbalance the loss of this element. With this view the problem of nationalization becomes one of determining what steps, if any, can be taken by a government to the advantage of public welfare, which will at the same time preserve and foster private initiative, exercised with the hope of reward, which seems alone to be capable of meeting the variable, elastic, and complex problems inherent in the development of a natural resource.

A first step toward a broad scientific attack on this problem would be the recognition of the fact that tariffs, taxes, conservation, international mineral questions, leasing laws, and various technical investigations of minerals are but parts of a great unit problem. With this recognition there should follow naturally an attempt to correlate and direct the many government agencies, legislative and administrative, now concerned with different aspects of the problem. Under present conditions, the various elements of the problem are considered by different groups of persons, without sufficient contacts or correlation to promise the development of a broad, underlying policy.

EFFECT OF OWNERSHIP LAWS ON EXPLORATION

The nature and the progress of exploration (and development) in different countries have been more or less related to the character of the mining laws.

Where the mineral resource has passed from government control into private ownership, exploration is a matter of commercial arrangement between the explorer and the owner. There is often some lag in exploration, especially where the lands are held in considerable blocks. The owner is often not inclined, or unable, to institute effective exploration himself; and even though he is willing to offer favorable exploration terms to others, the inducement is often less attractive than on government lands. For instance, it is stated that in England, due to the many requirements of law and custom, it takes on an average eight years, and in some cases even longer, to close a coal lease after the terms have been agreed upon. The slowness of exploration and development on the great land grants in the United States, and on the tracts of the large timber companies, also illustrates the retarding effect of private ownership. It is partly this situation that is making governments increasingly careful about parting with mineral ownership, and that is leading to the introduction of more or less coercive measures, either to regain control or to make it easier for the public to explore and develop minerals on privately owned lands. Under the great land grants to railroads in the United States it is becoming increasingly difficult to secure mineral patents from the government; and there has been litigation between government and grantees, as in the case of certain oil lands of the Southern Pacific Railway. The taxation in some states of mineral rights which have been reserved by large owners is indirectly resulting in appraisal of these rights by the owners and in efforts to utilize them. As long as mineral rights were not taxed independently of surface rights, they were often reserved in selling surface rights on the mere chance that mineral might be found in the future, and thereby general exploration and development were held back.

In the United States, minerals on the public domain have been open to exploration and acquirement with minimum restrictions, except for the considerable areas later withdrawn from entry. After long delay a part of these withdrawn lands are again open to private exploration, but not to fee ownership. Specified minerals—coal, oil, phosphates, and potash—may be explored for, and may be leased under certain restrictions as to amount and time of development. The effect of this act on exploration is yet to be proved; but since many of the lands have now been shown to be favorable for minerals which are in great demand, there is little doubt that exploration will be resumed on a large scale. On the whole, under the federal mining laws of the United States the individual prospector has maximum leeway,—and from the standpoint of development of resources this procedure probably has been justified.

In other countries where the mineral resources are owned by the government, there is in most cases considerable restriction, through licenses and other regulative measures, upon the activities of prospectors. This restriction, together with the fact that it is usually not possible to secure title to the land, but only to secure rights through rental or leasing, is to some extent a deterring influence on the penniless prospector. It does not follow that under these conditions exploration and development are absent. The charges imposed are light, and in the early stages require comparatively small contributions as evidence of good faith. It is to be remembered that exploration has become concentrated more and more into the hands of persons financially able to meet such conditions. Exploration is passing from the highly hazardous stage of individual effort into a systematic business with calculable returns.

USE OF GEOLOGY IN RELATION TO OWNERSHIP LAWS

The contacts between geology and laws relating to mineral ownership are many and varied; a few illustrative examples are offered.

Many difficulties arise from the loose use of mineral names in these laws. The laws governing location of mineral deposits in Cuba are so framed that iron ores may be located and claimed from the government either as "iron ores" or as "bog ores and yellow ochers." Some of the important ores of eastern Cuba, now being extensively used in the United States, came into litigation because rival claimants had overlapping claims under the two classifications. The wording of the law is of course ambiguous, and suggests that geologists did not have a hand in its framing. To establish title to these claims it was necessary to show whether these ores had been rightfully located as iron ores, or whether they should have been located as bog ores and yellow ochers. This involved an analysis of the geological conditions, to show that the ores are the result of normal weathering and concentration in place of the underlying rocks—an origin common to many iron ore deposits,—and that they do not have the characteristic origin of bog ores. In short, the question was settled on the scientific principles of origin of ores and of metamorphic geology.

The efforts of our federal government to frame and apply mining laws to public lands have involved extensive geological and mining surveys by the United States Geological Survey and the Bureau of Mines. The land classification work for this purpose by the Geological Survey has been of wide scope. The recently enacted leasing law, which opens up government lands for exploration of coal, oil, potash, and phosphate, requires carefully prepared geologic data for its proper administration.

State governments also have initiated surveys of an exploring nature for taxing and other public purposes (see pp. 306, 311).

In the United States there is a wide use of geologists as witnesses in litigation affecting "extralateral rights." The federal mining law gives the owner of the claim containing the "apex" or top of a mineral vein or lode the right to follow the vein down the dip, with certain limitations, even though this takes him on to adjacent properties under other ownership. Where two branches of a vein are followed down from separate claims, the owner of the oldest claim is entitled to the vein below the point of junction. The law was framed to validate a procedure more or less established by mining custom. It was obviously framed with a very simple and precise conception in mind—namely, a simple vein definitely and easily followed, without much interruption or contortion.

In nature, however, veins or lodes have a most astonishing variety of form and occurrence, making it difficult to frame a definition that is comprehensive and at the same time sufficiently precise for all cases. A commonly used definition of a vein or lode is a mineralized mass of rock which is followed for purposes of finding ore. The mineral matter may be continuous or discontinuous. There may be one definite wall, or two walls, or none at all. There may be associated gouges and altered or mineralized rock. The vein may consist of almost any combination of the elements of mineral matter, walls, gouges, and mineralized rock. Instead of being a simple tabular sheet, a vein may have almost any conceivable shape; it may consist of multiple strands of most complex relations; it may have branches and cross-over connections. It may be a more or less mineralized sedimentary formation with limits determined by original deposition. It is very often bent or folded, and even more often faulted; the faulting may be of great complexity, making it extremely difficult to follow the vein. The vein may be cut by other veins of different ages, which in places may be hard to distinguish one from another. Erosion working down on a complex vein displaced by faulting and folding may bring several parts of the same vein to the surface, developing what seem to be separate vein apices. Where there are many veins close together, it may be difficult to determine whether the entire mass should be considered a unit vein or lode (a "broad lode"), or whether each vein should be considered independently under the law.

The geologic aspects of these problems are obvious. There are few mining districts where the vein conditions are so simple that no geological problems are left to be solved with relation to extralateral rights. In the early stages of the mining, separate operations may be carried on for a considerable time in a district without mutual interference; but as mining is carried down the dip, what seemed to be separate veins may be found to be parts of the same vein or parts of a complex vein system, and separate mining organizations are thus brought into conflict. It then becomes necessary either to consolidate the ownerships or to go to the courts to see which claim has the extralateral rights. In either case, the geologist is called on to play a large part,—in the valuation of rights for the purpose of combination, or in litigation to settle apex rights. A geologic survey of the conditions is a prerequisite. In order to get the needed information for the courtroom, it may be necessary to go further, and to conduct extensive underground exploration under geologic direction. Some of the most intensive and complete geological surveys of mineral resources in existence have been done for litigation purposes. The study in these cases is not empirical, but goes into every conceivable scientific aspect of the situation which may throw any light on the underground conditions—the source of the ores, the nature and source of the solutions which deposited them, their paths of travel, the structural and metamorphic conditions, the mineralogical and chemical character of the ores and rocks, and even broader questions of geologic age. The many volumes of testimony which have accumulated during famous apex trials cover almost every phase of geology, and are important primary sources for the student of economic geology.

It is often argued that strictly scientific, impartial geologic work is impossible in connection with one of these trials, because the viewpoint is warped by the desire to win. The sharp contrast in the views of experts on the two sides is cited in evidence.

There is no denying the fact that the conditions of a trial tend toward a certain warp in scientific perspective. On the other hand, the very existence of competitive and opposing interests leads to the most intensive detailed study, and to complete disclosure of the facts. In most cases there are no substantial differences in the statements of scientific fact by reputable experts on the two sides, although there may be wide differences in the inferences drawn from these facts. The failure to note a fact, or any distortion or misstatement of a fact, is followed so quickly by correction or criticism from the other side, that the professional witness usually takes the utmost pains to make his statement of fact scientific and precise as far as his ability goes. Few scientific treatises in geology contain any more accurate accounts of mineral deposits than testimony in cases of this sort. If every student of geology, early in his career, could have a day on the witness stand on a geologic problem, under both direct and cross examination, he would learn once and for all the necessity for close and accurate thinking, the difference between a fact and an inference, and the difference between inductive study of facts and the subjective approach to a problem.

It is a common assumption that a witness called to testify on scientific matters is on a somewhat different basis from the eye-witness to an event or transaction. We are not sure that this assumption is justified. Seldom is it possible in mining operations to disclose the facts in three dimensions so completely that they may be empirically observed and platted by the layman. The grouping and presentation of the facts in adequate perspective require an analysis of the origin of the ores and rocks, the rock alterations, the structural systems, and other facts. No one ever saw the vein or lode in the process of formation. The true nature of the event and of its physical results must be inferred inductively from circumstantial evidence. If it be conceded that it is necessary and right to call an eye-witness to an event involved in litigation, it is equally necessary where there are no eye-witnesses to call the persons best qualified to interpret the circumstantial evidence.

It is to be remembered that apex cases are only one kind of a vast variety of cases affecting mineral resources. At one time or another, and in some connection or another, practically every geologist of considerable experience has found it necessary to testify on geologic matters in court. The wide interest attaching to certain spectacular apex cases has led in some quarters to hasty criticism of the participation of geologists therein, without apparent recognition of the fact that the criticism applies in principle to many other kinds of litigation and to practically all economic geologists. This criticism also fails to take cognizance of the fact that, for every case tried, there are many settled out of court through the advice and coöperation of geologists. While there may be in the geologic profession, as in others, a very few men whose testimony can be bought outright, in general it must be assumed that geologists will appear on the witness stand only when, after careful examination, they are satisfied that there is a legitimate point of view to be presented.

Geologists and engineers understand more clearly than almost any other group the extent to which the complexities of nature vary from the conditions indicated in the simple wording of the law of extralateral rights. Almost to a man, they favor either modification or repeal of the law. On the other hand, the law has been in force since 1872, it has been repeatedly interpreted and confirmed by the courts, and a vast body of property rights has been established under it. Lawyers see great legal difficulties in the way of its repeal or serious modification. Mining men for the most part are not primarily interested one way or another, unless there is potential application of the extralateral-rights provision to their particular properties. Of those who are thus interested, some hope to gain and some fear they may lose in the application of the law. The general public naturally has little direct interest in the problem. There is thus no effective public sentiment favoring the repeal or modification of the law. It seems likely that for some time to come the law, in spite of its recognized defects, must be applied, and the best geological effort must be directed toward reaching interpretations which come most near to meeting its intent. To refuse to lend geologic science to the aid of justice because the law was improperly framed is hardly a defensible position. Presumably it will never be possible to frame laws with such full knowledge of nature's facts as to eliminate the necessity for scientific advice in their interpretation.

It has been suggested that the courts, and not the litigants, should employ the geologists. The practical objection to this proposition lies in the difficulty encountered by the judge in the proper selection of geologists. On the assumption that the judge would select only men in whom he had confidence, it is not likely that he would override their conclusions. The outcome of the case, therefore, would be largely predetermined at the moment the selection of experts was made. It is to be doubted whether courts can have the knowledge of the scientific field and of the requirements of the situation necessary to make the wisest selection of men to interpret the given condition. The competitive element would be eliminated. From a judicial standpoint, there seems to be an equally good chance of getting at the best interpretation of the facts by listening to presentations from different standpoints, with the accompanying interplay of criticism and questioning.

Another practical objection to appointment of experts by the court is the limitation of court costs, which would make it impossible to secure the highest grade men. So far as these men are public employees, such as members of the federal or state geological surveys, this might be arranged. For others, it might be suggested that they should be willing to sacrifice their energy and time in the interests of justice; but as long as human nature and conditions are what they are, it is perhaps futile to argue this question.

If it is right to apply science to practical affairs, in other words, if the profession of economic geology is a legitimate one, it seems inevitable that the application must be in some part directed by the geologist himself, in order to avoid mistakes and confusion. The contention that the scientist must isolate himself in a rarified atmosphere to avoid contamination from a non-scientific, commercial, or legal atmosphere, seems to the writer practically untenable, if we recognize any obligation on the part of science to the practical conduct of human affairs. The fact that the geologist in making these applications may occasionally find himself in a non-scientific atmosphere may be deplored from the standpoint of maximum creativeness in science, and from this standpoint there may be reason for limitation of time given to this kind of work,—but to stay out entirely on this ground is to deny his obligation to make his science helpful to his fellows. The problem cannot be solved by staying out. It calls rather for an especial effort on the part of the scientist to establish and maintain his standards of science and ethics in the applied fields. Some doubtless fail in this effort. Others are strengthened scientifically and ethically, and contribute important aid in raising general standards. The principle of non-participation in such activities for fear of lowering scientific standards may make the geologist's problem easier, but at the expense of non-fulfillment of duties. Such a course has for its logical consequence an abandonment of the application of his science to untrained men without the ethical anchorage of scientific achievement. In short, there may be legitimate criticism of individual geologists for their methods and ethics in the applied field, and this is desirable as an aid to maintaining and improving standards; but it is not a logical step from this to the conclusion that, to avoid unfortunate incidents, economic geologists must cloister themselves and thus deny the very implication of their title.

II

LAWS RELATING TO EXTRACTION OF MINERAL RESOURCES

Under this heading come a wide variety of laws and regulations,—national, state, and local,—affecting the manner in which mineral resources shall be mined or quarried. Such laws may specify the number of shafts or outlets, the use of safety and prevention devices, miners' compensation and insurance, and many other features. Most of these laws are framed for the purpose of conserving human life and energy, but they directly affect the mining or extraction of the mineral resources themselves. Geology plays but little part in relation to such laws.

Where the government retains ownership and leases or rents the resources, there are often provisions regarding the manner of mining and the quality and quantity of the material to be mined, in the interests of efficient operation and conservation. The geologist is often called into consultation both in framing and in dealing with the infraction of such provisions. It may be noted that the control thus exercised on the operator by government ownership is very much the same as that often exercised by the private fee owner. It is not unusual for fee owners of mineral rights to maintain a geological staff in order to follow intelligently underground developments, to see that the best methods of exploration and mining are followed, and that ores are either extracted or left in accordance with the best conservational practice.

III

LAWS RELATING TO DISTRIBUTION AND TRANSPORTATION OF MINERAL RESOURCES

Under this heading come governmental regulations affecting directly or indirectly the transportation and the destination of mineral products. Transportation rates, tariffs, zoning, duties, and international trade agreements of all sorts have vital effects on distribution. In framing any of these measures for a mineral resource, it is desirable to know all about the character of the raw material, its physical occurrence and distribution, and the possibilities for future development. In adjusting the scientific naming and classification of mineral materials with the crude names and classifications used commercially—as in tariffs, in import and export laws, in reports of revenue collectors, in railway and ship rates, etc.—the geologic information is likewise necessary.

Heretofore, the formulation of measures concerning mineral distribution has often not been done on a scientific and impartial basis; but in recent years geologists have been called on more frequently for aid and advice, as a means of checking or verifying the special pleadings of the different industries. The rude disturbance of trade routes during the war brought home the necessity of basing control of distribution of mineral products on fundamental facts of geology and geography; thus it was that geologists had a considerable voice in the vast number of special measures taken for war purposes by such organizations as the Shipping Board, the War Trade Board, the War Industries Board, and other public organizations. The same was true in relation to the mineral resource questions at the Peace Conference. In the reconstructive measures of the future, a still larger use of scientific considerations may be looked for. Further suggestions as to the relation of geology to laws affecting distribution appear in the chapter on International Aspects (Chapter XVIII).

IV

OTHER RELATIONS OF GEOLOGY TO LAW

It is often assumed that the economic geologist is exclusively interested in mineral resources. However, there are varied applications of geology outside of the mineral resource field,—to many kinds of engineering and construction operations, to soils, to water resources, and to transportation,—any of which may develop legal problems requiring geologic service. A few illustrative cases follow.

The classification of mineral materials in contracts presents many difficulties. A contract for a railway cut, for a canal, or for any other kind of excavation may specify different prices for removing different mineral materials. Too often these are stated in extremely crude and arbitrary terms, such as *rock*, *hard rock*, *hardpan*, *earth*, *dirt*, etc., without regard to the actual variety of materials to be dealt with. When, therefore, in the case of the Chicago drainage canal, the contractor encountered a soft shale and claimed compensation for rock excavation, geologists played a considerable part in the extensive litigation that followed in the attempt to define the facts of nature in terms of a contract which did not recognize them. In a railway cut through glacial drift or till, a contractor came suddenly upon a mass of till which had been so thoroughly cemented in place as to have all the resistance of rock. Litigation was then necessary to decide whether this should be classified as dirt or rock.

Rock and dirt slides of all kinds, met with in open-pit mining, canals, and other excavations, present engineering problems with a geologic basis. The kinds of rocks, their strength, porosity, and moisture content, the effects of weathering, and the structural conditions must be determined in order to ascertain the cause of the slides, and are features which figure largely in litigation arising from troubles of this sort.

Both federal and state laws give the right to lateral and vertical support. When, therefore, adjacent or underlying excavations cause earth movements in a neighbor's property, litigation is likely to ensue and the geologist is likely to be called in. The long-wall method of coal mining, extensively practiced in certain parts of the United States, is slowly withdrawing support from the ground overlying the coal seams, resulting in damages to surface structures and in some cases to overlying mineral deposits. Extensive litigation has been the result, and the future seems to promise more of it. In certain metal-mining camps, where considerable amounts of materials have been mined to great depths, caves and cracking in the surface are reaching over unexpectedly wide areas, again threatening litigation.

The laws relating to the use of surface and underground waters touch the geologic conditions in many ways. The permanent lowering or raising of a water level through mining or damming may require a careful geological analysis of the underground conditions affecting the movements of ground-water. The use of streams for placer mining, as in California, has resulted in formulation of laws and in extensive litigation, again requiring analysis of geologic conditions.

In fact geologists, perhaps more than any other group, have come to realize how many and how varied are the ways in which people get into conflict in using the earth on which they live.

CHAPTER XVII

CONSERVATION OF MINERAL RESOURCES

THE PROBLEM

Conservation of mineral resources may be defined as an effort to strike a proper balance between the present and the future in the use of mineral raw materials.

Mineral resources have been used to some extent as far back as evidences of man go, but great drafts on our resources have come in comparatively recent years. The use of many minerals has started within only a few years, and for others the acceleration of production within the past two or three decades has been rapid (see pp. 63-64). In general, the use of mineral resources on a large scale may be said to have started within the lifetime of men still active in business. The wide use of power necessary to an industrial age, the development of metallurgy, the increasing size and complexity of demands for raw material, mean that the intensive development and use of our mineral resources is in its infancy, and is in many respects in an experimental stage.

As nations have awakened to their need of mineral raw materials and to the recent rapid depletion of these materials, they have been naturally led to inquire how long the reserves may last, and to consider prevention of waste and the more efficient use of materials, with a view to planning more prudently for future national supplies. The first inquiries seemed to reveal such shortage of mineral supplies as to call for immediate and almost drastic steps to prevent waste, and possibly even to limit the use of certain minerals in the interests of posterity.

More careful study of the problem, as might be expected, revealed new factors and greater complexity. The conservational idea has a wide sentimental appeal, but the formulation and application of specific plans meet many difficulties. In its practical aspects the problem is now a live one, the solution of which is requiring the attention of mining men, engineers, geologists, economists, and public officials. It is a question which is coming more and more into the field of actual professional practice of the economic geologist.

It is our purpose to indicate the general nature of the conservation problem. We may assume agreement to the desirability of preventing waste, of making a wise present use of mineral products, and of striking a proper balance between the present and future in their use. Nature has taken many long geologic periods to build up these reserves. We, of the present generation, in a sense hold them in trust; they are entailed to our successors. With this general thought in mind, how shall we proceed to formulate definite plans for conservation?

An initial step is obviously a careful taking of stock. With increasing knowledge of mineral resources, it is becoming apparent that early estimates of supplies were too low. Many of these estimates failed to take into account mining to great depths, and wide use of low-grade ores, rendered possible by improved methods; and especially they failed to put sufficient emphasis on the probabilities of new discoveries to replace exhausted supplies. Early predictions have already been upset in regard to a number of mineral resources. The recognition of the general fact that the world is far from explored in two dimensions, to say nothing of three, of the fact that known geologic conditions do not yet indicate definite limits to the possibilities of exploration for most mineral resources, and of the consequent fact that for a long time in the future, as in the past, discoveries of new mineral deposits will be roughly proportional to the effort and money spent in finding them,—which means, also, proportional to the demand,—makes it impossible, for most of the mineral resources, to set any definite limits on reserves. It is comparatively easy to measure known reserves; but a quantitative appraisal of the probable and possible reserves for the future is extremely difficult. Successive revisions of estimates have, with but few exceptions, progressively increased the total mineral supplies available. The result is that the time of exhaustion has been pushed far into the future for most of the important minerals, thus minimizing the urge for immediate and drastic conservational action, which followed naturally from early estimates of very limited supplies. For both coal and iron, supplies are now known for hundreds or even thousands of years. For oil and lead, on the other hand, the reserves now known have a life of comparatively few years, but the possibilities for successful exploration make it probable that their life will be greatly extended. Notwithstanding this tendency to lengthen the exhaustion period, the limits of mineral resource life are still small as compared with the life of the nation or of civilization,—and the fundamental desirability of conservation is not materially affected.

It is not easy to predict the rate of production for the future. At the present rate of coal production in the United States, the supplies to a depth of 6,000 feet might last 6,000 years; but if it be assumed that the recent *acceleration* of production will be continued indefinitely into the future, the result would be exhaustion of these supplies in less than 200 years. It is generally agreed that exhaustion will come sooner than 6,000 years, but will require more time than 200 years. The range between these figures offers wide opportunity for guessing. It is supposed that per capita consumption may not increase as fast in the future as in the past, that possibly an absorption point will be reached, and that there will be limits to transportation and distribution; but how to evaluate these factors no one knows. In the case of some of the metallic resources, such as iron, the fact that the world's stock on hand is constantly increasing—losses due to rusting, ship-wrecks, etc., being only a small fraction of the annual output—suggests that a point will be reached where new production will cease to accelerate at the present rate and may even decline. But again, the factors are so complex and many of them so little known, that no one can say how soon this point will be reached.

For the immediate future, there is little to be feared from shortage of mineral supplies in the ground. The difficulties are more likely to arise from the failure of means to extract and distribute these supplies fast enough to keep up with the startling acceleration in future demand indicated by the figures of recent years. The speed and magnitude of recent material developments in many lines cannot but raise question as to whether we have the ability to understand and coödinate the many huge, variable, and accelerating factors we have to deal with, or whether some of the lines of development may not get so far ahead of others as to cause serious disturbance of the whole material structure of civilization. Coal alone, which now constitutes a third of our railway tonnage, may with increased rate of production require two-thirds of present railway capacity. Will railway development keep up? It may be noted that national crises and failures in the past history of the world have seldom, if ever, been due to shortage of raw materials, or in fact to any failure of the material environment.

In its early stages the conservation movement in this country concerned itself principally with the raw material. Later there came the recognition of the fact that conservation of raw materials is closely bound up with the question of conservation of human energy. The two elements in the problem are much like the two major elements in mineral resource valuation (see pages 329-330). If in saving a dollar's worth of raw material, we spend two dollars worth of energy, it naturally raises question as to the wisdom of our procedure. It might be wiser in some cases to waste a certain amount of raw material because of the saving of time and effort. It might be better for posterity to have the product of our energy multiplied into raw material than to have the raw material itself. The valuation of these two major elements of conservation is again almost impossible of quantitative solution. We do not know what is the best result to be aimed for. We cannot foresee the requirements of the future nor the end toward which civilization is moving—or should move. The extravagance of the United States is often contrasted unfavorably with the thriftiness of Europe. When considered in relation to raw materials alone, there seems to be basis for this charge. When considered in relation to the product of human energy into raw materials, the conclusion may be far different; for the output per man in the industries related to mineral resources is far greater in the United States than in Europe. In the case of iron, it has been estimated that the output per man in the United States is two and one-half times as great as in the rest of the world. Which is best in the true interests of conservation, we are not yet able to see.

Our view of what is desirable in the way of conservation depends somewhat on the limitations imposed by self-interest or location. By devoting ourselves exclusively to one mineral resource, we might work out a conservation program very disadvantageous to the best use of some other mineral commodity. We might take steps to conserve chromite in the United States which would have a disastrous effect on the iron and steel industry. We might conserve coal by the substitution of oil, when the procedure is hardly warranted by the supplies of oil available. We might work out a program for the United States which would not be the best conservational plan for the world as a whole, and which would ultimately react to the disadvantage of the United States. The wisest and most intelligent use of mineral resources seems to call unquestionably for their consideration in their world relations, rather than for a narrow interpretation of local requirements.

DIFFERENCES BETWEEN PRIVATE AND PUBLIC EFFORTS IN CONSERVATION

It appears that a wide range of effective conservational practices has resulted solely from the effort to make more money through more efficient operations, and this is likely to be true in the future. Many improvements in mining, grading, sorting, concentration, and metallurgy of minerals, to yield larger financial returns, are coming naturally through private initiative, under the driving power of self-interest.

Another considerable group of conservational practices is possible only to governments or other public agencies. This group of practices on the whole requires some sacrifice of the immediate financial interest of the individual, in the interests of the community as a whole, or in the interests of posterity. In this group may be mentioned the compulsory use of methods of mining, sorting, and metallurgy which tend to conserve supplies but result in higher prices; the control of prices; the elimination or lowering of the so-called resource or royalty value (p. 375); and the removal of restrictions on private combination or coöperation, leading to more efficient methods, lessening of cost, and better distribution of the product; or, what might amount to the same thing, the acquirement by the government of the resources to be operated on this larger scale.

The most effective conservation measures yet in effect are the ones dictated by self-interest and instituted by private initiative. Governmental measures are not yet in effective operation. Illustrations of these two types of conservational effort are cited in relation to coal on later pages.

THE INTEREST RATE AS A GUIDE IN CONSERVATION

In striking a balance between the present and the future, economists have emphasized the importance of recognizing the interest rate as a guiding, if not a controlling consideration. It is obviously difficult for private capital to make investments of effort and money for the purpose of conservation which will not be returned with interest some time in the future. For the present, at least, this consideration furnishes the best guide to procedure in the field of private endeavor. So far as conservational measures, such as investment in an improved process of concentrating low-grade ores, promise return of capital and an adequate interest rate in the future, they are likely to be undertaken.

It is clear that governments are not so closely bound by this economic limitation. They can afford to carry their investments in raw materials and processes at a lower interest rate than the private investor. Their credit is better. Taxes do not figure so directly. They can balance losses in one field against gains in another. As a matter of insurance for the future of the nation, a government may feel justified in inaugurating conservational measures for a particular resource without hope of the interest return which would be necessary to the private investor. In appraising the iron ores of Lorraine taken over by France from Germany at the close of the war, the actual commercial value of these ores, as figured by the ordinary *ad valorem* method, was only ninety millions of dollars. It is clear, however, that to France as a nation the reserves were worth more. They could afford to pay more for them, and could afford to spend more money on conservational practice than under ordinary commercial limitations, because of the larger intangible and more or less sentimental interest.

The valuation of this larger interest, as a means of determining the limit to which conservational investments may be made, lies in the political field. It may be suggested, however, that a desirable first step in any governmental program of conservation is to ascertain the cost and the possibility of an adequate return of capital and interest. These determinations at least afford a definite point of departure, and a means for measuring the cost to the people of measures which are not directly self-supporting.

ANTI-CONSERVATIONAL EFFECTS OF WAR

Experience during the recent past indicates that the exploitation of mineral resources for war purposes is on the whole anti-conservational. It is true that the vast amount of war-time exploration and development, as well as the thoroughgoing investigations of the utilization of various minerals, have led to better knowledge of the mineral resources and their possibilities. It is also true that the war required a much more exhaustive census of mineral possibilities than ever before attempted. The immediate and direct effect of the war, however, was the intensive use of mineral resources without careful regard to cost, grade, or many other factors which determine their use in peace times. For instance, in ordinary times considerable quantities of high-phosphorus iron ores are mined; but, because of the fact that such ores require more time for conversion into steel, war-time practice concentrated on the higher-grade, low-phosphorus ores, resulting in an unbalanced production which in some cases amounted almost to robbing of ore deposits. In the case of coal, quantity was almost the only consideration; it was impossible to grade and distribute the coal to meet the specialized demands of industry. The results were a general lowering of the standards of metallurgical and other industrial practices, and increased cost. High-grade coals were used where lower-grade coals were desirable for the best results. In the making of steel, it is the custom to select the coal and coke with great care in regard to their content of phosphorus, sulphur, ash, and other constituents which affect the composition of the steel product; but during the war it became necessary to accept almost any kind of coal, with a resulting net loss in quantity and in grade of output.

For a considerable number of mineral resources, such as the ferro-alloys, foreign sources of supply were cut off during the war, requiring the development and use, at high cost, of low-grade scattered supplies in the United States. It was found possible to produce enough chromite in the United States for domestic requirements, but at two or three times the normal price of imported chromite. The grade was low and the loss in efficiency to the consuming interests was a high one. The extremely limited natural supplies were raided almost to the point of exhaustion.

With the post-war resumption of importation of minerals of this kind, producers naturally began a fight for a protective tariff, and the question is yet unsettled. The tariff, if enacted, would in most cases have to be a high one in order to permit the use of domestic supplies. The results would be a large increase in cost to other industries, decreased efficiency, and the early exhaustion of limited supplies in this country. Most of the mineral resources have been concentrated by nature in a comparatively few places in the world; and when the two elements of conservation are considered—the materials themselves and the human energy expended in obtaining and using them—it is clear that any measure which interferes with the natural distribution of the favored ores is anti-conservational from the world standpoint.

CONSERVATION OF COAL

In the sections on mineral resources, there are many casual references to conservation of specific minerals. Here we shall not go further than to introduce a brief discussion of the conservation of coal as illustrative of the general problem of conservation of mineral resources.

It has been estimated that the United States possesses, to a depth of 3,000 feet, in beds 14 inches or over, 3,538,554,000,000 tons of coal, and an additional reserve between 3,000 and 6,000 feet of 666,600,000,000 tons.[42] If all the unmined coal to a depth of 3,000 feet could be placed in one great cubic pile, the pile would be 18 miles long, 18 miles wide, and 18 miles high. Of the original amount of coal to this depth only about 0.4 of 1 per cent has been mined or wasted in mining. The wastage is estimated at about 50 per cent. If the annual production of coal were to remain the same as in recent years, the total life of the coal reserves (to a depth of 3,000 feet) would be between 4,000 and 6,000 years; but if the acceleration of production of recent years were to be maintained in the future, the life would be but little over 100 years, and the life of the highest-grade coal now being mined might not be over 50 years. All agree that the acceleration of production is not likely to continue indefinitely, which will mean that the life of coal reserves to 3,000 feet will be somewhere between the two extremes named. It seems clear that actual shortage of coal will not be felt for some hundreds of years;

but this period of years is short as compared with the probable life of the race.

MEASURES INTRODUCED OR PROPOSED TO CONSERVE COAL

The following list of measures for conservation of coal is taken from several sources. The exhaustive report of the British Coal Commission,[43] published in 1905, contains a considerable number of specific recommendations for conservation of the coal of Great Britain. The reports of the National Conservation Commission[44] of the United States, published in 1909, treat of the conservation of the coal of the United States and naturally follow some of the recommendations of the British report. The coal section of the National Conservation report was prepared by M. R. Campbell and E. W. Parker of the U. S. Geological Survey, and is contained in U. S. Geological Survey Bulletin 394. The recommendations there given are amplified and developed by Van Hise[45] in his book on Conservation, published in 1910. Since that time the subject has been discussed by Smith, Chance, Burrows, Haas,[46] and others, and certain additional conservational methods have been proposed. A considerable number of men have also discussed the sociologic and economic aspects of the question. The report of the Conservation Commission of Canada,[47] published in 1915, treats rather fully of the conservation of mineral resources.

It will suit our purpose, and avoid some repetition, if we group most of these recommendations without regard to authorship. In general, these recommendations can be grouped under the heads: (A) Methods of mining and preparation of coal; (B) Improvement of labor and living conditions at the mines; (C) Introduction or modification of laws to regulate or to remove certain restrictions on the coal industry; (D) Distribution and transportation of coal; (E) Utilization of coal; (F) Substitutes for coal as a source of power.

(A) Mining and preparation of coal. Under this heading may be included a large number of proposals which concern primarily the engineering treatment of the coal underground and in the mine plants. Some of the more important measures are:

1. Introduction of the long-wall system of mining in places where the conditions allow it, in order to minimize the waste underground.

2. Modification of the room-and-pillar system of mining, by which larger pillars are left while the mine advances, and are recovered in the retreat,—thereby recovering a larger percentage of coal than under the old system, where small, thin pillars were left, which failed and were permanently lost.

It has been argued that the great loss of coal by leaving it in pillars could be saved by using other material to support the roof; but an elementary calculation of the cost of this procedure shows that it is cheaper to use the coal. Chance[48] says:

The coal left as pillars to support the roof is thus utilized and performs a necessary and useful function, yet the principal part (perhaps two-thirds) of the 200,000,000 tons our friends the conservationists claim is wilfully and avoidably wasted every year is this coal that is left in pillars to support the roof. I think we can safely claim that this is not waste, but, on the contrary, is engineering efficiency of the highest type, in that it utilizes the cheapest and least valuable material available to support the roof and saves the whole labor cost of building supports of other materials. Investigation as to what becomes of that part of the 200,000,000 tons claimed as wasted, which is not utilized as pillars to support the roof, will disclose the fact that a very large portion is coal that is left in mine workings that are abandoned because the roof is unsafe and because a continuance of operation would result in injuries or loss of life. Coal left in the mines in order to conserve human lives cannot be classed as avoidable waste. A small part of the 200,000,000 tons is lost because it is intimately mixed with refuse and because the labor cost of recovering it and separating it from the refuse would be greater than its value.

3. Mining of shallow bituminous beds by means of the steam shovel. Progress has been made along this line in the last few years, and valuable deposits are thus mined which can be mined profitably by no other method.

4. New methods of filling mined-out spaces with sand, and new methods of mine survey and design. According to Haas[49]

the greatest advance in the question of method was the system of mine survey and design perfected in both the anthracite and bituminous fields. The relatively new method of filling old spaces with sand, etc., has also achieved success.

5. Use of methods by which coal is not left in the roof for the support where the roof is weak, and by which coal of inferior quality is not left in the roof.

6. Wider use of coal-cutting machines by which the wasting of thinner beds may be avoided.

7. Where conditions allow it, the working of the upper beds before the lower, in order not to destroy the upper ones by caving. The mining of a lower coal seam has often so broken up the overlying strata as to render it impossible to recover the upper coal seams contained therein. There are certain difficulties, however, in the way of this conservational measure. In some localities the seams are under separate ownership, and there is a resulting conflict of interests. Also, if the better coal seam happens to be below and the poorer seams above, market conditions may require that the lower seam be mined regardless of the destruction of the upper ones.

8. Elimination of coal barriers to mark the limits between properties. This involves more coöperation.

9. Improvement of mining machinery, power drills, etc.

10. Centralization of power stations, rather than the use of many small units.

11. Elimination of the wasting of slack or fine coal, through more careful methods of mining, through limitations on the excessive use of powder and larger use of wedges, through the abolition of laws for the payment of miners on a run-of-mine basis, and in the case of anthracite through recovery of the "silt" or dust caused by mining and sorting. It has been argued that the excessive use of powder ("shooting from the solid") means loss of coal, owing to the fact that it shatters the coal and makes a relatively large amount of slack, besides being accompanied by increased danger from fire and explosion and from weakening of the roof. Although the excessive use of powder makes a large amount of slack, it does not necessarily result in waste, for this fine coal is carefully saved and for certain purposes is as valuable as the lump coal. So far as the procedure endangers life, it is of course objectionable.

12. Better use of fine coal. It has been recommended that infirm and finely broken coal be washed and compressed, thus

avoiding the wasting of slack coal, which was formerly thrown away or burned. However, in recent years there has been comparatively little waste of this kind, for slack coal in general finds nearly as ready a market as lump coal and the use of slack is increasing. There has been much discussion also of the possibilities of using the coal waste on the ground to make power for electric transmission.

13. More careful attention to sorting and sizing of all grades of coal coming from the mine and to preparation of coals for special uses. On the other hand, some operators say that the ends of conservation will be best met by limiting the sorting and sizing now practiced. The large number of sizes now put on the market greatly increases the cost of production.

14. Wider use of the lower-grade fuels of the west, particularly with the aid of briquetting.

Progress in above methods. Methods of mining and preparation of coal have been improved. Campbell and Parker state:[50]

A much greater proportion of the product hoisted is now being sent to market in merchantable condition. Part of this is due to better and more systematic methods of handling, and part to the saving of small sizes which formerly went to the culm banks. The higher prices of coal and the development of methods for using these small sizes have also made it possible, through washing processes, to rework the small coal formerly thrown on the culm banks, and these are now furnishing several millions of tons of marketable coal annually.

In general there is increase in the percentage of recovery of coal. Whereas in the past the loss in mining was said by Campbell and Parker[51] to average 50 per cent, now an extraction of 70 to 90 per cent may be looked for.

Quoting from Smith and Lesher:[52]

Observation of the advances made in mining methods in the last decade or two affords slight warrant for belief in any charge of wasteful operation. As consumers of coal we might do well to imitate the economy now enforced by the producers in their engineering practice. In the northern anthracite field machine mining in extracting coal from 22- and 24-inch beds, and throughout the anthracite region the average recovery of coal in mining is 65 per cent., as against 40 per cent. only twenty years ago. Nor are the bituminous operators any less progressive in their conservation of the coal they mine.

In anthracite mining, powdered coal or "silt" has accumulated in stockpiles and in stream channels to many tens of millions of tons. It is estimated that this constitutes nearly 6 per cent of the coal mined. Significant progress has been made recently in the recovery and use of this silt as powdered fuel for local power purposes.

However, physical and commercial conditions do not in all cases allow of the full application of these new methods. Once a mine has been opened up on a certain plan, it is difficult to change it. As a whole the longer and better organized companies are better able to change than the smaller companies.

Conservation measures of the above kinds, as so far applied, have come mainly from private initiative based on self-interest,—though the coöperation of the government has been effective, particularly along educational and publicity lines.

(B) Improvement of labor and living conditions at the mines. Under this heading should be mentioned the improvement of housing, sanitation, and living conditions; improvements in the efficiency of labor, through making living conditions such as to attract a higher-grade labor supply and through educational means; the introduction of safety methods; the introduction of workmen's compensation and insurance; and other measures of a similar nature. All these measures as a class are sometimes grouped under the name of "welfare work."

Much thought and discussion have been devoted to the possibilities of improvement of labor and living conditions from the standpoint of conservation of human energy. In some quarters this subject has been treated as being independent of the physical conservation of mineral resources, and it has been the tendency to assume that conservation of human energy might be more or less inimical to conservation of mineral resources. Certain of the changes already introduced have undoubtedly increased the cost of mining; and, until there was a general increase in selling price, this increased cost may have had the effect of eliminating certain practices of mineral conservation which might otherwise have been possible. For instance, according to Smith and Lesher:[53]

The increased safety in the coal mines that has come through the combined efforts of the coal companies, the state inspectors, and the Federal Bureau of Mines necessarily involves some increase in cost of operation, but the few cents per ton thus added to the cost is a small price to pay for the satisfaction of having the stain of blood removed from the coal we buy. That form of social insurance which is now enforced through the workmen's compensation laws alone adds from 2 to 5 cents a ton to the cost of coal.

On the other hand, there can be no doubt that large advances have been made in welfare movements which were introduced for the purpose of insuring a steadier, better, and larger supply of labor, and that the general gain in efficiency of operation thereby obtained has absorbed a large part of the increased cost.

In general, conservation measures of this class have been developed coöperatively by private and public efforts, without important sacrifice of private interest. There is obviously room for much wider application of such measures, especially in some of the bituminous fields where conditions are still far from satisfactory.

(C) Introduction or modification of laws to regulate or to remove certain restrictions on the coal industry. It has been proposed:

1. To modify the laws so as to take care of situations where vertically superposed beds are owned by different parties, preventing the proper mining of the coal by either party.

2. To modify the laws so as to eliminate conflict in mining practice in cases where the coal is associated with oil and gas pools.

3. To allow larger ownership by companies utilizing the coal (now only 3 per cent owned by such companies).

4. To place restrictions on over-capitalization, which leads to wasteful mining in order to secure quick and large returns on large capital.

5. To remove restrictions on concentration of control. This means, as a corollary proposition, virtual restriction of competition. Concentration of control into comparatively few hands has undoubtedly favored conservation. It is easy to see that the stronger financial condition of the large companies makes it possible for them to take fuller advantage of modern methods of extraction, distribution, and marketing.

This proposal was especially urged for the bituminous coal industry before the war in order to avoid over-production and over-development. The very wide distribution of the bituminous coals, their enormous quantity, and their exceedingly diversified ownership had led to over-development of coal properties. Quoting from Smith and Lesher:[54]

In estimating the aggregate losses incurred by society by reason of the large number of mines not working at full capacity, the facts to be considered are that the capital invested in mine equipment asks a wage based on a year of 365 days of 24 hours, while labor's year averaged last year only 230 days in the anthracite mines and only 203 days in the bituminous mines, with only five to eight hours to the day.

These conditions prevented in some cases even the most modest introduction of better methods, or of changes that would enhance the average profits through a relatively short period of ten or fifteen years at the expense of the present year. It was necessary to get at the best of the coal available in the cheapest possible way, regardless of the losses of coal left in the ground.

To some extent the force of this argument was minimized by war and post-war conditions, but even yet development of coal mines is ahead of transportation and distribution.

6. To allow coöperation in the limitation of output, in the avoidance of cross freights, in gauging the market in advance, and in division of territory, all of which would allow cheaper mining and thus give larger leeway to conservational measures. This necessarily would be accompanied by government regulation. According to Van Hise,[55] who was active before the war in advocating this conservational measure, such a procedure

is neither regulated competition, nor regulated monopoly; but the retention of competition, the prohibition of monopoly, permission for coöperation and regulation of the latter. In

Chicago there cannot be one selling agency for the different coal companies which operate in Illinois, but there must be many selling agencies, and the coal of Pittsburgh must come into Illinois and the Illinois coal go toward Pittsburgh; every one of which things makes unnecessary costs, but all of which are inevitable under the extreme competitive system. Because of these facts it is necessary to waste the coal. If at the very same prices the different mines could coöperate in the limitation of the output, avoidance of cross freights, gauging the market in advance, and division of territory, they could mine their coal more cheaply, have a greater profit for themselves and conserve our resources.

To some extent the plan here advocated was put into effect during the war by the United States Coal Administration; but the conditions of this trial were so complicated by special war requirements, that the conservational advantages of unified control were not demonstrated.

7. To reduce the excessive royalties paid to fee owners. Smith and Lesher[56] have recently called attention to the relatively high resource cost in some of the coal fields, represented by the payment of royalties to fee owners. In the case of anthracite the payment averages 32 to 35 cents per ton, and exceptionally runs as high as a dollar per ton. For the bituminous coal the average resource cost is probably not much over five cents a ton. They suggest the possibility of lowering this cost by governmental regulation; and make an especially strong argument for not allowing the government-owned coal lands to go to private owners, who in the future, with the accumulation of interest on the investment, will feel justified in asking for a large "resource" return in the way of royalty.

If the resource cost could be lowered, further introduction of conservational methods by the operators would be possible without greatly increasing the cost to the public.

8. To require or allow, by government regulation, a raising of the price of coal to the consumer, thereby allowing wider application of conservational practices. Some of the increased recoveries of coal above noted have been made possible only by increase in the market price. If coöperation were permitted in the manner described in paragraph 6, the same results might be accomplished without increasing the price. Recent high prices caused by the war situation are reflected in the introduction of

many conservational changes which were not before possible. However, in some cases the demand for quick results under present conditions has an opposite effect, because of the desire to realize quick profits regardless of conservation.

9. The local conservation of coal at the expense of heavier drafts on coal of other parts of the world, by imposition of export taxes and preferential duties, has been discussed. While the effect of such a measure would doubtless be conservational from the standpoint of the United States, it is doubtful if it could be so regarded from the broader standpoint of world civilization. Under present world conditions such a step would be disastrous.

10. Government ownership has been proposed as a means of facilitating the introduction of conservation measures. In the United States there is yet no major movement in this direction. In England the question of nationalization of coal mines is an extremely live political problem (see pp. 343, 345-347).

Little progress has been made in conservation measures which involve legal enactments of the kinds above listed.

(D) Distribution, and transportation of coal. It has been argued that conservational results would ensue from:

1. Cheaper transportation.

2. Larger use of waterways.

3. Improvement in distribution of the product by partition of the market and by larger use of local coals. For effectiveness this proposition would have to include control of the agencies of distribution, in order to minimize excessive profits of middlemen.

4. Purchasing and storage of coal by consumers during the spring and summer months in anticipation of the winter requirements, in order to equalize the present highly fluctuating seasonal demands on the mines and railroads, and to eliminate the recurring shortages of coal in the winter months. This was particularly recommended by the United States Bituminous Coal Commission in a recent report.[57]

5. Where conditions allow it, conversion of coal into power at the mine and delivery of power rather than coal to consuming centers. This type of conservation is being put into practice on a large scale above Wheeling, on the Ohio River, where there has

recently been built a two hundred thousand kilowatt installation for steam-generated electric power. Some of the power will be delivered to Canton, Ohio, over fifty miles away. This plant uses local coal and the cost of coal is figured at two mills per kilowatt-hour.

Under this heading of distribution and transportation of coal, might be considered certain international relations. The international movements of coal are summarized in another place (pp. 115-117). Anything in the way of tariffs or trade agreements which would tend to interfere with or to limit the great natural international movements of coal—which in a free field are based on suitability of grade, cost, location, transportation, etc.—would be anti-conservational from the world's standpoint, although they might be of local and temporary advantage. For instance, the coal exported from England, which has heretofore dominated the international trade of the world, is of a high grade. American coal available for export is on the whole of considerably lower grade, being higher in volatile matter. Unless this coal is beneficiated at home, it can replace the English coal in the export field only at increased cost of transportation and lower efficiency in use. The time may come when it will be desirable to ship lower-grade coals long distances; but when the two factors of conservation are considered—the intrinsic qualities of the coal, and the efforts necessary to utilize it—it would seem to be conservational at this stage to ship to long distances only the coal which nature seems specially to have prepared for this purpose.

(E) Utilization of coal. Conservational proposals of this kind are:

1. Substitution of retort coke-ovens for beehive ovens, to save not only a larger quantity of coke but also valuable by-products (see pp. 118-119). Additional improvements in coking ovens may make possible the manufacture of some sort of coke from a much wider range of bituminous coals than can be used at present.

2. Larger use of smoke consumers and mechanical stokers.

3. Larger use of central heating plants, with higher efficiency than many local plants.

4. Substitution of gas engines for steam engines, and improvement of the steam engine.

5. Improvement in methods of smelting, leading to larger output of metal per ton of coke used. Also the development of electric smelting for certain metals.

6. More careful study and classification of the qualities of coals, in order to avoid use of higher-grade coals where inferior coals would serve the purpose.

7. More consumption at the collieries.

8. Larger use of powdered coal as fuel.

9. Improvement of force-draft furnaces.

10. Larger use of gas, a by-product of coal mining, and extraction of other by-products.

11. More efficient transformation of peat and coal into power and light.

12. The possible use of oil flotation to eliminate foreign mineral matter.

Most of the conservation measures above proposed have already been applied with good results, and with promise of large results for the future. The stimulus has come largely from self-interest. War conditions in some ways aided and in others hindered these developments. One of the conspicuous gains was the building of many by-product coke plants, under the necessity of securing the nitrates and hydrocarbons for munition and other purposes.

(F) Substitutes for coal as a source of power. Some of the more prominent measures along this line which have been discussed are:

1. Larger use of water power. This has sometimes been popularly assumed to be, at least potentially, a complete solution of the problem; but nevertheless it has its distinct limitations.

Water power has the advantages that its sources are not exhausted by use, and that the relatively greater initial cost of a hydro-electric plant is frequently more than compensated for by the saving in man power required and by the lower operating expense. However, the total amount of water power which can be developed on a commercial basis is rather closely limited, and much of the available power is so distributed geographically that it cannot be economically supplied to the industries which need it. Of the total water-power resources of the United States which have been estimated by the Geological Survey to be available for ultimate development, over 70 per cent is west of the Mississippi,—whereas over 70 per cent of the horse-power now installed in prime movers is east of the Mississippi. Electric power cannot at present be economically transmitted more than a few hundred miles. Furthermore, for many uses of coal, as in metallurgical and chemical processes which require the heat or reducing action of burning coal, and in its use as fuel for ships, hydro-electric power cannot be substituted. It seems clear that while the use of water power will increase, particularly as rising prices of coal make possible the development of new sites, it can never take the place of the mineral fuels in any large proportion.

For the immediate future, measures which have been suggested to extend the use of water power include: the more complete utilization of water powers already in use through more efficient machinery and methods; a certain degree of redistribution of industries, so that those requiring large amounts of power may be located in areas where water power is cheap and abundant; and the interconnection of hydro-electric plants so that their full capacity may be used. Some water powers which have been developed are not being fully utilized because the plants are not connected with distribution systems large enough to use all the power. During the war the United States Geological Survey, in coöperation with the Fuel Administration and the War Industries Board, collected the information required to prepare maps showing the locations and relations of power stations and transmission lines throughout the country. This survey of the situation showed many possibilities, which had before been but vaguely realized, of interconnections which would increase the efficiency of the plants.

2. Substitution of lower-grade coals—of bituminous for anthracite, and of low-grade bituminous for high-grade bituminous coals. Larger use of low-grade western coals. War and post-war conditions have shown Germany the way to a wide and effective use of its lignites. This has been accomplished by coöperation of the government and private interests. This vast improvement in methods of treatment and recovery of heating elements and by-products will doubtless have a widespread effect on utilization of lignites in other parts of the world.

3. Substitution of alcohol and natural gas, oil, oil shales, peats, etc., as a source of power. This merely concentrates the conservation problem more largely on these minerals, in some of which, at least, it is already considerably more acute than in the case of coal; it is not a solution of the problem, but merely a shifting of emphasis.

Business conditions have limited private enterprise in this class of measures, but some progress has been made. More rapid introduction of these measures would require sacrifice of private interest and probably may be accomplished only by application of public power.

Division of Responsibility Between Government and Private Interests in the Conservation of Coal

A review of the conservation measures above listed indicates that many of them are already in operation, and that the initiative for such measures has been largely supplied by private ownership endeavoring to advance its own interest. In this category are to be included most of the improvements in physical methods of mining, preparation, and utilization of coal, the use of substitutes for coal, the concentration of control into larger groups better able to introduce new methods, and the improvement of labor and living conditions; also, under recent conditions, the increase in selling price, allowing for a wider application of these measures. Another group of conservation proposals, which have not yet been put into substantial effect, are obviously beyond the power of private interests; and must be introduced, if at all, by the application of government power. These include the elimination of resource or royalty costs, the control of over-capitalization, the removal of restrictions on concentration of control, the granting of permission for coöperation among competitive units, the regulation of selling price minimums in order to insure during normal times the use of better physical practices, and the control of distribution. In short, it appears that there are two great spheres of conservational activity—one within the field of private endeavor, and the other possible only by collective action through the government. The principal advances thus far made have been in the field of private endeavor.

The government has aided greatly in the advancement of conservation measures arising within the field of private endeavor. One need only refer to many governmental investigations, to the spreading of information as to best methods, and to local compulsory requirements that the best practices be made uniform and that backward interests thereby be brought into line.

Recognition of the fact that there is a large body of sound conservational practice in the coal industry which falls within the range of self-interest seems essential in planning further changes in the direction of conservation. Conservational measures do not all require sacrifice of the individual to the public, nor of the present to the future generations. An exercise of public power is not in all cases essential to the advancement of conservation. The respective limits of the fields of public and private endeavor are not sharply defined, and vary from place to place and time to time, depending upon local conditions and special requirements.

In general, the sphere of private interest includes measures which will bring adequate commercial return. The interest rate is the limiting and controlling factor. When it is possible—by improvement of methods of mining, better planning, better preparation of coal, better transportation and distribution, or better utilization—to secure a larger average return on the investment, or to insure return through a longer period of years, self-interest naturally requires the introduction of such methods as rapidly as financial conditions allow. Even some of the improvements in labor and welfare conditions have been introduced in this way, with a view to securing a more permanent and more efficient labor supply, and thereby aiding the enterprise from the commercial standpoint.

Within the sphere of government activity lie the removal of unnecessary restrictions on private initiative, and such conservation measures as involve some sacrifice of individual returns—in other words, a reduction of the normal interest rate. Exercise of government power may be directly helpful within the field of private endeavor without materially sacrificing private interests; but beyond this point there are additional large possibilities of conservational activity which are clearly beyond the control of private interests. The introduction of any of these latter changes would evidently be so far-reaching in effect, and would require such broad readjustments not only within but without the mineral industry, that the necessity or desirability is not in all cases so clear as in the case of measures already introduced for private interest.

The most obviously helpful step possible to the government in the immediate future is to permit coöperative arrangements under private ownership,—which would make it possible to use common selling agencies, thereby reducing the cost of selling; to divide the territory to be served, thereby avoiding excessive cross freights; and to allot the output in proportion to the demand from various territories, thus eliminating excessive competition and over-production. All of these measures could be accomplished without detriment to the public if properly regulated by the government. The very large saving possible by this means would allow the introduction of conservational methods at the mines without raising the cost to the public.

War conditions required even more immediate and sweeping application of government power than above indicated, but conservational purposes were quite overshadowed by other considerations.

Where the mineral resources are already owned by the government, or can be acquired by the government, some of the troublesome factors in the problem are removed. In such cases it is possible to work out an intelligent plan for government control without the difficulties which arise in dealing with private ownership,—although, of course, new difficulties are introduced (see also pp. 345-347.)

The fact that there are conservational measures possible only to governments has been widely used as an argument for introducing government ownership or control. Recent vigorous demands for the nationalization of natural resources in Europe, and the increasing discussion of the subject in this country, may be regarded as phases of the conservation problem. It is not the purpose here to argue either for or against the drastic exercise of government power in the conservation of natural resources, but merely to call attention to the measures which are being discussed.

CONSERVATION OF MINERALS OTHER THAN COAL

The discussion of conservation as applied to specific minerals might be extended almost indefinitely; but perhaps enough has been said to indicate the general nature of the field. Before the war careful estimates of world supplies had been made for comparatively few minerals, although these included some of the most important, such as coal, oil, and iron. War conditions required a hasty estimate of world reserves of most of the mineral products. The reader interested in the problem will find an extremely interesting body of literature issued by the various governments on this subject. Of especial interest to the American reader will be the reports of the U. S. Geological Survey and of the Bureau of Mines.

In recent years there has been increasing recognition of the possibilities of conservational saving by concentration, refinement, and even manufacture of mineral commodities at or near the point of origin,—thus lessening the tonnage involved in transportation of the crude products. Limitations of fuel and other conditions often make this procedure difficult; but considerable progress is being made both through private initiative and, especially in international trade, through governmental regulations of great variety.

FOOTNOTES:

[42] Campbell, M. R., The coal fields of the United States: *Prof. Paper 100-A, U. S. Geol. Survey*, 1917, p. 24.

[43] Final report of the Royal Commission on coal supplies: House of Commons, London, vol. 16, 1905.

[44] Report of the National Conservation Commission: Senate Document No. 676, 60th Congress, 2d session, Govt. Printing Office, Washington, 1909.

[45] Van Hise, C. R., *The conservation of natural resources in the United States*: Macmillan Co., New York, 1910.

[46] Haas, Frank, The conservation of coal through the employment of better methods of mining: Abstract of paper presented to Pan-American Scientific Congress, Washington, Dec., 1915-Jan., 1916.

[47] Adams, Frank D., Our mineral resources and the problem of their proper conservation: *6th Ann. Rept., Commission of Conservation, Canada*, 1915, pp. 52-69.

[48] Chance, H. M., Address before the mine engineering class of the Pennsylvania State College, Quoted by F. W. Gray, The conservation of coal: *Bull. 47, Can. Mining Inst.*, 1916, p. 201.

[49] *Loc. cit.*

[50] Campbell, M. R., and Parker, E. W., Coal fields of the United States, Papers on the conservation of mineral resources: *Bull. 394, U. S. Geol. Survey*, 1909, p. 12.

[51] *Loc. cit.* p. 12.

[52] Smith, George Otis, and Lesher, C. E., The cost of coal: *Science*, vol. 44, 1916, p. 768.

[53] *Loc cit.*, pp. 768-769.

[54] *Loc. cit.*, p. 771.

[55] Van Hise, Charles R., *Coöperation in industry*, pp. 7-8, Address given before annual meeting of the National Lumber Manufacturers' Association, Chicago, Illinois, May 31, 1916.

[56] *Loc. cit.*, p. 767.

[57] Stabilization of the bituminous coal industry, Extracts from the award and recommendations of the United States Bituminous Coal Commission, Government Printing Office, Washington, 1920.

CHAPTER XVIII

INTERNATIONAL ASPECTS OF MINERAL RESOURCES

WORLD MOVEMENT OF MINERALS

Of the annual world production of minerals about two-thirds are used within the countries where the minerals are produced and one-third is shipped to other countries. In this chapter we are concerned primarily with the part which moves between countries. It may be assumed that the consumption within the countries of origin is a matter of national rather than international concern.

In pre-war times minerals constituted about 33 per cent[58] of the value of the total foreign trade of the United States, and 28 per cent of the foreign trade of Germany. Figures are not available to show the proportion of mineral tonnage to that of other commodities.

One of the several interesting facts in this world movement of minerals is that the movement of most of them shows a rather remarkable concentration. For instance, manganese moves from three principal sources to four or five consuming centers. Chromite moves from two principal sources; tungsten also from two. Even for certain commodities which are widely distributed and move in large amounts, the concentration of movement is rather marked; for instance, the world movement of coal is controlled by England, the United States, and Germany. In other words, although the world movement of mineral commodities is widespread and exhibits many complex features, most of the individual minerals follow two or three salient lines of movement. This means in general that for each mineral there are certain sources of limited geographic extent, which, because of location, grade, relation to transportation, cost—in short, all the factors that enter into availability—are drawn upon heavily for the world's chief demands. The convergence of these materials toward a few consuming centers indicates generally concentration of coal production necessary to smelting, high development of manufacturing, large per capita use, concentration of facilities, strong financial control, and, not least, a large element of enterprise which has taken advantage of more or less favorable conditions.

If a nation were fully supplied with mineral resources, without excess, the mineral problem might be almost exclusively domestic in its nature. But no country is so situated. For most of the mineral products the dominant supply is likely to be controlled by one or two nations, the other nations being correspondingly deficient and dependent. Even the United States, which is more nearly self-sustaining in mineral resources than any other country, is almost wholly dependent on other countries for certain mineral supplies; and in the case of minerals of which it has an excess it is dependent on other countries for markets. The view that the mineral resource problem is solely a local and national one, of no concern to outsiders, ignores this fundamental fact of distribution of raw materials.

Control of smelting facilities makes it possible for certain countries to exercise considerable influence over the production and distribution of minerals in other countries, and thus presents many difficult international questions. Even more difficult are the international problems created by the commercial ownership and control of minerals in the ground by nationals of other countries.

The national and international aspects of mineral resources are difficult to separate, so intimately do they react on each other. To some extent there may be conflict of interest between the two, but in the main the international questions may be logically approached from the standpoint of national self-interest; for, in the conduct of the national industry along broad and enlightened lines, world conditions must necessarily be considered. A clearer comprehension of the world mineral relations, and an understanding of our own opportunities and limitations in comparison with those of our neighbors, cannot but eliminate some of the unnecessary handicaps to the best use of the world mineral resources, and result in a lessening of causes of international discord.

A brief survey of the mineral conditions preceding, during, and following the war may serve as a convenient means of approach to a study of the present international aspects of the mineral problem.

Movement of Minerals Under Pre-war Conditions of International Trade

If the world pre-war movement of minerals is considered broadly, it may be regarded as conforming essentially to normal trade conditions of supply and demand. There have been barriers to overcome, such as tariffs and trade controls and monopolies of various kinds, but these barriers have not prevented the major movements between the best sources of supply and the principal consuming centers. These movements may be regarded as a more or less spontaneous internationalization of mineral resources by private enterprise. The aim of free trade or unrestricted commerce was equality of trade opportunities; but such conditions of unrestricted competition tended to concentrate trade in the hands of the strongest interests and to prevent equality of opportunity.

The efforts made to promote or hinder international mineral movements by tariffs, bonuses, embargoes, subsidies, transport control, patents, government management, financial pressure, and other means have been incited mainly by national or imperial self-interest, and have thus been to some extent inimical to an internationalization based on the principle of the greatest good to the greatest number. It may be supposed that, in any effort to attain supernational or international control, motives and measures based on national self-interest of the sort here mentioned will continue to play an important part.

Changes During the War

The war wrought fundamental changes in the world movement of minerals. The character and distribution of the demands changed. Customary sources of supply were cut off. Financial disturbances and ship shortage profoundly modified the nature, distribution, and extent of the world movement. Our domestic mineral industry was abruptly brought to a realization of its vital relations with international trade. To illustrate, the large movement of manganese from India and Russia to the United States was abruptly stopped, and we had to develop a source of supply in Brazil. The stoppage of pyrite importations from Spain as a means of saving ships required the development of pyrite and sulphur supplies in the United States. The export of oil from the United States to European countries was greatly stimulated, and the export to other countries was correspondingly decreased. The world movements of coal were vitally affected, principally by the limitation of the coal shipments from England and the United States to South America and the concentration of shipments to European countries. The closing of German coal supplies to nearby countries also had far-reaching consequences. The cutting off of the German potash left the world for the time being almost unsupplied with this vital fertilizing ingredient. The Chilean nitrates, on which the world had relied for fertilizer purposes, were diverted almost exclusively to the manufacture of powder. The total annual imports of mineral commodities into the United States were reduced by 1,200,000 tons. Our exports, though they continued in large volume, were mainly concentrated in Europe. The story of these disturbances in the world movement of minerals, though highly interesting, is too long to be told here.

Out of these sweeping and rapid changes in the world movement of mineral commodities there arose, partly as cause and partly as effect, international agreements for the allocation of minerals, as a means of insuring the proper proportions of supplies to the different countries for the most effective prosecution of the war. Inter-Allied purchasing committees in London and in Paris found it necessary to make an inter-Allied allocation of the output of Chilean nitrate, because the sum of the demands exceeded the total supply by a considerable fraction, and to agree on the distribution and prices of the world's supplies of tin, tungsten, and platinum. For many other commodities agreements of various sorts were made. For instance, the United States entered into an agreement with England and France for the purchase of iron ore and molybdenum from Scandinavia to keep it out of Germany. The United States and England agreed as to supplying Canada with ferromanganese. New problems of world allocation came up almost daily.

Another war change in mineral conditions, of a more permanent nature, was the liquidation of German ownership and control of minerals in allied countries, and in some cases even in neutral countries.

POST-WAR CONDITION OF THE MINERAL TRADE

The mineral industry has by no means reverted to its pre-war condition. The old movements have been only partially resumed, and new elements have entered. Shipping is still disturbed. Governments have been coöperating in various ways in the liquidation of government stocks of minerals. The German commercial control of minerals outside of its boundaries, as noted above, has been much weakened. The Reparations Committee created by the Peace Treaty has enormous powers over the use and distribution of the mineral resources of Germany, which directly and indirectly affect the mineral supplies of Europe and all the world. The terms of the Peace Treaty changed in fundamental ways the international channels of mineral movement.

The mineral situation of Europe is in such a state of chaos that the combined efforts of governments will be necessary for many years to bring order. This will be accomplished partly through the Reparations Committee, but may require other forms of coöperation. An international coal commission has already been formed to look after the distribution of coal through Europe. International coöperation in mineral distribution is not merely a theoretical possibility for the future,—it is now the outstanding fact with reference to the European situation.

The recognition of their dependence on neighbors for important mineral resources has led to earnest efforts on the part of nations to supply deficiencies. The great activity of the British government in acquiring oil is one example. The falling off of gold production the world over, together with the increased disparity between gold reserves and the currency issued against them, is causing serious consideration of government action to encourage the gold industry by financial measures tending to increase the profit of the miners (see pp. 224-225).

Before and since the war most countries of the globe, outside of England and the United States, have gone far in the exercise of the right of eminent domain over mineral resources within their own boundaries. Even in England the recent movement to nationalize the coal and oil resources is an indication of the general tendency. In the United States the movement has manifested itself thus far only in the increasing reluctance on the part of the government to part with mineral resources on the public domain,—as is clear from the terms of its new leasing law to cover oil, coal, gas, potash, and phosphates on public lands.

Before the war only the German government was clearly identified with private interests in international trade and in the acquirement of mineral reserves. Since the war all governments except that of the United States are taking an active part in these fields, both directly and in coöperation with private capital. The British government has taken a direct financial interest in certain companies, such for instance as the Anglo-Persian and Shell Oil Companies, and in some cases is actively interested in the acquirement of selling contracts. In England there is a wider use of voting trusts in controlling private companies, with the purpose of preventing the control from falling into alien hands. Government control of shipping in certain countries is involving various degrees of control of mineral movements. Also, through loans and bonds, mineral resources in certain countries have been tied up by the loaning governments. There has been wide extension of government control of minerals in mandatory territories and elsewhere through many new loans and regulations. These steps are in effect closing important parts of the world to private initiative, and particularly to nationals of other countries. Whether these activities of governments are economically desirable or not, they are the actual conditions, not theories.

If this situation continues, it raises the question whether our government will not be forced, in protection of its own mineral industries, also to take a direct part; for under present conditions, our importers and exporters find themselves dealing single-handed with governments or with private groups so closely identified with governments as to have much the same power. In matters of shipping, credits, exchange, tariffs, embargoes, and opportunity to acquire foreign reserves, the actual and potential disadvantage to American interests is obvious.

TENDENCIES TOWARD INTERNATIONAL COÖPERATION AND POSSIBILITY OF INTERNATIONAL CONTROL[59] OF MINERALS

Under the pre-war conditions, unrestricted competition in world trade by private enterprise had led to a certain kind of internationalization of mineral deposits based on natural conditions of availability. There is a natural tendency to work back as quickly as possible to this condition, but new elements have entered which seem to make it difficult for governments to keep their hands off. The participation of governments in world mineral trade, when not modified by international coöperation or some other higher form of control, seems to be having a tendency in the opposite direction—to be closing the doors of equal opportunity and preventing the natural world use of the world's resources.

These new conditions, together with others outlined in the preceding section, have made it necessary to pay more attention to the possibilities of international coöperation than ever before,—not as a restrictive measure, except temporarily in regard to the Central European powers,—but as a means of insuring open channels of movement for raw materials, and of insuring equal economic opportunities to all. Many of our mineral industries have already appealed to our government for coöperation and aid in their international dealings. Further, mineral industries in private hands in the various Allied countries have attempted to get together to arrange for private coöperation, and appealed to the Peace Conference for authority to do so. In certain cases the necessity for coöperative action became so apparent that pressure was brought to bear on the Peace Conference for the forming of some sort of international economic body which would make possible some of these steps. These movements were all dictated by considerations of self-interest, but self-interest broadened and educated by a knowledge of the world's situation.

Just as the increasing size of the units engaged in the mineral trade within national boundaries has led to discussion of the possibilities of government control in the interest of the public, so the increasing size of the units in the international mineral trade, the units in many cases being governments, is leading to discussion of the possibility of some international or supernational control in the interest of the world good. Just as national interest is the lengthened shadow of individual interest, so international interest may be regarded in some aspects as the lengthened shadow of national interest.

The general purpose of the suggested control is to minimize international friction; but more specifically it has been suggested that some sort of international coöperation is necessary in order to insure equality of opportunity among nations, both in supplies and in markets, and thereby to prevent the crowding of the weaker by the stronger nations. This is the gist of one of the famous fourteen points. The purpose might be accomplished by direct allocation of supplies or by control of tariffs and exchange.

One of the conditions which seems to require international coöperation is the exploitation of mineral deposits in backward countries. Unrestricted competition among nations in such exploitation has been an important cause of international controversy. It was planned at Peace Conference that the mineral resources in countries taken over by the great powers under mandatories should be developed and used in the interest of the group of nations, rather than for the special interest of the nation taking the mandatory. One of the natural functions of any international or supernational organization would be the adjustment and settlement of difficulties arising from this provision.

This topic brings up the question as to the right of any nation or group of nations to exert any force on weaker nations in the exploitation of mineral resources. On the principal of self-determination and of the complete freedom of action of nations, this procedure seems unjustified. On the other hand, whether rightly or wrongly, civilization has created great material demands which must be satisfied. The individuals, companies, and governments which use force to exploit resources in weaker countries are merely the agents in supplying the demand created by all of us. While their methods are often indefensible, the exploiters cannot be regarded merely as irresponsible buccaneers who are projecting themselves unnecessarily into somebody else's business. Whatever the sentimental and ethical aspects of the question, it seems almost inevitable that the demands of civilization will continue to require the exploitation of weaker countries; and in proportion as these countries are backward in coöperating, they must feel the world pressure. An agreement for international coöperation in such matters, therefore, is not to be regarded as merely a cold-blooded attempt to rob weaker nations,—but rather as a means of improving methods in satisfying the actually existing material demands of civilization. For illustration, the criticism of England's attempt to develop the oil industry of Mesopotamia and Persia has to a large extent confused the methods with the aim sought for. It is the writer's view that development of these resources is inevitable, and that criticism should not be directed toward nations and groups attempting to attain these results, but rather to the methods applied. For the purposes of this discussion, it is not necessary to go beyond the acceptance of the fact of demand, nor to argue the question as to whether the material demands of civilization should be curbed and progress restricted to matters of mind and human happiness.

METHODS OF INTERNATIONAL COÖPERATION

The first step in international consideration of minerals is obviously one of fact-finding. This became painfully evident during the great war, when the sudden cutting off of outside supplies and markets brought home the fact that the mineral question is only in part a domestic one. The average mining man had come to take the established marketing and commercial conditions more or less for granted, and had not looked into the underlying factors. There had been a tendency to assume that a kind Providence was in some manner looking after these elements in the situation. The nearest approach to Providence, as a matter of fact, was a small group of importers and exporters, possessing special knowledge of the international movements of certain commodities,—which knowledge was of unsuspected importance to the mineral industry. War conditions showed that neither the general public nor the mineral industry as a whole, much less the government, had even an elementary grasp of the important elements of the world mineral situation. The mobilizing of this information under high pressure, through the coöperation of government and private agencies, was an interesting and important feature in the complex activities back of the firing line. It is vastly to the credit of the men interested in the mineral industry in this country, and presumably also in other countries, that almost without exception they contributed their bits of knowledge to the common pool, even though these bits had been in a sense their private capital. Certain importers, who by their knowledge of international phases of the mineral situation had been able to exercise a profound influence on domestic markets, voluntarily sacrificed their own interest for the common cause and pointed out ways in which reductions of imports could be made.

The problems of the Peace Conference, and of other international agreements now pending, have required a still further systematizing of international information. One of the results has been the establishment of organizations of an international fact-finding character in our own and in certain other governments. In the chapters on the several minerals in this book, are summarized some of the salient features of the international situation developed by study of the kind indicated.

Knowledge of the physical facts of the world mineral situation is only a first step. Their interpretation and correlation, the study of the underlying principles, the formulation of the necessary international agreements and regulations, constitute even more difficult problems, which are far from solved.

There always has been some coöperation of governments in the mineral trade through the ordinary diplomatic channels. The question is now prominent whether, in view of the new conditions, it may not be necessary to develop better machinery—in the form of some international or supernational organization, possibly patterned on war procedure—in order to expedite the negotiations and to minimize possibilities of friction.

During the war, when the world demand exceeded the total world supply of certain commodities, such as nitrate and tin, international commissions were formed in order to make an equitable distribution of these minerals and prevent favored strong nations from taking too large a proportion of the total. This procedure presented no insurmountable difficulties. A canvass of the total supplies available and of the demands of the various countries ordinarily led to voluntary compromise in the allocation of supplies. Most of the regulations of these commissions were applied to mineral industries which were unable to meet the total demand. They were not tried out in cases where there were excess supplies; this process obviously would have been much more difficult, though perhaps not impossible.

The general success of international attempts to allocate mineral supplies during the war suggests the lines along which results might be accomplished during peace. The process is essentially a matter of getting at the facts, and then discussing the situation around a table,—thus eliminating the long delays and misunderstandings arising from the procedure through the older established diplomatic channels. How far such a procedure might be possible without the compelling common interest of war is debatable.

The great powers of the Reparations Committee, previously noted, and of the recently formed European coal commission, already indicate the general nature of the machinery for international control which might be exercised through a league of nations. It is not our purpose to argue for international control or for any specific plan of control, but rather to outline the problem. The question is not an academic one. Various kinds of international control are present facts, and the problem relates to the possibilities of more effective organization of existing agencies.

CONSERVATION IN ITS INTERNATIONAL RELATIONS

The interests of conservation, considering both its physical and its human energy phases (p. 362), seem to call for an international understanding in the use of mineral resources which will result in the minimum hindrance to their free movements along natural channels of trade. The essential fact of the concentration of mineral supplies in comparatively few world localities, and the fact that no nation is supplied with enough of all varieties of minerals, mean that artificial barriers to their distribution cannot but impose unnecessary handicaps on certain localities, which may be anti-conservational from a world standpoint. If the few countries possessing adequate supplies of high-grade ferro-alloy minerals, for instance, were to restrict their distribution by tariffs or other measures, the resulting cost to civilization through the handicapping of the steel industry would be a large one. Or if, for the general purpose of making the United States entirely self-supporting in regard to mineral supplies, sufficiently high import tariffs were imposed on these minerals to permit the use of the low-grade deposits in the United States, earlier exhaustion of the limited domestic supplies would follow, and in the meantime the cost to the domestic steel industry would be serious. Cost may be taken to represent the net result of human energy multiplied into raw material. The movement would therefore be anti-conservational. If each state in the United States were to start out to become entirely self-sustaining in regard to minerals, and by various regulations were able to prohibit the use of minerals brought in from without, or the export of its excess of minerals, the waste in effort and materials would be obvious. Nature has clearly marked out fields of specialization for different localities, and the effective use of mineral supplies is just as much a matter of specialization as the effective use of man's talents. If the United States, because of its vast copper deposits, is in a position to specialize in this line and to aid the world thereby, this should involve recognition of the fact that other countries are better able to specialize in other commodities,—thereby forming a basis for mutual exchange, which is desirable and necessary for world development.

This conservational argument against artificial barriers does not necessarily imply complete elimination of tariffs or other restricting or fostering measures. Within limits these may be necessary or desirable in order to maintain differences in the standard of living, or in order to permit the growth of infant industries; but to carry these measures to a point where they interfere with essential mineral movements determined by nature is obviously anti-conservational.

For some mineral commodities, international coöperation may prevent duplication in efforts and the development of excessive supplies in advance of the capacity of the world to use them. Partly because of lack of such coöperation, certain mineral commodities have been developed in such large quantities in various parts of the world that it may be many years before demand catches up with development. In the meantime, large and unnecessary interest charges are piling up. This financial loss measures the loss in effectiveness of collective human effort.

In the above discussion, little reference has been made to shortage of total world supplies as an argument for international coöperation. This is an argument often cited, and with some effectiveness during the war. It is the writer's view that this phase of the problem has been much exaggerated. Except for certain periods during the war, in considering the world as a whole adequate supplies of all mineral commodities have been available at all times. They have been developed as rapidly as needed, in some cases more rapidly; and geological conditions seem to indicate that this condition will continue for some time in the future, through national and individual effort. Combined efforts of governments seem hardly necessary as yet to accomplish this purpose. In fact, there is rather more danger of over-development, without due regard to the working of the interest rate, which might be prevented by international coöperation. The main problem now is not one of total supplies, but of their effective and equitable distribution.

EXPLORATION IN ITS INTERNATIONAL RELATIONS

When an explorer or prospector leaves his own country to discover and acquire minerals in other countries, with a view to exportation, it is reasonably obvious that he must first acquire a sound knowledge of at least some of the elements of international trade in minerals,—such as shipping facilities, rates, tariffs, attitude of the government toward ownership, toward export, etc. For example, the prospector for oil in foreign countries will not get very far without considering the recent steps taken by foreign governments, and mentioned on pp. 131-132.

The necessity of study of the international situation in conducting domestic exploration is not so generally recognized; and yet anyone today who confines his attention solely to the local physical facts of the situation, and who ignores international considerations, may find himself in difficulties. The investigation of international questions is not merely desirable from the standpoint of general information, but may be vital to the business or professional success of the explorer. For instance, he might take up the exploration and development in the United States of fertilizers and ferro-alloy minerals which are ordinarily imported; and without understanding the severe limitations imposed by the foreign situation, he might find himself with a property, sound from a physical standpoint, but financially a failure. It is comparatively easy, by running over the long list of mineral commodities used in the United States, to eliminate, on international grounds, a considerable number from the field inviting financial success, and to concentrate on others whose economic relations are sound. In the rapid changes during and since the war, the necessity for consideration of world conditions has been brought home at heavy expense to many business and professional men engaged in the mineral industry.

VALUATION IN ITS INTERNATIONAL RELATIONS

For mineral commodities of limited supply and steady demand, market conditions may be more or less taken for granted, and valuation may be based on local considerations. For a large number of mineral resources, however, the competitive market conditions are anything but stable, because of foreign competition. It is necessary not only to know the basis for this competition, but also to be able to follow intelligently its various changes. The value of many of our mineral deposits in recent years has varied widely with changes in the foreign situation.

RELATIVE POSITION OF THE UNITED STATES IN REGARD TO SUPPLIES OF MINERALS

The United States is more nearly self-sustaining in regard to mineral commodities as a whole than any other country on the globe. The following statement summarizes qualitatively our position:

1. Minerals of which our exportable surplus dominates the world situation:

Copper.

Petroleum has belonged in this class until recently. In the future imports will be required (see 5 following).

2. Minerals of which our exportable surplus constitutes an important but not a dominant factor in the world trade:

Cement.

Coal.

Iron and steel.

Phosphates.

Silver.

Sulphur.

Uranium and radium.

3. Minerals of which our exportable surplus is not an important factor in world trade. Small amounts of most of these minerals have been and will doubtless continue to be imported because of special grades, back-haul, or cheaper sources of foreign supply, but these imports are for the most part not essential as a source of supply:

Aluminum and bauxite.

Arsenic.

Artificial abrasives and emery (except Naxos emery).

Asphalt and bitumen.

Barite.

Bismuth.

Borax.

Bromine.

Building stone (except Italian marble).

Cadmium.

Feldspar.

Fluorspar.

Fuller's earth.

Gold.

Gypsum.

Lead.

Lime.

Magnesite.

Mineral paints (except umber, sienna, and ocher from France and Spain).

Molybdenum.

Pyrite.

Salt (except special classes).

Talc.

Titanium.

Tripoli and diatomaceous earth.

Zinc.

4. Minerals for which the United States must depend almost entirely on other countries:

Cobalt.

Nickel.

Platinum and metals of the platinum group.

Tin.

5. Minerals for which the United States will depend on foreign sources for a considerable fraction of the supply:

Antimony.

Asbestos.

Ball clay and kaolin.

Chalk.

Chromite.

Corundum.

Diamond dust and bort.

Garnet.

Graphite.

Grinding pebbles.

Manganese.

Mercury.

Mica.

Monazite.

Naxos emery.

Nitrates.

Petroleum (see below).

Potash.

Precious stones.

Pumice.

Tungsten.

Vanadium.

Zirconium.

In the past the production of petroleum in the United States has dominated the world petroleum situation; but domestic consumption has now overtaken production, and unless discoveries of oil come along at a rapid rate the domestic deficiency seems likely to increase, with corresponding increase in our dependence on foreign sources (see pp. 128-132).

Some of the minerals of this last class, such as potash, manganese, and chromite were developed under war conditions in the United States to such an extent as to materially lessen the demand for importation; but in normal times domestic sources can supply a considerable fraction of the demand only at high cost and with the aid of a protective tariff.

No attempt will be made here to present the detailed figures on which the above generalizations are based. In view of the present disturbed conditions of production and consumption, any judgment as to future demands or available surplus must take into account several factors which cannot be accurately measured,—such as financial control in foreign countries, possible tariffs, and foreign competition. For this reason the above statement should be regarded as only tentative, though it is the result of a rather exhaustive study of conditions in relation to the world control of shipping. The classes named overlap to some extent, and it is to be expected that some of the commodities placed in one class may in the near future be transferred to another.

In terms of value, the United States has a potential export surplus of minerals about twice as large as that of all the rest of the world put together. Countries which were neutral during the war have the remaining export surplus. Great Britain, France, and Italy have net import requirements considerably in excess of their exports. Germany has almost as large a deficit of minerals as the United States has a surplus.

From the above facts it is clear that, in any scheme of international control or coöperation, the United States would have by far the heaviest stake, and perhaps the most to lose by restriction. It seems equally clear that the preponderance of exportable surplus of minerals over necessary imports justifies the United States in taking a broad and liberal view of the importation of needed minerals. The war-time necessity of making our country as nearly self-sustaining as possible does not seem to obtain in peace times. To carry that principle to an extreme means not only the expensive use of low-grade domestic supplies, but the elimination of the imports which are so necessary to balance our export trade.

These facts also raise the question as to how far the United States is justified in exploiting the rest of the world to add to its already great preponderance of control,—as, for instance, in copper. Any further aggrandizement of our position in regard to such minerals may be directly at the expense of neighbors who are already far less well supplied than ourselves, and is to be justified only on the basis of adding to the world's supply for common use, and of lending our expert assistance to neighbors to make them more nearly self-supporting. To carry out our campaign in these cases without regard to the needs of other countries will obviously not hasten the ideal of a democratic world with equal opportunity for all. On the other hand, the great freedom allowed by our laws in regard to foreign commercial control of our minerals, as compared to the restriction on such control in other countries, suggests the desirability of exerting our pressure for the open door policy in all parts of the world, in the interest of desirable reciprocal relations.

In this connection there has been a tendency to criticise England's post-war activity in securing oil reserves for the future. Self-interest has clearly dictated the necessity for improving England's weak position in regard to this vital energy resource. The success of this movement obviously means a lessening of the future preponderance of the United States in the oil industry, and calls for increased activity on the part of the United States in maintaining the desirable leading position it has long held. From the writer's viewpoint, however, the fair success of a rival does not call for criticism of motives. If there is any just criticism, it applies to methods (see pp. 390-391).

Whatever action may be taken by the United States in regard to international mineral questions, it is clear that the war has brought this country into such world relations that it has become imperative for us to study and understand the world mineral situation much more comprehensively than before,—in the interest not only of intelligent management of our own industries, but of far-sighted handling of international relations. Under the stress of war the government, especially through the Geological Survey, the Bureau of Mines, and the several war boards, found it necessary to use extraordinary efforts to obtain even elementary information on the international features of mineral trade. Much progress has been made, but only a start. The geologist or engineer who fails to follow these investigations may be caught napping in the economic phases of his work.

THE COAL AND IRON SITUATION OF WESTERN EUROPE UNDER THE TERMS OF THE PEACE TREATY

A mineral problem of special international importance at the present time relates to the disposition of the coal and iron resources of Germany. Germany's coal and iron have been the basis for its commanding position in industry and commerce. In fact, its development of these resources has been probably the most vital element in the European economic situation. The terms of the Peace Treaty in regard to these commodities have far-reaching consequences, not only for Germany but for all Europe, and indirectly, for the world.

Germany (Westphalia) outclasses all other European sources in grades of metallurgical coal, in quantities produced, and in cheapness of production. Both France and Belgium must continue to be dependent on this source for important parts of the coking coal for metallurgical purposes, notwithstanding France's acquisition of the Saar Basin, which produces mainly non-coking coal, and the development of new reserves in Belgium. Germany's command of coal is wrecked in several ways. The French take over full and absolute possession of the coal of the Saar Basin, though Germany has the right to repurchase it at the end of fifteen years, in case this territory then elects for union with Germany. The coal of Upper Silesia, with a production of about 23 per cent of the total of all German hard coal, is to be ceded to Poland, subject, however, to plebiscite. Germany undertakes to deliver to France each year, for not to exceed ten years, an amount of coal equal to the difference between the annual pre-war production of the French coal mines destroyed as a result of the war, and the production of the mines of the same area during the years in question,—such delivery not to exceed 20,000,000 tons in any one year of the first five, nor 8,000,000 in any one year of the succeeding five years. In addition, Germany agrees to deliver coal, or its equivalent in coke, as follows: to France 7,000,000 tons annually for ten years; to Belgium 8,000,000 tons annually for ten years; to Italy an annual quantity rising by annual increments from 4,500,000 tons in 1919-20 to 8,500,000 tons in each of the six years 1923-24 to 1928-29; and to Luxemburg, if required, a quantity of coal equal to the pre-war annual consumption of German coal in Luxemburg.

The total pre-war coal production of Germany in 1913 was 191,500,000 tons. The diminution of production due to loss of territory in Alsace-Lorraine, in the Saar Basin, and in Upper Silesia amounts to about 61,000,000 tons. The further required annual distribution of coal to France, Italy, Belgium, and Luxemburg amounts to about 40,000,000 tons. This leaves about 90,000,000 tons for Germany's domestic use, as compared with a pre-war domestic use of 139,000,000 tons. Even then, these calculations make no allowance for coal to be used in export trade to neutrals or other countries, some part of which seems vital to Germany's trade. They make no allowance for the deterioration of plant and machinery in the mines, which will delay resumption of coal production. They make no allowance for the diminution in working hours and the lack of transportation. In short, unless there is a miraculous recovery and development of Germany's coal industry, impossible conditions have been imposed. Some recognition of this fact appears in the great powers to adjust terms which have been vested in the Reparations Committee. Successive revisions of requirements by the Reparations Committee have already reduced the direct contributions of coal from Germany nearly fifty per cent. The entire European coal situation is in a state of chaos. It was found necessary in 1918 to appoint a Coal Commission under international control, to attempt to allocate and distribute supplies. It seems inevitable that the physical facts of the situation will prevail, and that the control of the Allies will resolve itself into efforts to distribute and coördinate supplies so as to keep the European machinery going, more or less regardless of the terms of the Peace Treaty.

One of the important outcomes of this situation has been the recent rapid development of German lignite production, based on newly worked-out methods of treatment and utilization.

By taking over Alsace-Lorraine, France acquires about 70 per cent of the iron ore reserves and annual production of Germany. This production was in minor part smelted locally,—the larger part moving down the Rhine to the vicinity of the Ruhr coal fields, and Ruhr coal coming back for the smelting in Lorraine. This great channel of balanced exchange of commodities has been determined by nature, and is not likely to be permanently affected by political changes. For the time being, however, the drawing of a political boundary across this trade route hinders the full resumption of the trade. Self-interest will require both Germany and France to keep these routes open. France requires German coal to supply the local smelters near the iron fields, and German markets for the excess production of iron ore. On the other hand, Germany's great smelting district in the Ruhr Basin is largely dependent on the Lorraine iron ore, and the movement of this iron ore requires coal from down the Rhine as a balance.

The intelligent handling of this great coal and iron problem is of far-reaching consequence to the mineral industries of the world.

CONCLUSION

In the foregoing discussion it is not our purpose to argue for any specific national or international plan or procedure, but rather to show something of the nature of the problem,—and particularly to show that intelligent and broadened self-interest requires a definite national policy in regard to world mineral questions. Realization of this fact is a long step toward the solution of the international problems. No geologist, engineer, or business man is safe, in the normal conduct of his affairs, without some attention to these matters.

It is our purpose further to bring home the fact that international coöperation in the mineral field is not merely an academic possibility, but that in many important ways it is actually in existence. The terms of the Peace Treaty alone have far-reaching consequences to the explorer or mining man in all parts of the world. The modifications of these terms, which are inevitable in the future, will not be of less consequence. It is necessary not only to know what these are, but to aid in their intelligent formulation.

LITERATURE

A vast new literature on the subject of international mineral relations has sprung into existence during and following the war, and anyone may easily familiarize himself with the essentials of the situation. Some of the international features are noted in the discussion of mineral resources in this book. For fuller discussion, the reader is especially referred to the following sources:

The reports of the United States Geological Survey. Note especially *World Atlas of Commercial Geology*, 1921.

The reports of the United States Bureau of Mines.

Political and commercial geology, edited by J. E. Spurr, McGraw-Hill Book Co., New York, 1920.

Strategy of minerals, edited by George Otis Smith, D. Appleton and Co., New York, 1919.

Coal, iron and war, by E. C. Eckel, Henry Holt and Company, New York, 1920.

The iron and associated industries of Lorraine, the Sarre district, Luxemburg, and Belgium, by Alfred H. Brooks and Morris F. LaCroix, Bull. 703 U. S. Geological Survey, 1920.

The Lorraine iron field and the war, by Alfred H. Brooks, Eng. and Min. Journ., vol. 109, 1920, pp. 1065-1069.

Munitions Resources Commission of Canada, final report, 1920.

FOOTNOTES:

[58] Umpleby, Joseph B., *Strategy of minerals—The position of the United States among the nations*: D. Appleton and Co., New York, 1919, p. 286.

[59] Control is here used in a very general sense to cover activities ranging from regulation to management and ownership. The context will indicate in most cases that the word is used in the sense of regulation when referring to governmental relationships.

CHAPTER XIX

GEOLOGY AND WAR

GEOLOGY BEHIND THE FRONT

The experience of the great war disclosed many military applications of geology. The acquirement and mobilization of mineral resources for military purposes was a vital necessity. In view of the many references to this application of geology in other parts of this volume, we shall go into the subject in this chapter no further than to summarize some of the larger results.

As a consequence of the war-time breakdown in international commercial exchange, the actual and potential mineral reserves of nations were more intensively studied and appraised than ever before, with the view of making nations and belligerent groups self-sustaining. This work involved a comprehensive investigation of the requirements and uses for minerals, and thus led to a clearer understanding of the human relations of mineral resources. It required also, almost for the first time, a recognition of the nature and magnitude of international movements of minerals, of the underlying reasons for such movements, and of the vital inter-relation between domestic and foreign mineral production. The domestic mineral industries learned that market requirements are based on ascertainable factors and that they do not just happen. Large new mineral reserves were developed. Metallurgical practices were adapted to domestic supplies, thus adding to available resources. Better ways were found to use the products. Some of these developments ceased at the end of the war, but important advances had been made which were not lost. One of the advances of permanent value was the increased attention to better sampling and standardization of mineral products, as a means of competition with standardized foreign products. For instance, the organization of the Southern Graphite Association made it possible to guarantee much more uniform supplies from this field, and thereby to insure a broader and more stable market. Such movements allow the use of heterogeneous mineral supplies in a manner which is distinctly conservational, both in regard to mineral reserves and to the human energy factors involved. In another war the possibilities and methods of meeting requirements for war minerals will be better understood.

In these activities, geologists had a not inconsiderable part. The U. S. Bureau of Mines, the U. S. Geological Survey, state geological surveys, and many other technical organizations, public and private, turned their attention to these questions. One of the special developments was the organization by the Shipping Board of a geologic and engineering committee whose duty it was to study and recommend changes in the imports and exports of mineral commodities, with a view to releasing much-needed ship tonnage. This committee was also officially connected with the War Industries Board and the War Trade Board. It utilized the existing government and state mineral organizations in collecting its information. Over a million tons of mineral shipping not necessary for war purposes were eliminated. This work involved also a close study of the possibilities of domestic production to supply the deficiencies caused by reduction of foreign imports.

Other special geological committees were created for a variety of war purposes. In the early stages of the war a War Minerals Committee, made up of representatives of government and state organizations and of the American Institute of Mining Engineers, made an excellent preliminary survey of mineral conditions. A Joint Mineral Information Board[60] was created at Washington, composed of representatives of more than twenty government departments which were in one way or another concerned with minerals. It was surprising, even to those more or less familiar with the situation, to find how widely mineral questions ramified through government departments. For instance, the Department of Agriculture had men specially engaged in relation to mineral fertilizers and arsenic. Sulphur and other mineral supplies were occupying the attention of the War Department. Mica and other minerals received special attention from the Navy Department. The Tariff Board, the Federal Trade Commission, the Commerce Department, even the Department of State, had men who were specializing on certain mineral questions. All these departments had delegates on the Joint Mineral Information Board, in which connection they met weekly to exchange information for the purpose of getting better coördination and less duplication.

The National Academy of Sciences established a geologic committee, with representatives from the U. S. Geological Survey, the state geological surveys, the Geological Society of America, and other organizations. This committee did useful work in correlating geological activities, mainly outside of Washington, and in coöperation with the War Department kept in touch with the geologic work being done at the front.

While the activities of geologists for government, state, and private organizations were for the most part in relation to mineral resource questions, this was by no means the total contribution. The U. S. Geological Survey and other organizations, in coöperation with the War Department, did a large amount of topographic and geologic mapping of the eastern areas for coast-defense purposes. This work involved consideration of the topography for strategic purposes, as well as the stock-taking of mineral resources—including road materials and water supplies. The revision of Geological Survey folios, with these requirements in mind, brought results which should be of practical use in peace time. Studies were likewise made of cantonment areas, with reference to water supplies and to surface and sub-surface conditions.

Many geologists were engaged in the military camps at home and abroad, and in connection with the Student Army Training Corps at the universities, in teaching the elements of map making, map interpretation, water supply, rock and soil conditions in relation to trenching, and other phases of geology in their relation to military operations. The textbook on Military Geology,[61] prepared in coöperation by a dozen or more geologists for use in the courses of the Student Army Training Corps, is an admirable text on several phases of applied geology. The name of the book is perhaps now unfortunate, because most of it is quite as well adapted to peace conditions as to those of war. There is no textbook of applied geology which covers certain phases of the work in a more effective and modern way. The topics treated in this book are rocks, rock weathering, streams, lakes and swamps, water supply, land forms, map reading and map interpretation, and economic relations and economic uses of minerals. Another book,[62] on land forms in France, prepared from a physiographic standpoint, was a highly useful general survey of topographic features and was widely used by officers and others.

GEOLOGY AT THE FRONT[63]

Perhaps the most spectacular and the best known use of geology in the war was at and near the front. This use reached its earliest and highest development in the German army, but later was applied effectively by the British and British Colonial armies, and by the American Expeditionary Force.

One of the first intimations to the American public of the use of geology at the front appeared in the publication of German censorship rules in 1918,—when, among the prohibitions, there was one forbidding public reference to the use of earth sciences in military operations. A leading American paper noted this item and speculated at some length editorially as to what it meant.

It was discovered that geologists to the number of perhaps a hundred and fifty were used by the Germans to prepare and interpret maps of the front for the use of officers. Features represented on these maps included topography; the kinds of rocks and their distribution; their usefulness as road and cement materials; their adaptability for trench digging, and the kinds and shapes of trenches possible in the different rocks; the manner in which material thrown out in trenching would lie under weathering; the ground-water conditions, and particularly the depth below the surface of the water table at different times of the year and in different rocks and soils; the relation of the ground-water to possibilities of trench digging; water supplies for drinking purposes; the behavior of the rocks under explosives, and the resistance of the ground to shell-penetration; the underground geological conditions bearing on tunnelling and underground mines; and the electrical conductivity of rocks of different types, presumably in connection with sound-detection devices and groundings of electric circuits. Some of the captured German maps were models of applied geology. They contained condensed summaries of most of the features above named, together with appropriate sketches and sections. During the Argonne offensive by the American army the captured German lines disclosed geologic stations at frequent intervals, each with a full equipment of maps relating to that part of the front. From these stations schools of instruction had been conducted for the officers in the adjacent parts of the front.

The British efforts were along similar lines, although they came late in the war, under the leadership of an Australian geologist. Their efforts were especially useful in connection with the large amount of tunnelling and mining done on the British front. Among the many unexpected and special uses of geology might be cited the microscopical identification of raw materials used in the German cement. It became necessary for certain purposes to know where these came from. The microscope disclosed a certain volcanic rock known to be found in only one locality. In the Palestine campaign, the knowledge of sources of road material and water supply based on geologic data was an important element in the advance over this arid region. Wells were drilled and water pipes laid in accordance with prearranged plans.

In spite of the fact that the usefulness of geology had been clearly indicated by the experience of the German and British armies, the American Expeditionary Force was slow to avail itself in large measure of this tool; but after some delay a geologic service was started on somewhat similar lines under the efficient leadership of Lieutenant-Colonel Alfred H. Brooks, Director of the Division of Alaskan Resources in the U. S. Geological Survey. The work was organized in September, 1917, and during the succeeding ten months included only two officers and one clerk. For the last two months preceding the armistice there was an average of four geologic officers on the General Staff, in addition to geologists attached to engineering units engaged in road building and cement making, and plans had been approved for a considerable enlargement of the geologic force. The work was devoted to the collection and presentation of geologic data relating to (1) field works; (2) water supply; and (3) road material. Of these the first two received the most attention. Maps were prepared, based somewhat on the German model, for the French defenses of the Vosges and Lorraine sectors, and for the German defenses of the St. Mihiel, Pont-a-Mousson, and Vosges sectors. Water supply reports covered nearly 15,000 square kilometers. The following description of the formations, taken from the legend of one of the geologic maps, shows the nature of the data collected:

Silt, clay and mud, with some limestone gravel, usually more or less saturated, except during dry season (June to September), in many places subject to flooding. Surface usually soft except during Summer. These deposits are ½ to 2 meters thick in the small valleys, and 2 to 3 meters in the —— Valleys. Unfavorable to all field works on account of ground-water and floods, and not thick enough for cave shelters.

Silts with some clay and fine sands and locally some fine gravel and rock débris. These deposits occur principally on summits and slopes, and are probably from 1 to 2 meters thick. Even during dry season (June to September) they retain moisture and afford rather soft ground. In wet season the formation is very soft and often muddy. In many places water occurs along bottom of these deposits. Favorable for trenches, but which require complete revetment, and ample provision for drainage, not thick enough for cave shelters; cut and cover most practical type of shelter.

Clay at surface with clay shales below. This deposit occurs in flats and is usually saturated for a depth of 1 to 2-½ meters, during wet season, for most of the year the surface is soft, but in part dries out in Summer. Deep trenches usually impossible, and even shallow trenches likely to be filled with water; defensive works will be principally parapets revetted on both sides. Cave shelter construction usually impracticable, unless means be provided for sinking through saturated surface zone into the dry ground underneath. Cut and cover usually the most practical type of shelter in this formation.

Clay at surface with calcareous clay shale and some thin limestone layers below. This formation occurs in low rounded hills; surface saturated during wet weather, but terrain permits of natural drainage, and dries out during Summer; during wet season (October to May) the surface zone is more or less saturated, and ground may be muddy to a depth of a meter or more, ground-water level usually within two or three meters of surface. Trench construction easy, but requires complete revetment, and ample provision for surface drainage. Cave shelters can be constructed in this formation where the slope is sufficient to permit of drainage tunnels. The depth to ground-water level should always be determined by test shafts or bore holes in advance of dugout construction.

Surface formation usually clay 1 to 2 meters in depth; below this is soft clay shales or soft limestone. Surface usually fairly well drained, and fairly hard ground. In general, favorable for trenches and locally favorable for cave shelters. In some localities underground water prevents cave shelter construction. The presence or absence of underground water should always be determined by test shafts or bore holes in advance of dugout construction.

Surface formation consisting of weathered zone ½ to 1-½ meters thick, made up of clay with limestone fragments and broken rock. Below is compact limestone formation. The surface of this formation is usually fairly hard, and well drained except in wettest season. Trenches built in it require little revetting; very favorable for cave shelters, but requires hard rock excavation. Some thin beds of clay occur in some of the limestone, and at these a water bearing horizon will be found. Where a limestone formation rests on clay as near —— a line of springs or seepages is usually found. Such localities should be avoided, or the field works placed above the line of springs or seepages. This formation is best developed in the plateau west of ——. Here it is covered by only a thin layer of soil, hard rock being close to the surface.

The limestones afford the only rock within the quadrangle which can be used for road metal.

Quarries (in part abandoned).

Limestone gravel pits.

Locus of springs and seepages. These should be avoided as far as possible in the location of field works, especially of dugouts. Field works should be placed above the lines of springs.

The water supply maps with accompanying engineer field notes are models of concise description of water supply conditions, with specific directions for procedure under different conditions. A few paragraphs taken from these notes are as follows:

Ground overlying rock, such as limestone, compact sandstone, granites, etc., which are usually fractured, is from the standpoint of underground water, most favorable for siting of field works. Clay shales and clay hold both surface and underground water, and are, therefore, unfavorable for field works. The contact between hard rocks resting on clay or clay shales is almost invariably water bearing, and should be avoided in locating field works.

At localities where impervious formations (clay, etc.) occur at or near the surface, they hold the water and form a superficial zone of saturation. This condition makes trench construction and maintenance difficult, and cave shelters can usually only be made by providing means of sinking through the saturated zone. The surface saturated zone often dries out in summer.

In pervious, or almost pervious rocks, the zone of saturation, or ground-water level, lies at much lower depth, and may permit of the construction of field works as well as cave shelters above it.

Underground water bearing horizons and water bearing faults should be avoided in locating field works.

Wherever there is any uncertainty about the underground water conditions, test shafts or bore holes should always be made in advance of the construction of extensive deep works.

EFFECT OF THE WAR ON THE SCIENCE OF ECONOMIC GEOLOGY

In general, the war required an intensive application of geology along lines already pretty well established under peace conditions. Much was done to make the application more direct and effective, and a vast amount of geologic information was mobilized. The general result was a quickened appreciation of the possibilities of the use of geology for practical purposes. Perhaps the most important single result was a wider recognition of the real relations of mineral resources to human activities, and of the international phases of the problem. More specifically, there was a most careful stock-taking of mineral resources and a consideration of the "why" of their commercial use. Many new resources were found, as well as new ways to utilize them.

FOOTNOTES:

[60] Now known as Economic Liaison Committee.

[61] *Military geology and topography*, Herbert E. Gregory, Editor. Prepared and issued under the auspices of Division of Geology and Geography, National Research Council, Yale Univ. Press, New Haven, 1918.

[62] Davis, W. M., *Handbook of Northern France*, Harvard Univ. Press, Cambridge, 1918.

[63] For more detailed description of this subject the reader is referred to The use of geology on the Western Front, by Alfred H. Brooks, *Prof. Paper 128-D, U. S. Geol. Survey*, 1920.

CHAPTER XX

GEOLOGY AND ENGINEERING CONSTRUCTION

Economic applications of geology are by no means confined to mineral resources (including water and soils). The earth is used by the human race in many other ways. Human habitations and constructions rest on it and penetrate it. It is the basis for transportation, both by land and water. Its water powers are used. In these various relations the applications of geology are too numerous to classify, much less to describe. While only a few of these activities have in the past required the participation of geologists, the growing size of the operations and increasing efficiency in their planning and execution are multiplying the calls for geologic advice. The nature of such applications of geology may be briefly indicated.[64]

FOUNDATIONS

The foundations of modern structures such as heavy buildings, especially in untried localities, require much more careful consideration of the substrata than was necessary for lighter structures. In planning such foundations, it is necessary to know the kinds of rocks to be excavated, their supporting strength, their structures, the difficulties which are likely to be caused by water, and other geologic features. Failure to give proper attention to these factors has led to some disastrous results.

The planning of foundations and abutments of bridges requires similar geologic knowledge. In addition, there must be considered certain physiographic factors affecting the nature and variation of stream flow and the migration of shore lines.

SURFACE WATERS

Construction of great modern dams is preceded by a careful analysis of sub-surface conditions, in regard to both the rocks and the water. It is necessary to know the supporting strength of the rocks in relation to the weight of the dam; to know whether the rocks will allow leakage around or beneath the dam; and to know whether there are any zones of weakness in the rocks which will allow shearing of foundations under the weight of the dam in combination with the pressure of the ponded water. It is necessary to know whether the valley is a rock valley or whether it is partially filled with rock débris; if the latter, how deep this débris is, and its behavior under load and in a saturated condition. Here again physiographic factors are of vital importance, both in relation to the history of development of the valley, and to questions of stream flow and reservoir storage.[65]

Construction of dams is only an item in the long list of engineering activities related to surface waters. River and harbor improvements of a vast range likewise involve geologic factors. Problems of wave action, shore currents, shifting of shores, erosion, and sedimentation, which are of great importance in such operations, have long occupied the attention of the geologist. They belong especially in the branch of the science known as physiography.

Geology in relation to underground water supplies is discussed in Chapter V.

TUNNELS

The digging of tunnels for transportation purposes, for aqueducts, and for sewage disposal requires careful analysis of geologic conditions in regard to both the rocks and the underground water. Knowledge of these conditions is necessary in planning the work, in inviting bids, and in making bids. It is necessary during the progress of the work. Too often in the past disastrous consequences, both physical and financial, have resulted from lack of consideration of elemental geologic conditions.

The building of the great New York aqueducts and subways through highly complex crystalline rocks has been under the closest geological advice and supervision. The detailed study of the geology of Manhattan Island through a long series of years has resulted in an understanding of the rocks and their structures which has been of great practical use. In the aqueduct construction the kinds of rock to be encountered in the different sections, their water content, their hardness, their joints and faults, were all platted and planned for, and actual excavation proved the accuracy of the forecasts. An interesting phase of this work was the tunneling under the Hudson at points where the pre-glacial rock channel was buried to a depth of nearly a thousand feet by glacial and river deposits,—this work requiring a close study of the physiographic history of the river.

SLIDES

Slides of earth and rock materials, both of the creeping and sudden types, have often been regarded as acts of Providence,—but studies of the geologic factors have in many cases disclosed preventable causes. A considerable geologic literature has sprung up with reference to rock slides, which is of practical use in excavation work of many kinds.

The cause of such movements is gravity. The softer, unconsolidated rock materials yield of course more readily than the harder ones, but even strong rocks are often unable to withstand the pull of gravity. The relative weakness of rock masses on a large scale was graphically shown by Chamberlin and Salisbury,[66] in a calculation indicating that a mass of average hard rock a mile thick, domed to the curvature of the earth, can support a layer of only about ten feet of its own material. The structural geologist, through his study of folds, faults, and rock flowage, comes to regard rocks essentially as failing structures.

Disturbances of equilibrium, resulting in rock movements under gravity, may be caused by local loading, either natural or artificial. Natural loading may be due to unusual rainfall, or raising of water level, or increased barometric pressure. Artificial loading may come from construction of heavy buildings or dams. Movement may also result from excavation, which takes away lateral support—and such excavation again may be caused by natural processes of erosion or by artificial processes involved in construction. Movement may be caused by mere change in the moisture content of rocks, or by alterations of their mineral and chemical character, affecting their resistance to gravity. In still other cases, earthquakes are the initiating cause of movement.

In unconsolidated rocks, a frequent cause of movement is the presence of wet and slippery clay layers. The identification and draining of these clay layers may eliminate this cause. In certain sands, on the other hand, water may actually act as a cement and tend to increase the strength of the rock. Planes of weakness in the rock, such as bedding, joints, and cleavage, are also likely to localize movement.

Earth materials, and even fairly hard rocks, may creep under gravity at an astonishingly low angle. The angle from the horizontal at which loose material will stand on a horizontal base without sliding is called the angle of rest or repose. It is often between 30° and 35°, but there is wide variation from this figure, depending on the shapes and sizes of the particles and on other conditions. It has been suggested that even the slight differences in elevation of continents and sea bottoms may, during long geologic eras, have caused a creep of continental masses in a seaward direction.

In problems relating to slides, the geologist is concerned in determining the kinds of rocks, their space relations, their structures and textures, their metamorphic changes, their water content and the nature of the water movement, their strength, both under tension and compression, and other factors.

In the digging of the Panama Canal, a geological staff was employed in the study of the rock and earth formations to be met. However, had more attention been paid to geologic questions in the planning stages, this great undertaking, so thoroughly worked out from a purely engineering standpoint, would have avoided certain mistakes due to lack of understanding of the geological conditions. It is a curious fact that in these early stages no strength tests of rocks were made, and that no thorough detailed study was made of the geologic factors affecting slides and their prevention. It was only after the slides had become serious that the geological aspects of the subject were intensively considered. The results of the geologic study, therefore, are useful only for preventive measures for the future and for other undertakings.

One of the interesting features of this investigation was the discovery that certain soft rock formations were rendered weaker rather than stronger by the draining off of the water. It had been more or less assumed that the water had acted as a lubricant rather than as a cement.

SUBSIDENCE

Not the least important application of geology to slides is in relation to deep mining operations. While the mining geologist has been principally engaged in exploration and development of ores, he is now beginning to be called in to interpret the great earth movements caused by the sinking of the ground over mining openings. For instance, the long-wall method of coal mining has resulted in a slow progressive subsidence of the overlying rock, affecting overlying mineral beds and surface structures over great areas. Detailed studies have been made of this movement, in order to ascertain its relation to the strength and structure of the rocks, its relation to the nature of the excavation, its speed of transmission, and the possible methods of prevention. German scientists have perhaps gone further with this kind of study than anyone else. In an elaborate investigation of subsidence over a coal mine in Illinois,[67] unusually complete data were obtained as to the nature, direction, and speed of the transmission of strains through large rock masses, and as to their effect in producing secondary rock structures.

RAILWAY BUILDING

In railway building, the planning and estimation of cuts and fills is now receiving geologic consideration, in order to make sure that no geologic condition has been overlooked which will affect costs, the stability of the road, or the accurate formulation of contracts. The location of best sources of supply for ballast is also a geologic problem (see pp. 90-91).

The physiographic phases of geology also are finding important applications to railroad building. The physiographer studies the surface forms with a trained eye, which sees them not as lawless or heterogeneous units but as parts of a topographic system, and he is able to eliminate much unnecessary work in the location of trial routes. Further study of some of the older railroads from this standpoint has led to considerable improvements. Physiographic study has also been applied to railway bridge construction, in the appraisal of the difficulties in surmounting stream barriers. A still broader use of physiography or geography, not popularly understood, is illustrated in the case of certain transcontinental railroads, in the study of the probable future development of the territory to be served—many features of which can be predicted with some accuracy from a study of the rocks, soils, topography, conditions of transportation, and natural conditions favoring localization of cities. The location of new towns in some cases has been based on this kind of preliminary study.

In locating an Alaskan railway close to the end of a momentarily quiescent glacier, troubles were not long in appearing, due to the fact that the glacier was really not as stable as it seemed to the layman. A specialist on glaciers, knowing their behavior, their relations to precipitation, their relations to earthquakes, the speed of their movement, and the periodicity of their movement, was ultimately called into consultation on the location of the railroad.

ROAD BUILDING

Road building in recent years has become a stupendous engineering undertaking, which is requiring geologic aid to locate nearby sources of supply for road materials. A considerable number of geologists are now devoting their attention to this work. It relates not only to the hard-rock geology but to the gravel and surface geology. Certain northern states are using specialists in glacial geology to aid in locating proper supplies of sand and gravel.

GEOLOGY IN ENGINEERING COURSES

Many engineering courses include elementary geologic studies, in recognition of the close relationship between geology and engineering. Men so trained, though not geologists, have been responsible for many applications of geology to engineering. With the increasing size and importance of operations, calling for more specialization, the professional geologist is now being called in to a larger extent than formerly. A logical trend also is the acquirement of more engineering training on the part of the geologist, for the purpose of pursuing these applications of his science.

FOOTNOTES:

[64] Excellent texts on this subject may be found in *Military Geology and Topography*, Herbert E. Gregory, Editor, prepared and issued under the auspices of Division of Geology and Geography, National Research Council, Yale Univ. Press, New Haven, 1918, and *Engineering Geology*, by H. Ries and T. L. Watson, Wiley and Sons, New York, 2d ed., 1915.

[65] Atwood, W. W., Relation of landslides and glacial deposits to reservoir sites in the San Juan mountains, Colorado: *Bull. 685, U. S. Geol. Survey*, 1918.

[66] Chamberlin, T. C., and Salisbury, R. D., *Geology*, vol. 1, 1904, pp. 555-556.

[67] Schultz, Robert S., Jr., *Bull. Am. Inst. Mining and Metallurgical Engrs.* In preparation.

CHAPTER XXI

THE TRAINING, OPPORTUNITIES, AND ETHICS OF THE ECONOMIC GEOLOGIST

Economic geology is now an established and well-recognized profession, but there is yet nothing approaching a standardized course of study leading to a degree in economic geology. There are as many different kinds of training as there are institutions in which geology is taught. Within an institution, also, it is seldom that any two persons take exactly the same groups of geologic studies. This situation allows wide latitude of training to meet ever changing requirements, but in other respects it is not so desirable.

PURE VERSUS APPLIED SCIENCE

In no institution are all the applied branches of geology taught. There is constant pressure for the introduction of more applied courses; this seems to be the tendency of the times. The economic geologist, fresh from vivid experiences in his special field, is often insistent that a new course be introduced to cover his particular specialty. Any attempt, however, to put into a college course a considerable fraction of the applied phases of geology would mean the crowding out of more essential basic studies. To yield wholly to such pressure would in fact soon develop an impossible situation; for, on the basis of time alone, it would be quite impossible to give courses on all of the applied subjects in a training period of reasonable length.

On the other hand, the failure to introduce a fair proportion of applied geology, on the ground that the function of the college is to teach pure science and that in some way economic applications are non-scientific, seems to the writer an equally objectionable procedure,—because it does not take into account the unavoidable human relations of the science, which vivify and give point and direction to scientific work. The development of science in economic directions does not necessarily mean incursion into less scientific or non-scientific fields. It is true that many of the economic applications of geology are so new and so constantly changing that they are not yet fully organized on a scientific basis; but this fact is merely an indication of the lag of science, and not of the absence of possibilities of developing science in such directions. There is today a considerable tendency among geologists of an academic type, whose lives have been spent in purely scientific investigation and teaching, to assume that anything different from the field of their activities is in some manner non-scientific, and therefore less worthy. Many economic geologists have been made to feel this criticism, even though seldom expressed openly. For the good of geologic science, this tendency seems to the writer extremely unfortunate. The young man entering the field of economic geology should be made to understand that his is the highest scientific opportunity; and that if parts of his field are not yet fully organized, the greater is his own opportunity to participate in the constructive work to be done.

Under war requirements many geologists were called upon to extend their efforts to bordering fields of endeavor. In some quarters these activities were regarded as non-scientific, and as subtracting from efficiency in purely geological work,—and yet out of this combined effort came a wider comprehension of new scientific fields, between the established sciences and between sciences and human needs. It is inevitable that in the future these fields, now imperfectly charted, will be occupied and developed, perhaps not by the men who are already well established in their particular fields of endeavor, but by coming scientists. In this light, it was a privilege for geologists to participate in the discovery and charting activities of the war.

Still another attempt to discriminate between scientific and non-scientific phases of geologic effort has been the assumption by certain scientific organizations with reference to standards of admission,—that work done for practical purposes may be regarded as scientific only if it leads to advancement of the science through the publication of the results. There is by no means any general agreement as to the validity of this distinction. On this basis, some of the most effective scientific work which is translated directly into use for the benefit of civilization is ruled out as science, because it is expressed on a typewritten rather than on a printed page.

While applied phases of the geologist's work may be truly scientific in the broader sense, it is undoubtedly easy in this field to drift into empirical methods, and to emphasize facility and skill at the expense of original scientific thought. The practice of geology then becomes an art rather than a science. This remark is pertinent also to much of non-applied geologic work in recent years. A considerable proportion of this empirical facility is desirable and necessary in the routine collection of data and in their description; but where, as is often the case, the geologist's absorption in such work minimizes the use of his constructive faculties, it does not aid greatly in the advancement of science.

Geology is by no means the only science in which there has been controversy as to the relative merits of the so-called pure and applied phases; but as one of the youngest sciences, which heretofore has been pursued mainly from the standpoint of "pure science," it is now, perhaps more than any other science, in the transition stage to a wider viewpoint. In the past there was doubt about the extension of chemistry toward the fields of physics and engineering, and of physics toward the fields of chemistry and engineering, and of both physics and chemistry toward purely economic applications; but out of these fields have grown the great sciences of physical chemistry, chemical engineering, and others,—and few would be rash enough to attempt to draw a line between the pure and applied science, or between the scientific and non-scientific phases of this work. This general tendency means a broadening of science and not its deterioration.

COURSE OF STUDY SUGGESTED

There are almost as many opinions on desirable training for economic geology as there are geologists, and the writer's view cannot be taken as representing any widely accepted standard. On the basis of his own experience, however, both in teaching and in field practice, he would lay emphasis on the fundamental branches both of geology and of the allied sciences,—general geology, stratigraphy, paleontology, physiography, sedimentation, mineralogy, petrology, structural and metamorphic geology, physics, chemistry, mathematics, and biology. After these are covered, as much attention should be given to economic applications as time permits. The time allowance for training, at a maximum, is not sufficient to cover both pure and applied science. Subsequent experience will supply the deficiencies in applied knowledge, but will not make up for lack of study of basic principles.

It is safe advice to a student wishing to prepare for economic geology that there is no royal road to success; that his best chance lies in the effort to make himself a scientist, even though he cover only a narrow field; that if he is successful in this, opportunities for economic applications will almost inevitably follow. To devote attention from the start merely to practical and commercial features, rather than to scientific principles, brings the student at once into competition with mining engineers, business men, accountants, and others, who are often able to handle the purely empirical features of an economic or practical kind better than the geologist. In the long run the economic geologist succeeds because he knows the fundamentals of his science, and not because he has mere facility in the empirical economic phases of his work. Of course there are exceptions to this statement,—there are men with a highly developed business sense who are successful in spite of inadequate scientific training, but such success should be regarded as a business and not a professional success.

Geology is sometimes described as the application of other sciences to the earth. This statement might be made even broader, and geology described as the application of all knowledge to the earth. In the writer's experience, the best results on the whole have been obtained from students who, before entering geology, have had a broad general education or have followed intensively some other line of study. Whether this study has been the ancient languages, law, engineering, economics, or other sciences, the results have usually been good if the early training has been sound. To start in geology without some such background, and without the resulting power of a well-trained mind, is to start with a handicap in the long race to the highest professional success. It follows, then, that intensive study of geology should in most cases not begin until late in the undergraduate course, and preferably not until the graduate years. Two or three years of graduate work may then suffice to launch the geologist on his career, but so great is the field, and so rapid the growth of knowledge within it, that there is no termination to his study. It is not enough to settle back comfortably on empirical practice based solely on previously acquired knowledge. Each problem develops new scientific aspects. It is this ever renewing interest which is one of the great charms of the science.

However, whether the student has a general training in geology, a specialized knowledge of certain branches, or takes it up incidentally in connection with engineering and other sciences, he will find opportunities for economic applications. The frequent success of the mining engineer in the geological phases of his work is an indication that even a comparatively small amount of geological knowledge is useful.

The writer is inclined to emphasize also the desirability of what might be called the quantitative approach to the subject,—that is, of training in mathematics and laboratory practice, which gives the student facility in treating geologic problems concretely and in quantitative terms. Geology is passing from the descriptive and qualitative stages to a more precise basis. For this reason the combination of geology with engineering often proves a desirable one. It is not uncommon for the student trained solely in the humanities and other non-quantitative subjects to have difficulty in acquiring habits of mind which lead to sufficient precision in the application of his science. He may have a good grasp of general principles and be able to express himself well, but he is handicapped in securing definite results. This does not necessarily mean that a large amount of time should be given to study of quantitative methods; exact habit of mind is more important in the early stages than expert facility with methods.

The teacher of economic geology finds his data so voluminous that it is difficult to present all the essential facts and yet leave sufficient time for discussion of general principles or for drill in their constructive application. It is difficult to lay down any rule as a guide to the proper division of effort; but from the writer's point of view, it is a mistake to attempt to crowd into a course too many facts. At best they cannot all be given; and in the attempt to do so, the student is brought into a passive and receptive attitude, requiring maximum use of his memory and minimum use of his reasoning power. Presentation of a few fundamental facts, combined with vigorous discussion tending to develop the student's ability to use these facts, and particularly tending to develop a constructive habit of investigation, seems to be the most profitable use of time during the course of training. The acquirement of facts and details will come fast enough in actual practice.

The variety, amount, and complexity of the data available in geology tend in themselves toward generalizations in teaching—toward the deductive rather than the inductive method. A certain amount of generalization is desirable, but its over-emphasis develops bad habits of mind on the part of the student, and requires radical readjustment of his ideas in subsequent field investigations. To retain a proper emphasis on inductive methods, it is necessary to limit the amount of data presented. Good results have been obtained by using the "case system," now common in the teaching of law—that is, by starting with a specific fact or situation as a basis for developing principles.

Another advantage in the restriction of data is the opportunity thus afforded for spending more time in the study of original reports rather than of the short textbook summaries. The student thus learns where the best primary sources of information are, how to find them, and how to extract essentials from them.

FIELD WORK

Field work is an essential part of any course of geologic training. Not only should it be taken at every opportunity during the regular school year, but no summer should be allowed to pass without geologic practice in the field. Opportunities for such work are offered in the summer field courses given by various institutions. In recent years it has usually been possible, also, for the student with elementary training to take part in summer geological survey work for state, national, or private organizations. In fact, after two or three years of geologic training, it is comparatively easy for the student to earn at such intervals during the year a fair fraction of his year's expenses.

The ideal arrangement, from the writer's viewpoint, would be about an equal division of time between indoor and outdoor study. The alternation from one to the other supplies a much needed corrective to clear thinking. It is impossible to bring all the subject materials into the classroom and laboratory; such study must inevitably be more or less deductive and generalized. If the student at frequent intervals is not able to acquire and renew a mental picture of field conditions, there is likely to be a faulty perspective even in regard to principles, and a considerable gap between the theoretical and applied phases of his knowledge. It may be possible in the classroom, for instance, to discuss faults in great detail with the aid of maps, diagrams, and pictures; and yet it is extremely difficult to get a real three-dimensional conception of the problems without actually standing on the ground.

Specialization in Studies

With the increasing size and efficiency of human operations has come an inevitable tendency to specialization. Where, in the past, the necessary geologic work might be passably done by the mining engineer, the local superintendent or operator, it is now being intrusted to specialists. Even within the more strictly engineering phases of the mining engineer's work, there is the same tendency toward specialization; his work is being divided up among the electrical engineers, the mechanical engineers, the hydraulic engineers, and others. The opportunities for geologic work, therefore, are distinctly in the direction of specialization. The student in determining the field he shall enter needs to take this fact into account and to prepare accordingly, but not at the sacrifice of the broad basal training. Only a small part of the specialization can be accomplished in college. The remainder will come with experience.

In the future there is likely to be increasing specialization among the different educational institutions in the phases of applied geology which are taught. Geographic location has a good deal to do with this tendency. Where an institution is located near a coal or oil field, it is likely, as a matter of course, to specialize to some extent in the application of geology to these resources. Or, the specialization may arise from the fact that the teachers have had special training in certain phases of applied geology, and such training naturally and properly determines the emphasis to be placed. Courses in engineering geology are finding a natural development in the leading engineering colleges.

In view of the fact that it is impossible for any one institution to cover all phases of applied geology, because of lack of time, and in view of the fact that even if this were attempted the results would be very unequal, because of the varied experience of teachers or because of geographic location, it would seem wise definitely to recognize these limitations and for each institution to play up the work it can do best. With freedom of migration among universities, a student by moving from place to place can thus secure any combination of specialized courses which best fits his requirements.

A Degree of Economic Geology

There has been some agitation in recent years for standardization of courses in economic geology, and for the granting of a special degree in evidence of the completion of such a course. The principal argument for this procedure is that it would tend to insure a better average of training and would draw a line between worthy geologists and a host of ill-trained pseudo-geologists. The earth is so accessible, and its use so varied, that geology is handicapped perhaps more than any other science by persons who really have no valid claim to a scientific title.

The writer doubts whether a special degree in economic geology would go far toward improving this situation. Even if the courses were the same in different institutions, the manner of treatment and the ability of the teachers would be so varied that in the future, as in the past, anyone inquiring into the real standing of a geologist would be likely to consider his individual training rather than the degree attached to his name. There would be no guarantee that institutions not qualified to give the degree might not do so. However, the principal objection in the writer's mind to a degree of economic geology is the assumption that it is possible for anybody, in the present stage of knowledge, to formulate a standardized course adequate or best to meet the varied requirements. Considering the breadth and the variety of the field, any such attempt at standardization would have to be highly arbitrary. Once established, it would be a hindrance to the natural development of new courses to meet the ever changing requirements. When, if ever, the science of economic geology becomes fully organized, a standardized course may be possible. In the present stage of the science, more elasticity is required than seems to be possible in any of the courses proposed.

One of the purposes of the introduction of a degree of economic geology, to separate the sheep from the goats, may be accomplished in another way,—namely, by the establishment and maintenance of high standards of admission and high aims on the part of the various professional societies having to do with geology and mining. If this is done, membership in such societies may be regarded as evidence of sound training and achievement. To some extent this procedure may relieve the pressure on universities for uniformity of courses and degrees, leaving them free to develop in such manner as seems best. Scientific organizations, overlooking the entire field, are in a position to take into account the greatest variety of factors of training and experience in selecting their members. Failure of any university course to make men eligible for such recognition will obviously react on the course in a desirable way.

THE OPPORTUNITIES OF THE ECONOMIC GEOLOGIST

It has been the aim in this book to present a general view of the fields of activity of the economic geologist; and the list of chapter headings in itself summarizes the variety of his opportunities. The rapidly increasing use of earth materials promises far greater calls for geologic aid in the future than in the past. The profession is in its infancy.

Opportunities for employment are ordinarily found in three main directions—in educational institutions, in the federal and state geological surveys, and in private organizations. Connection with the United States Geological Survey excludes participation in private work, and in recent years even in teaching. In the state surveys there is ordinarily more latitude in this regard. In the educational institutions, it is rather the common procedure for the instructor to secure his field practice and experience through private agencies, or through part time connection with state surveys,—an arrangement with advantages to all concerned. The educational institution secures the benefit of the field experience which it cannot afford to provide, and is enabled to hold geologists at salaries far below their earning capacity. The geologist gains by the opportunity to alternate between office and field study, and to correct his perspective by the constant checking of theory with field conditions. The combination tends to keep the clearly scientific and the applied phases in a proper relative proportion; it minimizes the danger of drifting into purely empirical field methods on the one hand, and of losing touch with actualities on the other. Geologists devoting their attention solely to field work often complain that they do not have time to digest and correlate their results, nor to keep up with what others are doing. On the other hand, geologists without current field practice are likely to develop too strongly along subjective, deductive, and theoretical lines. The teacher gains in freshness and force in the presentation of his subject in the classroom, and the very effort necessary for presentation requires better analysis and coördination of his field observations. The private or state organization gains in this combination by drawing on the general and varied knowledge which has necessarily been accumulated for teaching and investigative purposes.

Temperament and circumstances will determine in which of these directions the student will turn. However, in view of the present natural tendency to be attracted by the large financial rewards in the commercial field, it may not be out of place to emphasize the fact that these rewards are perhaps more likely to be gained through perfected training and experience in state and national surveys and in educational institutions, than through early concentration in the commercial field. In any case, the financial side will take care of itself when sufficient knowledge and proficiency have been attained in any branch of the science.

The world is the geologist's laboratory; it is the only limit to his activities. The frontiers are near at hand, both physically and intellectually. There are few fields so attractive from the scientific standpoint. There are few in which the successful prosecution of the science can be of so much direct benefit to civilization and can yield such large financial rewards. If, in addition, the opportunities for travel and adventure are taken into account, what profession promises a more interesting and useful life?

So far we have discussed geology as a profession. It has proved its value also as a training for administrative and other public careers. The profession contributes its full share of men to these activities. The practice of geology deals with a wide variety of factors, and requires the constant exercise of judgment in balancing, correlating, and integrating these factors in order to reach sound conclusions. This objective treatment of complex situations is valuable training for the handling of human affairs.

ETHICS OF THE ECONOMIC GEOLOGIST

Ethical questions involved in the practice of economic geology have called out much discussion, and, in some cases, marked differences of opinion among men equally desirous of doing the right thing. In the plain choice between right and wrong, there is of course no difference of opinion. Unfortunately in many of the questions which arise the alternatives are not so clearly labeled.

The lure of discovery and quick returns always has, and doubtless always will, draw into the field large numbers of persons without sound ethical anchorage or standards. Fortunately, these are not the persons in control of the mineral industries; they are mere incidents in the great and stable business built up by legitimate demands for raw materials.

The view is sometimes expressed that the geologist should hold himself aloof from the business or applied phases of his profession, because of the danger of being tainted with commercialism. This argument would apply to the engineer as well as to the geologist. To carry such a procedure through to its logical conclusion would mean substantially the withdrawal of scientific aid from industry,—which, to the writer, is hardly a debatable question. Circumstances are trending inevitably to the larger use of geologic science in the commercial field. The problems of ethics cannot be solved by staying out. The economic geologist is rather called upon to do his part in raising the standards of ethics in that part of the field in which he has influence. This he can do by careful appraisal of all the conditions relating to a problem which he is asked to take up, and by refusing to act where questionable ethical standards are apparent or suspected. He must understand fully the purposes for which his report is to be used; merely as a matter of professional self-interest, there is no other course open to him. In a field in which there is so much danger from loose ethical conceptions, the premium on rigid honesty and nice appreciation of professional ethics is proportionately higher. The extreme care taken in this matter by acknowledged leaders in the profession of economic geology should be carefully considered by the young man entering the profession. There is a reason.

In other chapters reference is made to certain special ethical questions, such as the use of geology in mining litigation (pp.

349-355), and the necessity of the geologist's recognizing his own limitations (pp. 92-94), but no attempt has been made to cover the variety of such questions that may come up. It is safe to assume that no special ethical code can be made sufficiently comprehensive, detailed, and elastic to cover all the contingencies which are likely to be met in the practice of economic geology; nor is it likely that any such code, if attempted, would be any improvement on the spirit of the Golden Rule. Simple decency and common sense in their broader implications are essential to the practice of the profession.

Made in the USA
Monee, IL
05 November 2026

67808168R00270